Wilson

ORGANIC CHEMISTRY: AN INTERMEDIATE TEXT

TOPICS IN ORGANIC CHEMISTRY
A Series of Advanced Textbooks

SERIES EDITORS

Kendall N. Houk, UCLA
G. Marc Loudon, Purdue University

▼

ORGANIC STRUCTURE ANALYSIS
Phillip Crews, Marcel O'Neil-Johnson, and Jamie Rodriguez

ORGANIC CHEMISTRY: AN INTERMEDIATE TEXT
Robert V. Hoffman

ORGANIC CHEMISTRY: AN INTERMEDIATE TEXT

ROBERT V. HOFFMAN
New Mexico State University

New York   Oxford
Oxford University Press
1997

Oxford University Press

Oxford   New York
Athens   Auckland   Bangkok
Bogota   Bombay   Buenos Aires   Calcutta
Cape Town   Dar es Salaam   Delhi
Florence   Hong Kong   Istanbul   Karachi
Kuala Lumpur   Madras   Madrid   Melbourne
Mexico City   Nairobi   Paris   Singapore
Taipei   Tokyo   Toronto

and associated companies in
Berlin   Ibadan

Published by Oxford University Press, Inc.
198 Madison Avenue, New York, New York 10016

Oxford is a registered trademark of Oxford University Press

Library of Congress Cataloging-in-Publication Data

Hoffman, Robert V.
Organic chemistry : an intermediate text / Robert V. Hoffman.
p.   cm. — (Topics in organic chemistry)
Includes bibliographical references and index.
ISBN 0-19-509618-5
1. Chemistry, Organic.   I. Title.   II. Series.
QD251.2.H58   1997
547–dc20   96-17671

1   3   5   7   9   8   6   4   2

Printed in the United States of America
on acid-free paper

CONTENTS

PREFACE

This text was inspired by two observations. The first is that many entering graduate students took organic chemistry as sophomores but have since had little exposure to organic chemistry in a formal sense. Because of this time lapse in their organic preparation, they often have difficulty performing well when placed directly into mainstream graduate-level organic courses. What is much more effective is to first place them in a course which will bring them back up to speed in basic organic chemistry and at the same time introduce many of the advanced topics which are crucial to understanding current advances in the field. A course well suited for this purpose is a one-semester advanced organic course at the senior undergraduate/beginning graduate level. Most departments have such a course, but textbook selection for this course is problematic. If one of the standard advanced texts is used, only a small part is actually covered and students are not prepared to master the complexities, whereas an undergraduate text often fails to push the students to the next level. Consequently there is a real need for a one-semester text which reviews basic principles in addition to presenting the ideas that are currently of great importance in organic chemistry. This book was written to fill this need.

A second observation instrumental in shaping the approach of this text was made during group discussions of the organic faculty and students. One common exercise in these discussions is to work through some practice cumulative exam problems. It is very common for the students to analyze the question in terms of reactions and transformations and try to arrive at a solution based on the question as written. On the other hand, it is very common for the faculty to ask very simple questions first: "What is the oxidation change?" "What is the pK_a of the acid and what is the base?" and "What stereochemical changes occur?" are typical. It is clear that more experienced organic chemists begin from a very basic point of view and progress to a more complex solution, whereas novice organic chemists tend to jump in at a much more difficult level. It thus appears very important to initially emphasize the basic principles on which organic chemistry depends, and then progress to more specialized topics, all the while emphasizing their relationship to the basic principles. This book utilizes this organizational approach.

The result is a textbook designed for a one-semester advanced organic chemistry course. First and foremost it is a textbook and not a reference book. There is plenty of material to fill a semester, but it is not comprehensive in its coverage. Topics were chosen to provide a basic and well-rounded discussion of ideas important in modern organic chemistry and to provide students with the necessary tools to succeed in more specialized advanced courses. It is a book to be taught from; thus instructors should take the opportunity to include special or favorite topics at appropriate points. References to alternative textbooks and literature reviews of the subjects are included so that students can go to the library and get a different explanation. Students are thus encouraged to do library work as a means to independently gain insight and understanding. Finally, there are abundant problems included at the end of each chapter so that students can practice applying what they are learning. Working problems is the single most effective way to learn and organize the large amount of information that is encountered in organic chemistry, so numerous practice problems at all levels of difficulty are available.

The goal of this text is to provide senior undergraduate students with the organic background required to move on successfully in their careers. For beginning graduate students who lack this background it provides a succinct yet rigorous preparation for advanced organic courses.

Las Cruces, New Mexico   R.V.H.
June 1996

ORGANIC CHEMISTRY: AN INTERMEDIATE TEXT

FUNCTIONAL GROUPS AND
CHEMICAL BONDING

Functional Groups

There are now more than 12 million known compounds, of which more than 80 percent are organic. In order to make sense of 9 million organic compounds, and to be able to manipulate them and make new compounds, there must be some system of organization whereby organic compounds can be categorized by a particular property or group of properties. A natural method utilized by early practitioners was to group organic compounds by the reactions they underwent. Thus there developed a large variety of qualitative tests, called classification tests, which could be used to systematically categorize the reactivity of a compound and thus allow it to be grouped with others of similar chemical reactivity. These tests are still very useful to practicing organic chemists and collectively are known as organic qualitative analysis.

Classification tests are used to distinguish organic compounds and segregate them into different functional classes based on their chemical properties. Originally, a group of compounds that showed similar chemical behavior based on the classification tests were named for a property or behavior (e.g., acids from *acer*, meaning sour, and aromatic compounds from the odors). With the evolution of the science of chemistry and the development of more modern views of atoms and molecules, a different definition of functional classes is possible. The behavior of organic compounds is now organized into patterns that are based on recurrent groups of atoms—*functional groups*. The sites in molecules at which chemical reactions occur are localized at the functional groups in the molecule; the rest of the molecule is the same after the reaction as before. Thus instead of thinking of the whole molecule in terms of its chemical reactivity, it is necessary only to recognize what functional group or groups are present in the molecule. It is then possible to predict the chemical behavior of the molecule based on the known chemistry of the functional groups it contains.

This turns out to be a huge simplification. Since the numbers of functional groups are relatively small, it is possible to classify a very large number of individual compounds using a relatively small number of functional groups. The first step to enlightenment in organic chemistry is therefore to realize the key role that functional groups play in simplifying the subject, and the second step is to learn the functional groups by name, structure, and formula. Although a great number of them may have already been encountered in the introductory organic course, it is helpful to review them. Table 1.1 is a list of the most common functional groups. Although many other functional groups are not shown, those in Table 1.1 are present in the vast majority of compounds. Notice that not all functional groups contain only carbon atoms (e.g., the nitro group and the carbodiimide groups), and some functional groups differ at atoms other than carbon (compare the nitro and nitroso groups and the sulfoxide and sulfone groups). Since functional groups are reference points for predicting and understanding the reactions of individual organic molecules, it is very important to be able to recognize these common ones (and others that might be encountered in the future). It is also useful to learn normal structural abbreviations used to indicate functional groups that are present in chemical structures. The abbreviations in Table 1.2 correspond to the groups that are shown in Table 1.1.

A major reason that the behavior of organic compounds can be generalized in terms of the functional groups they contain is that the bonds holding a given functional group together are the same regardless of the compound which contains that functional group. The four compounds shown in Figure 1.1 all contain the carboxylic acid functional group which is highlighted within the boxes. Thus all four contain the bonding pattern characteristic of the —COOH functional group which is *independent of the bonds found in rest of the molecule!*

Since most organic reactions involve the conversion of one functional group to another, it follows that most organic reactions simply involve bond changes involving functional groups. If one knows the bonds found in the reactant functional group and the bonds found in the product functional group, one automatically knows what bonding changes are required to effect the desired chemical change. Thus, in addition to being able to recognize functional groups, it is also important to be able to describe the numbers and types of bonds found in functional groups.

Bonds in functional groups can first be described by Lewis structures, which are merely formalisms for denoting numbers of shared and unshared electron pairs, formal charges, and types of bonds (numbers of shared pairs, single, double, and triple). Chemistry students learn to write Lewis structures in virtually all of their early chemistry courses. How to write Lewis structures will not be reviewed here, but knowing the correct Lewis structures for molecules and functional groups in molecules is an indispensable first step in being able to describe the structure and bonding of functional groups.

The next level of insight into functional groups comes from the translation of Lewis structures into more accurate bonding descriptions based on modern bonding theories. Structural details, including geometries, are also evident from the proper description of the bonding in the functional group. The ideas of structure and bonding currently in use had their origins in the late 1920s. It is again beyond the scope of this book to trace the developments which were seminal in the development of current theories; however, early studies were all rooted in the quest to understand and be able to describe the behavior of electrons in atoms. The development of quantum mechanics and the particle-wave duality of the electron and the

TABLE 1.1 Common Functional Groups

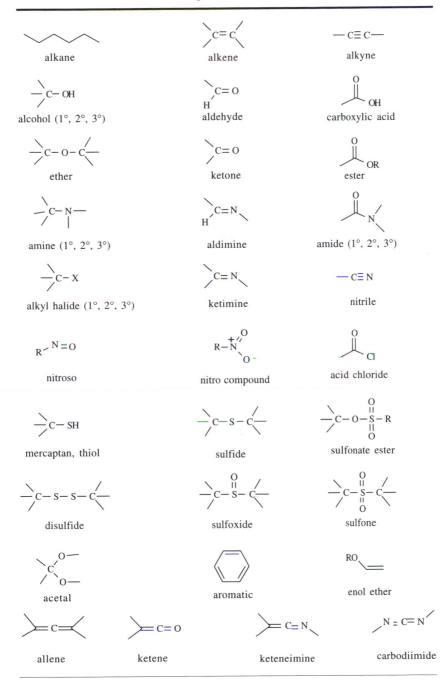

alkane	alkene	alkyne
alcohol (1°, 2°, 3°)	aldehyde	carboxylic acid
ether	ketone	ester
amine (1°, 2°, 3°)	aldimine	amide (1°, 2°, 3°)
alkyl halide (1°, 2°, 3°)	ketimine	nitrile
nitroso	nitro compound	acid chloride
mercaptan, thiol	sulfide	sulfonate ester
disulfide	sulfoxide	sulfone
acetal	aromatic	enol ether
allene	ketene	keteneimine carbodiimide

TABLE 1.2 Common Functional Group Abbreviations

Group	Abbreviation
alkane	R
alkene	$R_2C{=}CR_2$
alkyne	$RC{\equiv}CR$
alcohol	ROH
aldehyde	RCHO
carboxylic acid	RCO_2H
ether	ROR
ketone	RC(O)R
ester	RCO_2R
amine	RNH_2,R_2NH,R_3N
aldimine	RHC=R
amide	$RC(O)NH2,RC(O)NHR,RC(O)NR_2$
alkyl halide	RX
ketimine	$R2C{=}NR$
nitrile	RCN
nitroso	RNO
nitro	RNO2
acid chloride	RC(O)Cl
mercaptan, thiol	RSH
sulfide	RSR
sulfonate ester	RO_3S
disulfide	RSSR
sulfoxide	RS(O)R
sulfone	RSO_2R
acetal	$(RO)_2CR_2$
aromatic	Ar
enol ether	$ROCH{=}CR_2$
allene	$R_2C{=}C{=}CR_2$
ketene	$R_2C{=}C{=}O$
keteneimine	$R_2C{=}C{=}NR$
carbodiimine	$R_2N{=}C{=}NR_2$

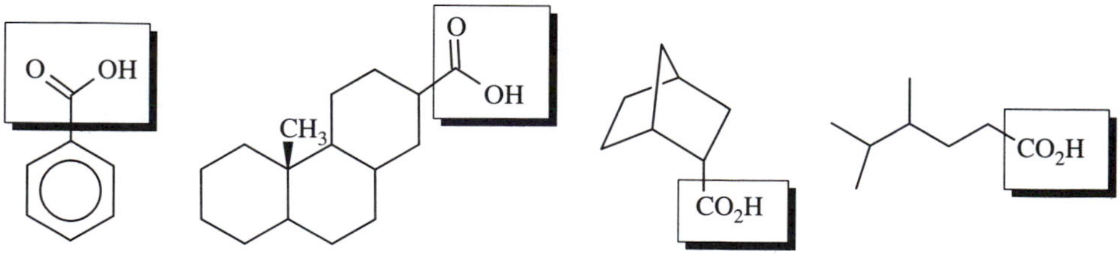

FIGURE 1.1

uncertainty principle led to mathematical descriptions of the behavior of electrons in the electric field of the nucleus. The solutions of those equations resulted in a new conceptual framework for understanding chemical bonding.

Orbitals

The theory suggests that electrons around the nucleus occupy orbitals, which are regions of space with a nonzero electron population. Instead of thinking of where an electron *is,* it is more correct to think about where the electron is *likely to be.* Orbitals are thus regions of space where an electron is more likely to be found. These regions of space with a significant electron population (orbitals) have shape, size (distance from the nucleus), and energy. Familiar examples of s, p, and d atomic orbitals are shown in Figure 1.2 The most common elements present in organic compounds are first-row elements (C, H, N, O); 1s, 2s, and 2p atomic orbitals are therefore most commonly encountered. The concept of atomic orbitals (AOs) was a breakthrough in understanding the properties of atoms.

In molecules, the problem of understanding the interactions of electrons with the nuclei is more complicated because there are more nuclei and more electrons that interact. Imagine, however, the situation that occurs when two nuclei approach one another. If the two atoms come close enough together, an AO of one atom that contains a single electron will occupy to some extent *the same* region of space as an AO of a second atom that contains a single electron. When those atomic orbitals overlap, an electron from one atom shares a region of space with the electron from the other atom. When such an event occurs, each electron is no longer influenced by just one nucleus but by two. This requires a new mathematical description of the behavior of electrons influenced by two nuclei. Again the solution to those equations defines a new region of space where there is a high probability of finding *both* electrons. Furthermore, only two electrons can occupy any particular region of space. This new region of space is called a molecular orbital (MO). The electrons in the MO are of lower energy than when they were in their separate AOs, and the lowered energy gives rise to a chemical bond between the atoms. This process is shown in Figure 1.3. In other words, chemical bonds result from the overlap of singly occupied AOs to give a doubly occupied MO in which each electron of the pair interacts with both nuclei. Because each of the electrons interacts with two nuclei, they are more tightly bound (i.e., of lower energy) than they were in the separated atoms, and they are more likely to be found between the two nuclei.

Bonding Schemes

Bond formation between atoms occurs primarily to enable each atom to achieve an inert gas electron configuration in the valence level (a valence octet for all elements except hydrogen, which requires only two electrons to achieve the configuration of He). An atom can achieve an inert gas electronic configuration by giving up electrons, by accepting electrons, or by sharing electrons with another atom. An ionic bond is formed when one atom gives up one or more electrons to reach an octet electronic configuration (as a positively charged ion) and a second atom accepts one or more electrons to reach an octet electronic configuration (as a

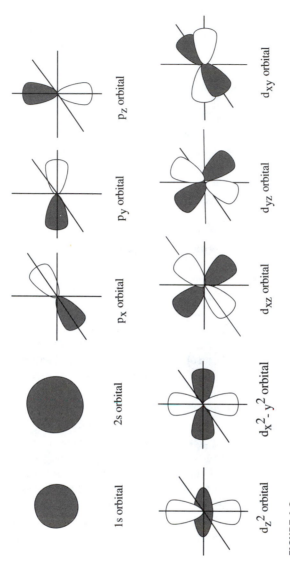

1s orbital

2s orbital

p_x orbital

p_y orbital

p_z orbital

d_{z^2} orbital

$d_{x^2 - y^2}$ orbital

d_{xz} orbital

d_{yz} orbital

d_{xy} orbital

FIGURE 1.2

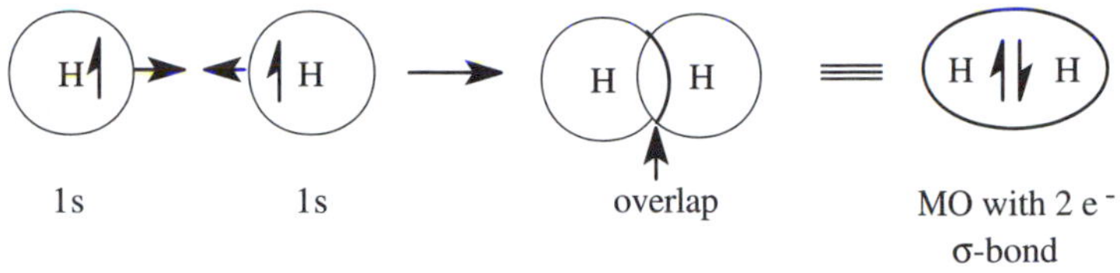

FIGURE 1.3

negatively charged ion). For example, the reaction of a cesium atom with a chlorine atom occurs by the transfer of an electron from the cesium atom to the chlorine atom. By doing so, both cesium and chlorine have reached a valence octet electron configuration. The cesium atom has been converted to a positively charged cesium ion with the octet electronic configuration of xenon, and the chlorine has been converted to a negatively charged chloride ion with the octet electronic configuration of argon. The "bond" between cesium and chlorine is due to the electrostatic attraction of the cesium and chloride ions:

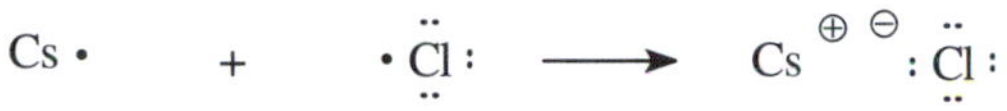

$$Cs \cdot \quad + \quad \cdot \ddot{\underset{..}{Cl}} : \quad \longrightarrow \quad Cs^{\oplus} \; {}^{\ominus} \; : \ddot{\underset{..}{Cl}} :$$

The reaction of potassium metal with *tert*-butanol gives an ionic bond between the *tert*-butoxy anion and a potassium cation by transfer of electrons from potassium to the hydroxyl functional group. Hydrogen is evolved as a by-product.

$$\underset{|}{\overset{|}{-}} -\ddot{\underset{..}{O}}H \;+\; K\cdot \;\longrightarrow\; \underset{|}{\overset{|}{-}} -\ddot{\underset{..}{O}} : K^+ \;+\; H\text{-}H$$

(Based on functional group behavior, any other alcohol is predicted to react with potassium in the same way—and it does!)

Most bonds in organic molecules, however, are covalent bonds in which electrons are shared between two atoms. Sharing electrons is a way to enable each atom of the bonded pair to reach an octet electronic configuration without having to give up or gain an electron. Covalent bonds are formed by the overlap of singly occupied AOs to form new MOs that contain a pair of electrons. Each atom in essence gains an electron by sharing. The reaction of a chlorine atom with a fluorine atom occurs by the overlap of a singly occupied 3p orbital of chlorine with a singly occupied 2p orbital of fluorine to give a bond between the two atoms that contains two electrons. The type of bond formed is called a σ *bond* because the region of greatest electron density falls on the internuclear axis. This is shown both by using Lewis structures and orbital pictures.

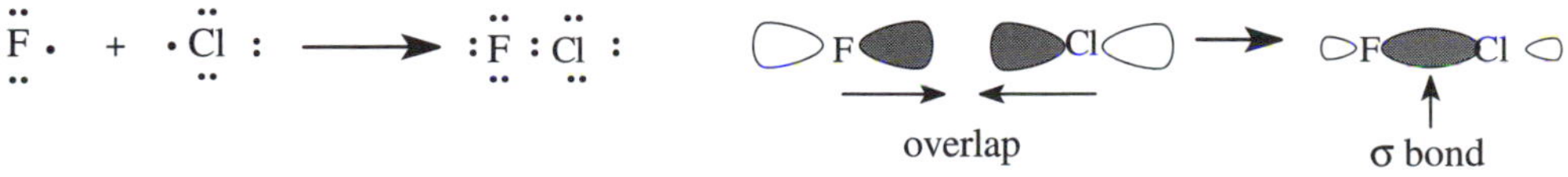

$$\ddot{\underset{..}{F}}\cdot \;+\; \cdot \ddot{\underset{..}{Cl}} : \;\longrightarrow\; : \ddot{\underset{..}{F}} : \ddot{\underset{..}{Cl}} :$$

This simple picture is adequate for many diatomic molecules with univalent atoms, but it is not sufficient to describe the bonding in most polyatomic molecules. In addition to electron sharing to reach octet electronic configurations, other considerations require some modification of the picture: for example, the number of bonds to an atom, the number of electron pairs that are shared between two bonded atoms, and repulsion energies that are present between electron pairs. These factors are ameliorated by the combination of valence shell atomic orbitals (2s and 2p's) to form hybrid AOs. These hybrid AOs overlap with AOs of other atoms in the usual fashion to form covalent bonds. Hybrid AOs have energies, shapes, and geometries which are intermediate between the atomic orbitals from which they are formed. Hybridization of atomic orbitals is an outgrowth of bond formation that enables atoms to derive the greatest amount of bond energy from electron sharing and to allow bonded atoms to achieve octet electronic configurations.

If four single bonds or electron pairs originate from a single atom, the s orbital and the three p orbitals of the valence shell combine to form four equivalent sp^3-hybrid orbitals that are then used in bond formation to other atoms, as in Figure 1.4. Depending on the number of electrons in the valence shell of the atom, these sp^3-hybrid orbitals can contain either a single, unpaired electron, which can be shared with another atom by overlap and bond formation, or an unshared pair of electrons, which is normally not involved in bond formation. Thus alkanes, which have all single bonds, have carbon atoms that are sp^3-hybridized. For example, methane has four single C—H bonds originating at carbon and these bonds are σ bonds produced by the overlap of four sp^3-hybrid orbitals of carbon with four 1s atomic orbitals of 4 hydrogen to give four sp^3– 1s σ bonds from carbon to hydrogen. The geometry of the four equivalent sp^3-hybrid orbitals (and hence the compound produced by overlap with these orbitals) is tetrahedral. Thus methane has four equivalent C—H σ bonds which point toward the corners of a regular tetrahedron and have H—C—H bond angles of 109.5°.

In a similar fashion each carbon of propane is sp^3-hybridized and tetrahedral since each carbon has four single bonds to other atoms originating from it. For example, the central carbon of propane has two equivalent sp^3–1s C—H σ bonds and two equivalent sp^3–sp^3 C—C σ bonds. (Note that sp^3 orbitals from one carbon can overlap with sp^3 orbitals from another carbon to produce carbon–carbon bonds.) The geometry is very close to tetrahedral, but the C—C—C bond angle is slightly larger (111°) to accommodate the bigger CH_3 groups (see Figure 1.5).

Other elements can also be sp^3-hybridized. The only requirement is that they have a combination of four single bonds or electron pairs originating from a single element. Ammonia, which has three N—H bonds and a lone pair on nitrogen, is thus sp^3-hybridized and has three equivalent sp^3–1s N—H σ bonds and a lone pair which occupies an sp^3-hybrid orbital. The

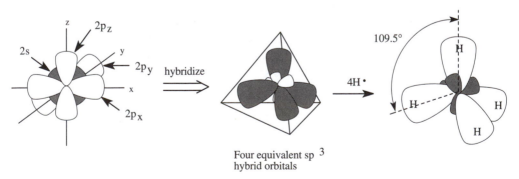

Four equivalent sp^3
hybrid orbitals

FIGURE 1.4

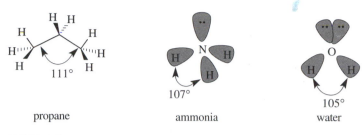

FIGURE 1.5

geometry is close to tetrahedral with an H—N—H bond angle of 107° (Figure 1.5). Other amines also have sp³-hybridized nitrogen and are close to a tetrahedral geometry around the nitrogen atom.

The oxygen atom in the water molecule has two bonds and two lone pairs, so it too is sp³-hybridized. There are two equivalent sp³–1s O—H σ bonds and two lone pairs occupying sp³-hybridized orbitals. Electron–electron repulsions of the lone pairs cause greater distortions from a true tetrahedral geometry so that the H—O—H bond angle is 105°. Other singly bonded oxygen functional groups such as alcohols, ethers, and acetals have sp³-hybridized oxygens and nearly tetrahedral geometries.

Second-row elements such as silicon, phosphorus, and sulfur can also have sp³-hybridization of the valence shell orbitals. In these second-row elements, however, the 3s and 3p AOs hybridize to form the sp³-hybrid orbitals. Tetramethylsilane, the standard reference for nmr spectra, has tetrahedral geometry because of the sp³-hybridization of the 3s and 3p valence shell orbitals of silicon (see Figure 1.6). Dimethyl sulfone has nearly tetrahedral bond angles because the sulfur is sp³-hybridized. Although formal charges are present, the two bonds to oxygen can be thought to arise by the overlap of a filled sp³ orbital on sulfur with an unfilled sp³ orbital on oxygen. The resulting σ bond is called a coordinate covalent, or dative, bond because both of the shared electrons in the bond come from only one of the bonded elements.

When two pairs of electrons are shared between two elements, a different bonding arrangement is required to enable the atoms to reach valence octet electron configurations. Because of the Pauli exclusion principle, only one σ bond is possible between any two atoms because only one pair of electrons can occupy the space along the internuclear axis. The second pair of electrons that is shared by the two atoms must therefore be located in space someplace other than along the internuclear axis. The second pair of shared electrons is located in a different type of covalent bond, a π *bond,* which has electron density found on either side of the internuclear axis. π Bonding results from the parallel overlap (or sideways overlap) of atomic p orbitals.

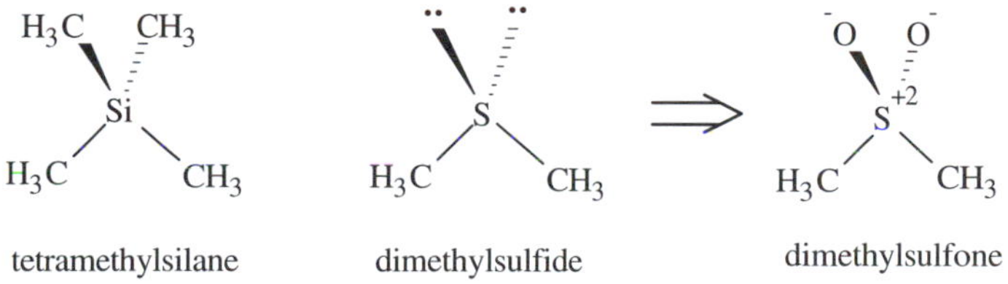

tetramethylsilane   dimethylsulfide   dimethylsulfone

FIGURE 1.6

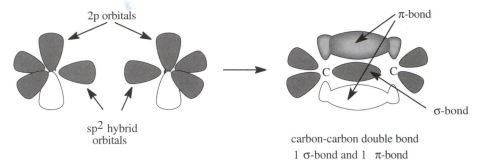

FIGURE 1.7

To accommodate the need for a singly occupied p orbital available for the formation of a π bond, hybridization of the valence AOs takes place between the s orbital and two of the three p orbitals. Hybridization of one s and two p AOs produces three equivalent sp²-hybrid AOs, and a p orbital remains unhybridized in order to produce a π bond (see Figure 1.7).

   This bonding scheme permits two pairs of electrons to be shared between two atoms so that each pair occupies a different region of space and does not violate the Pauli exclusion principle. Since only two p orbitals are used in the hybridization, and since they are orthogonal and define a plane, the sp²-hybridized carbon is planar with bond angles of 120°. The remaining p orbital, which is left unhybridized to form the π bond, is perpendicular to the molecular plane. Once formed, the π bond keeps the entire system rigid and planar, because rotation of one end of the π-bonded system relative to the other end requires that the π bond be broken.

   Elements other than carbon are also sp²-hybridized if they share two electron pairs with another atom. Thus imines have sp²-hybridized nitrogen (and carbon) to account for formation

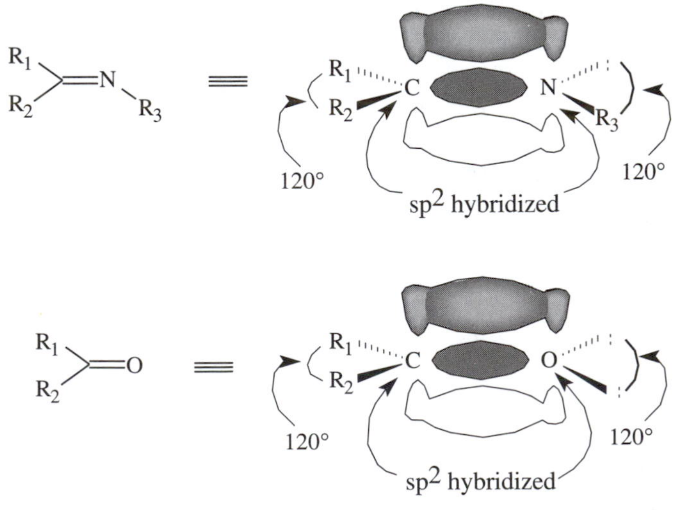

FIGURE 1.8

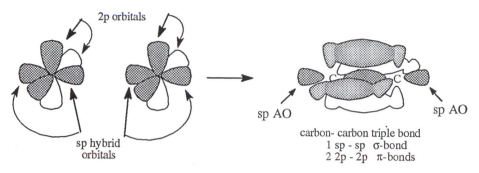

2p orbitals

sp AO

sp AO

sp hybrid orbitals

carbon- carbon triple bond
1 sp - sp σ-bond
2 2p - 2p π-bonds

FIGURE 1.9

of the C=N double bond (see Figure 1.8, top). The lone pair on nitrogen occupies an sp^2-hybrid orbital. The bond angles are all 120° around both carbon and nitrogen since both are sp^2-hybridized. Similar considerations hold for the oxygen atom of carbonyl groups of all kinds (Figure 1.8, bottom). Both unshared pairs of electrons on oxygen occupy sp^2 orbitals. The interorbital angle is 120°, as expected for trigonal hybridization.

The sharing of three pairs of electrons between two atoms can be accomplished by extrapolation of the above considerations. That is, since there can be only one σ bond connecting the atoms, the other two pairs of shared electrons must be in two different π bonds, each of which is formed by the parallel overlap of a p orbital. Furthermore, the π bonds must be mutually orthogonal so as not to violate the Pauli exclusion principle. Hybridization of one s orbital and one p orbital gives two equivalent sp-hybrid AOs which are linearly opposite one another (Figure 1.9). The two remaining p orbitals, which are mutually orthogonal, are used to produce two orthogonal π bonds. The geometry of triply bonded systems is thus linear about the triple bond.

Similar considerations apply to the triply bonded nitrogen found in nitriles. The sp-hybridized carbon and nitrogen atoms form an sp–sp σ bond and two 2p–2p π bonds between carbon and nitrogen. The unshared pair on nitrogen occupies an sp-hybrid orbital, as shown in Figure 1.10.

Another instance requiring sp-hybridization occurs in molecules with cumulated double bonds such as allenes, ketenes, and carbodiimides. The end atoms of the cumulated units are sp^2-hybridized because each shares two electron pairs with another element (the central carbon) and there are a σ bond and a π bond. The structure, however, requires that two π bonds originate from the central carbon: one going toward one end of the cumulated system, and one going toward the other end. Thus two 2p atomic orbitals are required for π bonding from the central carbon, and sp-hybridization is therefore appropriate. Consequently, the geometry is

R—C≡N: ⟹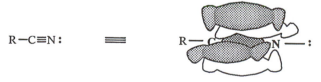

FIGURE 1.10

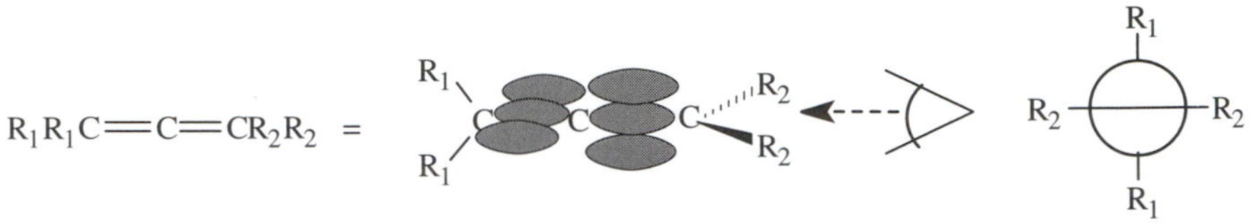

$$R_1R_1C{=\!\!=\!\!=}C{=\!\!=\!\!=}CR_2R_2 \;=\;$$

FIGURE 1.11

linear at the middle atom and trigonal at the end atoms (see Figure 1.11). A further consequence of the orthogonal π bonds is that planar bonds originating at the end carbons lie in two
orthogonal planes with a dihedral angle of $90°$.

Besides providing a theoretical framework by which the structure, geometry, and octet
structure of bonded elements can be explained and understood, the concept of hybridization
also predicts the ordering of stabilities and energies of bonds and the energy of lone pairs
of electrons in hybrid orbitals. Because s atomic orbitals are of lower energy than p atomic
orbitals, hybrid orbitals with a greater proportion of s character should be more stable and
thus form stronger bonds. Unshared pairs of electrons in hybrid orbitals with greater s character should also be of lower energy (more stable). As the s-character percentage of hybrid
orbitals increases from sp^3 (25% s character) to sp^2 (33% s character) to sp (50% s character), the strength of bonds formed by overlap with those orbitals increases in a parallel fashion.
For example, the bond dissociation energies of primary C—H bonds have been measured and
fall in the order predicted by the s-character percentage of the hybrid orbitals on carbon: sp^3
C—H, 98 kcal/mol; sp^2 C—H, 103 kcal/mol; and sp C—H, 125 kcal/mol. Electron pairs are
more stable in orbitals with more s character; thus the acidities of primary C—H bonds are
found to be sp^3 C—H, $pK_a = 50$; sp^2 C—H, $pK_a = 44$; sp C—H, $pK_a = 25$. This is true
because the anions formed by proton removal yield carbanions that have the negative charge
in sp^3, sp^2, and sp orbitals, respectively. Because the lone pair is more stable in an orbital of
greater s character, the anion formed by removal of an sp C—H proton is more stable (and
hence the proton more easily removed) than the anion formed by removal of an sp^2 C—H proton, which is in turn more stable than the anion formed by removal of an sp^3 C—H proton.
Other examples of the effects of greater s character in orbitals are encountered routinely.

The concept of hybridization of atomic orbitals to give new hybrid atomic orbitals involved
in the bonding patterns of atoms is a useful and practical way to describe the way in which
functional groups are constructed. It provides insight into the structure as well as the geometry
and electron distribution in functional groups and molecules in which they are found. It can
also be used to predict reactivity patterns of functional groups based on these considerations.

Antibonding Orbitals

The overlap of atomic orbitals to give a new molecular orbital (MO) in which an electron pair
is shared by the interacting atoms is illustrated in Figure 1.3. The new MO, which contains the
shared electron pair, is of lower energy than the atomic orbitals from which it was produced
by overlap. This energy change (ΔE) is illustrated in Figure 1.12. (N represents the nucleus
of some element in the bond formation process.) The ΔE is related closely to the bond energy

FIGURE 1.12 Energy changes that occur during the overlap of atomic orbitals to form covalent bonds.

of the bond produced. The same model holds irrespective of the type of AOs which overlap (simple AOs or hybrid AOs) or the type of bond formed (σ or π).

Although this model is easy to visualize and understand, it is actually only half of the story. When atomic orbitals interact, the number of new MOs produced from that interaction must equal the number of AOs which initially interact. Furthermore, for each MO produced which is lower in energy than the energy of the interacting atomic orbitals, another will be produced which is *higher* in energy by the same amount (Figure 1.13). Thus when two half-filled atomic orbitals interact, two MOs will be produced, one of lower energy which will contain the electron pair, termed the *bonding MO,* and a second that is of higher energy and unfilled, termed the *antibonding MO.*

For each bond in a molecule which is described by the overlap of AOs, there will be a bonding MO of lower energy that, when filled with an electron pair, gives rise to a stable bond between elements. There will also be an antibonding MO which is of higher energy and thus unfilled. Antibonding orbitals correspond to the situation in which nuclei are moved to within bonding distance of one another but there is *no* electron sharing—in fact, the electrons and nuclei actually repel one another. This electronic and nuclear repulsion is what increases the energy of the antibonding level. Because the bonding MO is filled and the antibonding MO is unfilled, the system is at a lower net energy than the individual AOs, and bond formation takes place. This occurs for both σ bonds and π bonds; the antibonding orbitals are indicated by an asterisk in Figure 1.14. Overlap of an sp^3 AO on a carbon with a 1s AO on a hydrogen thus gives a σ-bonding MO that is filled with two electrons and an unfilled, higher-energy antibonding MO termed a $\sigma*$ MO. Likewise, overlap of two 2p AOs on carbon gives a π MO which contains a shared pair of e^- and a $\pi*$ MO which is a higher-energy, unfilled π-antibonding orbital (see Figure 1.15).

Thus far it would appear that antibonding orbitals are real orbitals, but they seem to be merely mathematical artifacts since they are unfilled and thus do not enter into bonding or

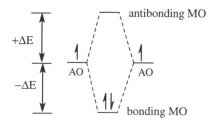

FIGURE 1.13

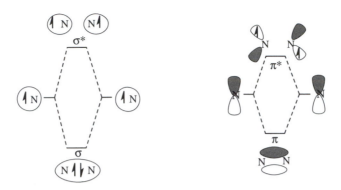

FIGURE 1.14

energy considerations. For ground state molecules, this is actually true: all of the electrons are found in bonding orbitals. Why, then, should we even concern ourselves with their existence? The answer lies in the realization that antibonding orbitals are still, in fact, orbitals. They are regions of space that electrons *could* occupy. In ground state molecules, electrons fill the lower-energy bonding orbitals. Suppose, however, one wished to take an electron out of a bonding orbital and move it to a higher level. Where would it go? Or suppose one wished to add electrons to a molecule which already had its bonding orbitals filled. Where would the electrons go? Suppose an electron-rich reagent were to donate electrons to a molecule. Where would the electrons go?

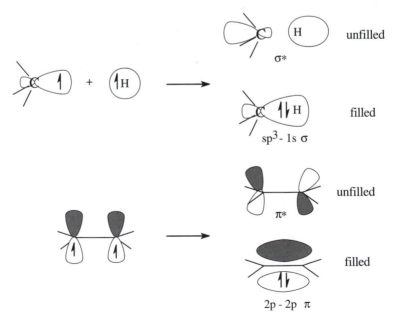

FIGURE 1.15

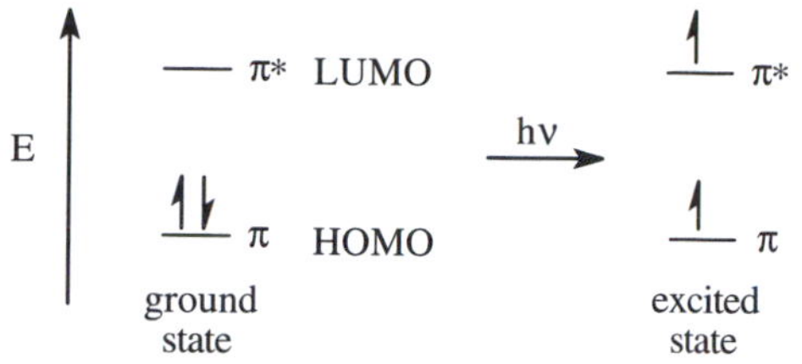

$$
\begin{array}{ccc}
\underline{} & \pi* \;\; \text{LUMO} & \underline{\uparrow} \;\; \pi* \\[2mm]
& \xrightarrow{\;\; h\nu \;\;} & \\[2mm]
\underline{\uparrow\downarrow} & \pi \;\;\; \text{HOMO} & \underline{\uparrow} \;\; \pi \\[2mm]
\text{ground} & & \text{excited} \\
\text{state} & & \text{state}
\end{array}
$$

FIGURE 1.16

The answer to all these questions is antibonding orbitals. Although they are of high energy, antibonding orbitals are unfilled and can accept electrons from several sources if sufficient energy is available to promote electrons into the antibonding energy level. Absorption of light energy can cause an electron to be promoted from the highest occupied molecular orbital (HOMO), which is usually a bonding MO, to the lowest unoccupied molecular orbital (LUMO), which is normally an antibonding molecular orbital. For example, if an olefin which contains a carbon–carbon π bond is exposed to ultraviolet light of the correct frequency (and hence energy), the molecule can absorb the energy of the light by promoting a π electron from the bonding MO into the antibonding MO (Figure 1.16). This new electronic state is termed an *excited state* and is higher in energy than the initial electron-paired state, called the *ground state*. (The electron spins can be paired in the singlet excited state or unpaired in the triplet excited state.) Excited states of molecules are high-energy states, which are much more reactive than ground states and can be described in terms of the population of antibonding orbitals. Consequently, all photochemical reactions which occur by the reactions of excited state species are intimately dependent on the existence of and population of antibonding orbitals.

The reduction of organic molecules by the addition of electrons can take place by chemical reagents or at the surface of electrodes. In either case electrons are added to the organic compound, thus reducing it. Electrons cannot go just anywhere; they must go into an unfilled orbital. During a reduction, then, electrons are injected into the LUMO of the molecule, which is normally an antibonding orbital. Population of the antibonding orbital raises the total energy of the molecule, and subsequent reactions follow. The electrochemical reduction of alkyl bromides schematized in Figure 1.17 illustrates the process well. An electron is added into the $\sigma*$ orbital of the carbon–bromine bond, which is the LUMO of a saturated alkyl bromide. Population of the antibonding orbital raises the energy of the molecule and weakens the carbon–bromine bond, which then dissociates to give a bromide ion and a carbon-centered free radical with an

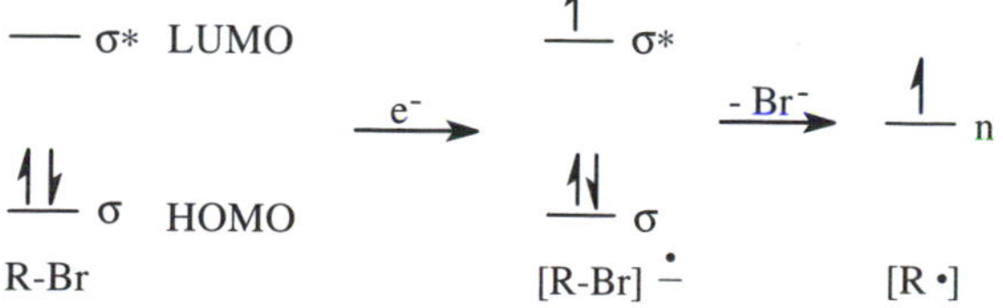

$$
\begin{array}{ccccc}
\underline{} \;\; \sigma* \;\; \text{LUMO} & & \underline{\uparrow} \;\; \sigma* & & \\[2mm]
& \xrightarrow{\;\; e^- \;\;} & & \xrightarrow{\;\; -\,Br^- \;\;} & \underline{\uparrow}_{n} \\[2mm]
\underline{\uparrow\downarrow} \;\; \sigma \;\; \text{HOMO} & & \underline{\uparrow\downarrow} \;\; \sigma & & \\[2mm]
\text{R-Br} & & [\text{R-Br}]\,\overset{\bullet}{}{}^{-} & & [\text{R}\,\bullet]
\end{array}
$$

FIGURE 1.17

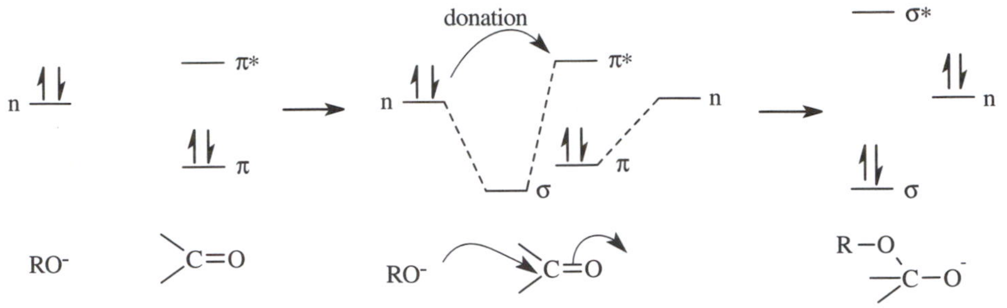

FIGURE 1.18

unpaired electron in a hybrid atomic orbital (nonbonded energy level). Almost all dissolving metal and electrochemical reductions follow this same general sequence. An electron is donated into an antibonding orbital, the energy of the molecule is raised, and chemical change ensues.

When a nucleophile attacks an electrophile, it donates a pair of electrons to the electrophile. Electron donation must take place by an overlap interaction between a filled orbital on the nucleophile that contains the electron pair to be donated and an unfilled orbital (LUMO) on the electrophile that is usually an antibonding orbital. Population of the LUMO by electron donation raises the energy of the system, leading to bonding change and new bond formation. Addition of an alkoxide to a ketone is a typical example of the process. The electron pair to be donated is in a hybrid AO and therefore is at a nonbonding energy level (n). Overlap with the $\pi*$ orbital of the carbonyl group starts to populate the $\pi*$ orbital. This weakens the π bond, the carbon–oxygen π bond of the carbonyl group is broken, and a new, lower-energy σ bond is formed between oxygen of the alkoxide and the carbonyl carbon. The electrons of the π bond end up in a nonbonding atomic orbital on oxygen in the product. This process is shown schematically in Figure 1.18.

Nucleophilic additions and substitutions, the most widespread of all organic reactions, all have the same general orbital requirements. An orbital containing an electron pair of the nucleophile overlaps with an antibonding orbital of the electrophile, which leads to population of the antibonding level. This raises the energy of the system, and bond and electron reorganization follows to give products. The electron pair must be able to be donated (i.e., not tightly bound or of higher energy), and the antibonding orbital must be of sufficiently low energy to ensure effective overlap.

Thus one can see that, although antibonding orbitals are not a major factor in describing the bonding of ground state molecules, they can play a pivotal role in the reactions of molecules. It is therefore important to keep in mind the existence of antibonding orbitals and their ability to accept electrons and control the reactivity of molecules.

Conjugated π Systems

The same principles of overlap hold when more than two p atomic orbitals overlap to form π systems. First, the number of MOs produced will be the same as the number of p orbitals which interact. Thus for the allyl system, in which three contiguous p orbitals interact, three

MOs will be produced from the interaction of three 2p atomic orbitals. For the butadienyl system, in which four contiguous p orbitals interact, four MOs will result, and so on:

allyl   butadienyl

Second, the energy distribution of the MOs will be disposed symmetrically about the energy of the atomic orbitals before they interact (nonbonded energy level). For example, if one MO is of lower energy by $-\Delta E$ because of overlap, there must be an antibonding MO raised to higher energy $(+\Delta E)$. MOs which are lower in energy than the nonbonding energy are bonding MOs $(-\Delta E)$, those which are higher in energy than the nonbonding energy are antibonding MOs $(+\Delta E)$, and those at the same energy as the nonbonding energy are nonbonding MOs $(\Delta E = 0)$.

For the allyl system, which has three MOs from the overlap of three 2p atomic orbitals, one MO will be lowered in energy $(-\Delta E)$, so another MO will be raised by the same amount. The remaining MO must stay at the nonbonding level $(\Delta E = 0)$ to maintain energy symmetry around the nonbonding level.

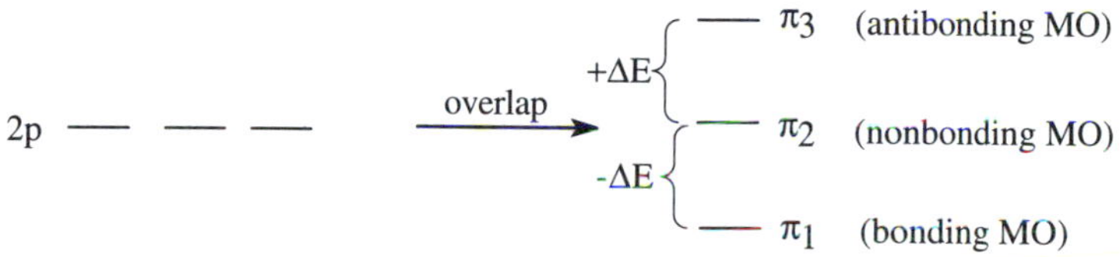

What is interesting is that this overlap model allows one to construct the orbital diagram without being concerned with electrons. The MOs produced by the interaction of atomic orbitals can each hold two paired electrons, and these can be filled in depending on the number of electrons present in the π system. Thus the bonding diagrams for the allyl cation, allyl radical, and allyl anion can be constructed by merely filling the orbitals with the number of π electrons present in these species (two, three, and four π electrons, respectively). Figure 1.19 also demonstrates that all three intermediates in the allyl system are stabilized because each contains two electrons in the π_1-bonding MO.

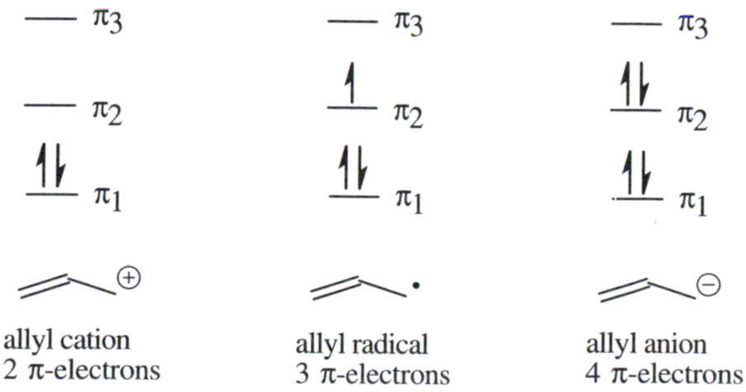

allyl cation
2 π-electrons

allyl radical
3 π-electrons

allyl anion
4 π-electrons

FIGURE 1.19

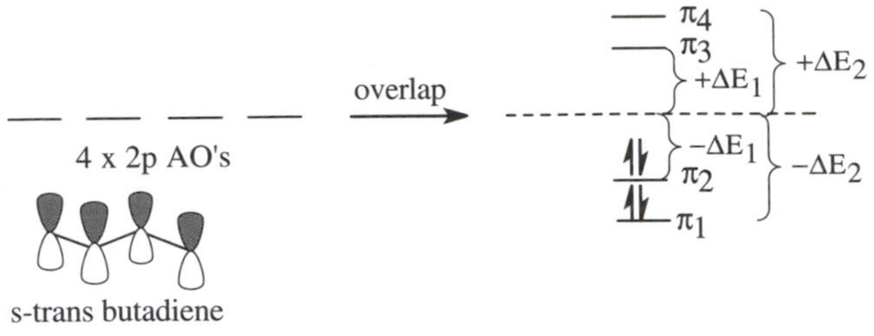

s-trans butadiene

FIGURE 1.20

Two of the four MOs of the butadienyl system are at lower energy than the nonbonded energy level ($-\Delta E_1$, $-\Delta E_2$), and two are at higher energy than the nonbonded energy level ($+\Delta E_1$, $+\Delta E_2$). The four π electrons of butadiene fill the two bonding MOs and give a stable molecule (Figure 1.20). It should also be obvious that butadienyl species with less than, or more than, four π electrons should be significantly less stable than butadiene itself. Removal of an electron requires energy because the electron would have to come from a relatively stable bonding MO. Addition of an electron to the butadienyl π system requires that it be put into an antibonding MO, which is also energetically unfavorable.

A great many π systems have been examined by this approach and the orbital diagrams understood. As mentioned before, the antibonding orbitals are often unfilled in the ground state, but play an important part in the excited states and reactions of these compounds.

Aromaticity

A special type of orbital interaction occurs when a conjugated π system is in a ring. The π system of benzene is a classic example of this behavior. In benzene, the carbons of the six-membered ring are sp²-hybridized, so each has a singly filled 2p orbital to interact with the others of the conjugated system. The six 2p orbitals interact, giving rise to six new MOs: three bonding MOs and three antibonding MOs. Because of the symmetry properties of the six-membered ring, the six MOs are distributed energetically as shown in Figure 1.21. The six available π electrons completely fill the bonding levels, leading to an enhanced stability of the π system, termed *aromatic stabilization*, or *aromaticity*.

This "extra" stability of benzene and other aromatic compounds is a well-known phenomenon. In fact, aromaticity was first described as chemical stability (unreactivity) toward

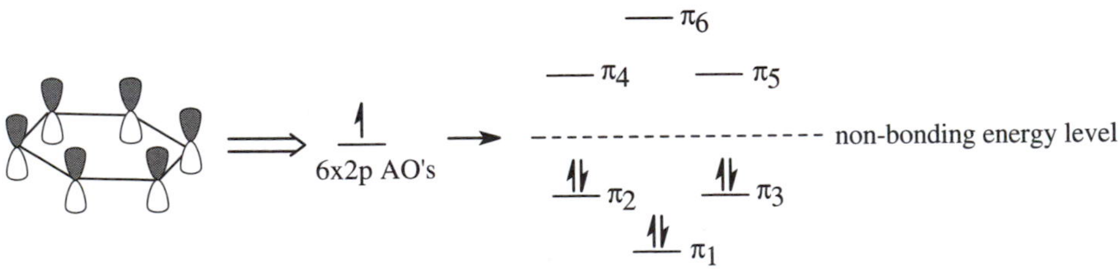

FIGURE 1.21

reagents that normally attacked double bonds and π systems. Moreover, it was known that reagents which attack the aromatic ring give substitution products in which the aromatic ring is retained; the same reagents usually give addition products with typical double bonds and conjugated π systems. Since stability refers to energy level, aromaticity was later defined as the energy difference between an aromatic π system and a model π system in which there is no aromatic stabilization. The aromatic stabilization of benzene was taken as the difference between the heat of hydrogenation of benzene ($\Delta H_{hyd} = -49.8$ kcal/mol) and the heat of hydrogenation of the hypothetical molecule cyclohexatriene ($\Delta H_{hyd} = -85.8$ kcal/mol), which has three noninteracting double bonds in a six-membered ring. (The heat of hydrogenation of cyclohexatriene was estimated as being three times the heat of hydrogenation of cyclohexene.) Since both benzene and cyclohexatriene give cyclohexane upon hydrogenation, a difference in the heats of hydrogenation must be due to a difference in the energies of the starting materials (see Figure 1.22). This difference, which amounts to 36 kcal/mole, is termed the *resonance energy* (R.E.) of benzene; it corresponds to the extra stability of benzene due to aromatic stabilization. The same approach can be used to estimate the resonance energy of other aromatic molecules.

Benzene and the hypothetical model compound are physically distinct in that benzene has equal bond lengths and bond angles and is planar, whereas the hypothetical model would have localized bonds and unequal bond lengths (double bonds are shorter than single bonds). Thus the resonance energy determination is only as good as the model system that is used.

Aromaticity was found to be a general property of many (but not all) cyclic, conjugated π systems. Moreover, it was found that aromaticity in molecules can be predicted by Huckel's rule. The structural requirements implicit in Huckel's rule are that there be $4n + 2$ (n an integer) π electrons in a cyclic, conjugated π system (see Figure 1.23). Obviously, benzene, which has six π electrons ($4n + 2$, $n = 1$) in a conjugated π system, is aromatic. Huckel's rule also predicts that molecules such as cyclodecapentaene, with $4n + 2 = 10$ ($n = 2$), and [18]-annulene, $4n + 2 = 18$ ($n = 4$), should be aromatic, have equal bond lengths, and be planar—and they are. On the other hand, cyclobutadiene and cyclooctatetraene, which do not have $4n + 2$ π electrons, are not aromatic. In fact, these molecules, which contain $4n$ π electrons, are *less* stable than the planar model compounds and are termed *antiaromatic*. Both of these molecules adopt shapes that *minimize* interactions of the π orbitals.

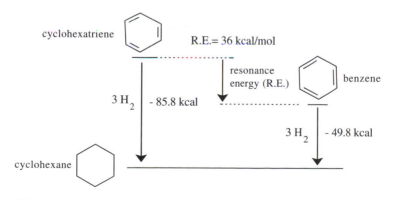

cyclohexatriene   R.E.= 36 kcal/mol

resonance energy (R.E.)   benzene

$3 H_2$ - 85.8 kcal

$3 H_2$ - 49.8 kcal

cyclohexane

FIGURE 1.22

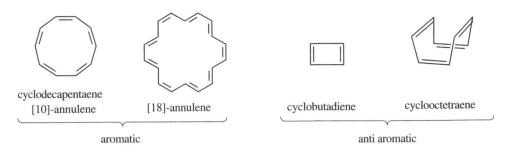

cyclodecapentaene
[10]-annulene

[18]-annulene

cyclobutadiene

cyclooctetraene

aromatic

anti aromatic

FIGURE 1.23

Cyclobutadiene is an antiaromatic, $4n = 4$ ($n = 1$), system, and it is quite unstable and can be observed only at very low temperatures. Although it must be planar (accounting for its instability), it distorts to a rectangular geometry with unequal bond lengths to minimize π bond interactions. Cyclooctatetraene is an antiaromatic, $4n = 8$ ($n = 2$), system that adopts a boat shape so that the π bonds are orthogonal and cannot interact.

Huckel's rule is more than an operational way to identify aromatic molecules. Its origins are in molecular orbital theory and its applicability is general, regardless of ring size or charge. In terms of Huckel's rule, the requirement for aromatic stabilization is that there be a cyclic system with all atoms having a p orbital available for interaction. The array of MOs produced from this interaction is populated by the total number of electrons present in the interacting p orbitals. If that number of electrons is $4n + 2$, the molecule will have aromatic stabilization. It turns out that the preceding requirements lead to a situation in which the bonding MOs are completely filled, the nonbonding orbitals are either completely filled or completely empty, and antibonding levels are unfilled.

As noted in the MO description of benzene, there are three bonding MOs that are filled by the six electrons of the π system. As another example, the tropilium ion (Figure 1.24) is known to be aromatic. The interaction of seven 2p orbitals leads to an MO array with three bonding MOs and four antibonding MOs. The six electrons fill the bonding MOs and give an aromatic system, irrespective of the fact that, to do so, one of the seven interacting p orbitals must be unfilled, leading to a net positive charge on the delocalized aromatic ion.

It is also clear why cyclooctatetraene is not aromatic. The interaction of eight contiguous 2p atomic orbitals in a planar ring gives rise to an MO array which has three occupied bonding MOs and two nonbonding MOs which are degenerate and thus singly occupied (Figure 1.25). Since this is an unstable bonding situation, the molecule distorts to the shape of a boat so that interactions are avoided and four isolated π bonds can form. It is clear that by either removing two electrons (resulting in 6 π electrons) or adding two electrons (resulting in 10 π electrons),

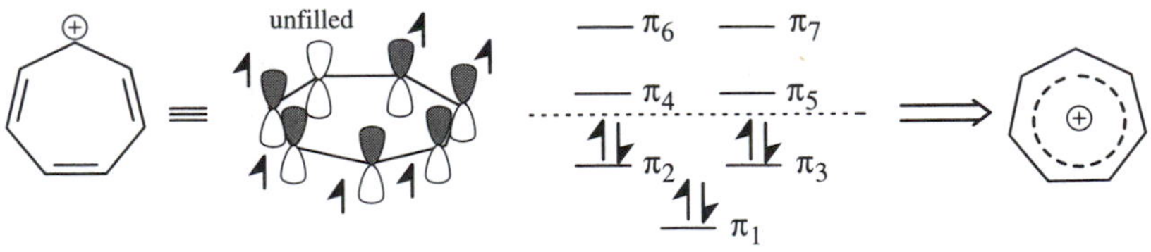

FIGURE 1.24

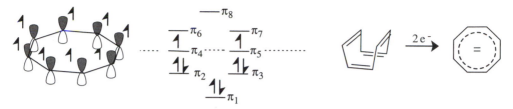

FIGURE 1.25

one could reach an aromatic system. It turns out that cyclooctatetraene is easily reduced by the addition of two electrons, which fill the nonbonding MOs and give a planar, aromatic dianion. Examples of simple aromatic molecules and ions which have been studied are shown in Figure 1.26.

Other elements can also participate in the formation of aromatic species. Furan, pyrrole, and thiophene are all aromatic molecules because, if the heteroatom is sp^2-hybridized, a doubly occupied p orbital interacts with the carbon 2p orbitals to give an MO array which contains 6 π electrons and is aromatic (Figure 1.27). Note that in the development of the MO diagram for these systems, the identity of the heteroatom is not important; it is important only in determining the magnitude of the aromatic stabilization.

The added stability of an aromatic system is a significant energetic feature of molecules. Reactions which occur with the formation of an aromatic system are generally facile, whereas reactions in which an aromatic system is disrupted are generally very difficult. Thus aromaticity can dramatically influence the reactivity of compounds and should be kept in mind.

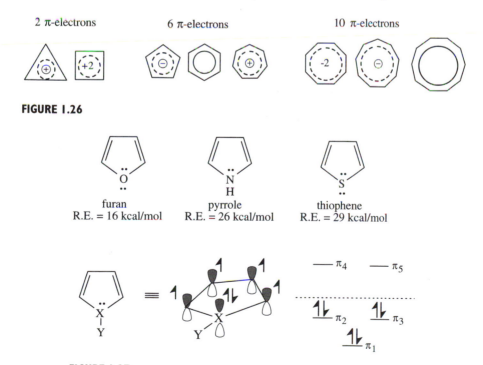

2 π-electrons   6 π-electrons   10 π-electrons

FIGURE 1.26

furan
R.E. = 16 kcal/mol

pyrrole
R.E. = 26 kcal/mol

thiophene
R.E. = 29 kcal/mol

FIGURE 1.27

Bibliography

A very nice pictorial presentation and discussion of overlap and bonding is found in Chapters 1 and 2 in L. G. Wade, *Organic Chemistry*, 2d ed., Prentice Hall, Englewood Cliffs, NJ, 1991.

A discussion of bonding for each of the functional groups when each is first introduced is found in M. A. Fox and J. K. Whitesell, *Organic Chemistry*, Jones & Bartlett, Boston, 1994.

An excellent presentation of overlap and bonding is found in Chapter 1 of P. H. Lowery and K. S. Richardson, *Mechanism and Theory in Organic Chemistry*, 3d ed., Harper & Row, New York, 1987.

An advanced discussion of bonding theory is found in H. E. Zimmerman, *Quantum Mechanics for Organic Chemists*, Academic Press, New York, 1975.

Problems

1. Excluding alkyl groups, name and point out the functional groups in the following molecules:

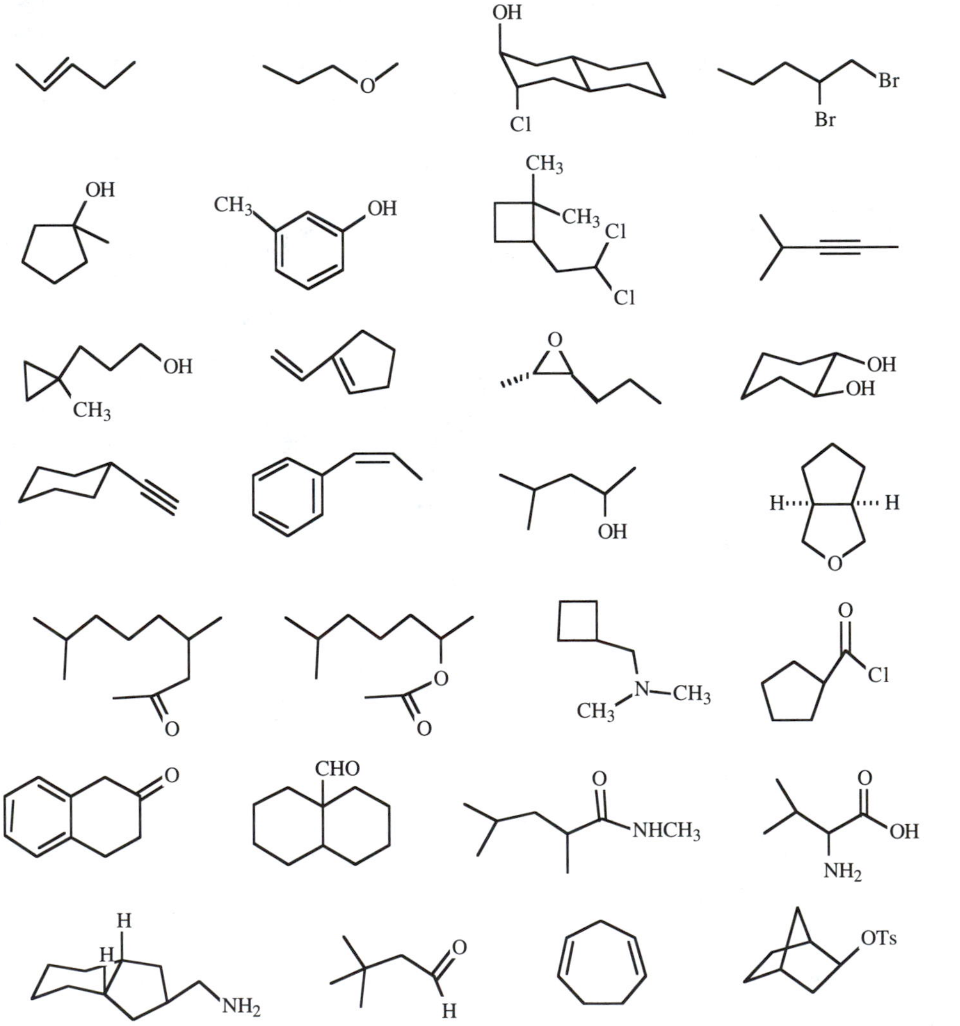

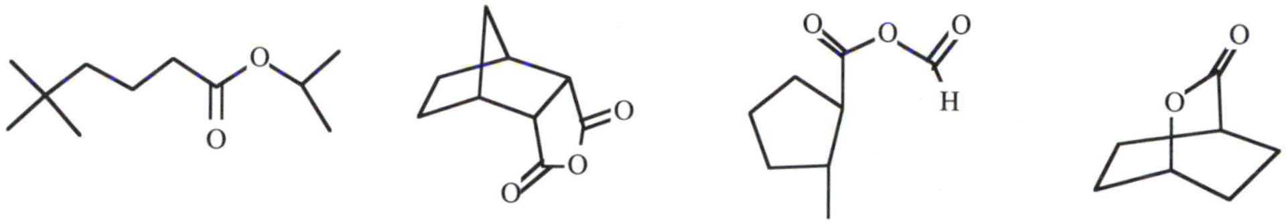

2. Give the bonding scheme (orbitals, etc.) and geometry for each of the following functional groups:

(**a**) alkyl nitrile (use R for alkyl group)

(**b**) alkyl azide (use R for alkyl group)

(**c**) nitro alkane (use R for alkyl group)

(**d**) *N*-methyl pyrrole (it is aromatic)

3. For the following compounds, give the approximate bond angles around the atoms indicated by an arrow.

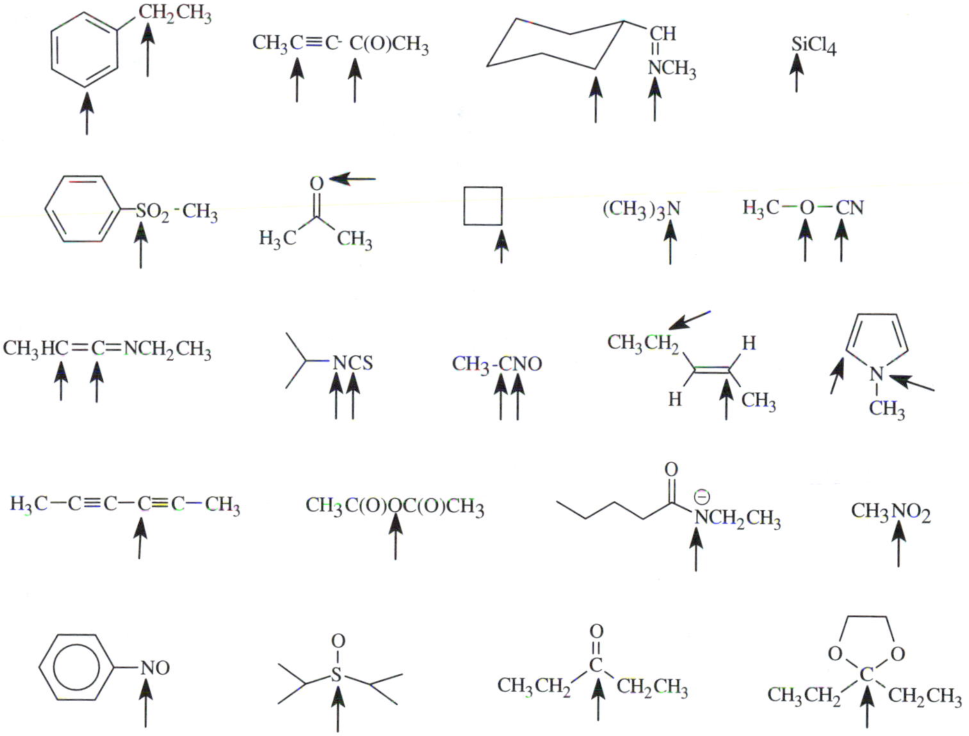

4. For the following compounds, add all lone pairs of electrons to the structures and then specify the type of orbital in which they are located:

CH_3OCH_3 CH_3CHO CH_3O^- $CH_3C{\equiv}O^+$

$CH_3CH{=}NCH_3$ $CH_3CH_2NH_2$ $CH_2{=}N^-$ CH_3CN

CH_3F CH_3Cl CH_3Br

CH_3CNO CH_3NC CH_3SCN CH_3NCS

5. On the basis of electronic structure and orbital energies, supply predictions for the following and explain your answer.
(*a*) Which will be more nucleophilic toward methyl iodide?

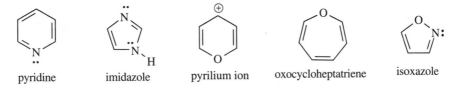

or

(*b*) Which will be more basic?

$CH_3CH_2-\overset{..}{\underset{..}{O}}H$ or   [structure: H_3C with carbonyl $\overset{..}{\underset{..}{O}}$ and chain]

(*c*) Which anion will be more stable?

$CH_3CH{=}CH^{\ominus}$ or $CH_3C{\equiv}C^{\ominus}$

6. Which of the following compounds or ions are aromatic? Draw orbital diagrams to demonstrate why.

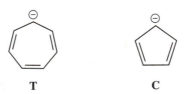

pyridine   imidazole   pyrilium ion   oxocycloheptatriene   isoxazole

7. Consider the tropanyl anion **T** and the cyclopentadienyl anion **C**. Which one is more stable and why? Predict the structure of each based on your analysis.

T   **C**

OXIDATION STATES OF ORGANIC

COMPOUNDS

Oxidation Levels

Besides bonding patterns, functional groups also vary with respect to the oxidation states of carbon in those functional groups. Thus another way to classify functional groups is by the carbon oxidation level. Correspondingly, organic reactions can be categorized as to whether an oxidation, a reduction, or no change in oxidation level occurs in the organic reactants as they become products. This is a very useful distinction because the reagents used in a given transformation must be compatible with the oxidation change that occurs in the reaction. It is important to remember this fundamental truth: no oxidation can occur without a corresponding reduction, and no reduction can occur without a corresponding oxidation. As a consequence, if a transformation of an organic compound involves a change in its oxidation level, the reagents necessary to cause that change must be able to undergo the complementary change in oxidation level. For an oxidation to take place, an oxidizing agent, which gets reduced in the process, is required. Similarly, a reducing agent that gets oxidized is needed for the reduction of an organic compound. Reagents can thus be categorized on the basis of their oxidizing or reducing properties. If no change in oxidation state occurs during a chemical reaction, reagents used to effect the transformation should undergo no oxidation or reduction. Moreover, if a reagent is not normally an oxidizing agent, it is not easily reduced and cannot be used to oxidize something else. Conversely, if a reagent is not normally a reducing agent, it is not easily oxidized and cannot be used to reduce something else.

Oxidation is defined as the loss of electrons. This concept is very straightforward for metal ions. Thus the change $Mg \rightarrow Mg^{2+}$ is an oxidation because magnesium has lost two electrons in going from the element to the positive ion. Similarly, oxidation of Cu^+ involves loss of an electron from Cu^+ to give the Cu^{2+} species.

Reduction is defined as the gain of electrons. The conversion of Ag^+ to Ag involves a gain of an electron by the silver ion, so the silver is reduced. Likewise the permanganate ion MnO_4^- has manganese [VII], but MnO_2 manganese dioxide has manganese [IV]. Thus the gain of three electrons by manganese causes a reduction from the oxidation level of +7 to +4.

Because organic compounds have an overwhelming preponderance of covalent bonds, changes in the oxidation state of carbon are not so easily determined by inspection as they are for metal ions. The definition of oxidation and reduction for organic compounds is the same as it is for metal ions (i.e., gain or loss of electrons), but the oxidation state of a carbon atom is determined by the types of covalent bonds originating from it. Rules have been developed that assign numerical values for the contributions of atoms covalently bonded to a particular carbon. Summation of the contributions of its covalently bonded substituents gives the oxidation state of that carbon in a molecule. The oxidation levels of various carbons can then be compared just as +2 or +3 oxidation states in metal ions can be compared.

These rules are simple and are summarized as follows:

1. Bonds to hydrogen or other elements more electropositive than carbon contribute −1 to the oxidation level.

2. Bonds to other carbon atoms contribute 0 to the oxidation level.

3. Bonds to oxygen or other elements more electronegative than carbon contribute +1 to the oxidation level.

4. Multiple bonds to an element count as multiple single bonds to that element. That is, the carbon–oxygen double bond of the carbonyl group (C=O) is oxidatively equivalent to a carbon atom with two single bonds to oxygen (−O−C−O−).

5. A pair of electrons on carbon contribute −1 to the oxidation level.

6. A positive charge on carbon contributes +1 to the oxidation level.

Given these contributions, the oxidation level of a given carbon can be determined by adding together the contributions of the four attached bonds.

Oxidation States in Alkanes

Considering the various alkanes shown in Figure 2.1, one sees that carbon atoms in alkanes can have several different oxidation levels. Oxidation levels can range from −4 for methane and −3 for the carbon atom of methyl groups all the way to 0 for the quaternary carbon of

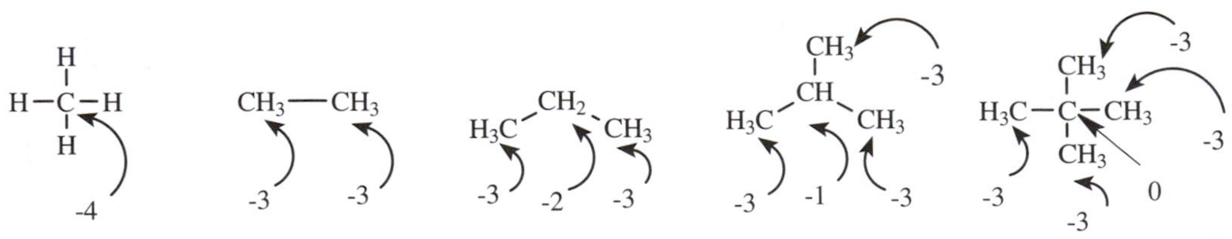

FIGURE 2.1

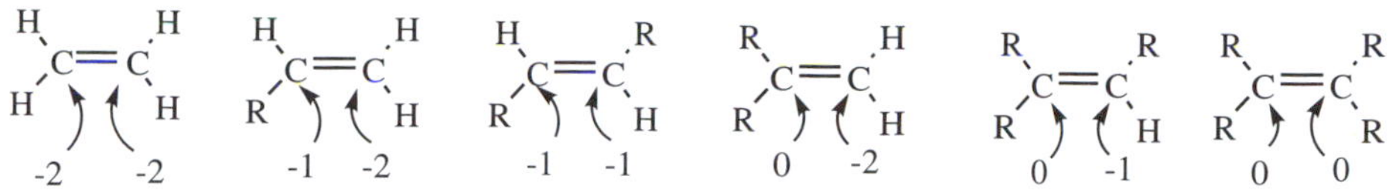

$$\begin{array}{cccccc}
\underset{-2\quad-2}{\overset{H\quad H}{\text{C}=\text{C}}} & \underset{-1\quad-2}{\overset{H\quad H}{\text{C}=\text{C}}} & \underset{-1\quad-1}{\overset{H\quad R}{\text{C}=\text{C}}} & \underset{0\quad-2}{\overset{R\quad H}{\text{C}=\text{C}}} & \underset{0\quad-1}{\overset{R\quad R}{\text{C}=\text{C}}} & \underset{0\quad0}{\overset{R\quad R}{\text{C}=\text{C}}}
\end{array}$$

FIGURE 2.2

neopentane. In spite of the several oxidation levels possible in alkanes, the functional group approach tells us that all are saturated alkanes and thus have the same functional equivalency and similar reactivity patterns. This conclusion is derived from the fact that all the carbons in alkanes have four single bonds originating from them and those σ bonds go to either carbon or hydrogen. Thus one cannot automatically assign a molecule to a functional class based solely on a certain oxidation level of the carbons that it contains.

Oxidation States in Alkenes

For alkenes, several carbon oxidation levels are again possible. Furthermore, *both* carbon atoms must be considered as part of the same alkene functional group. Although the total oxidation level can go from -4 for ethylene (as the sum of the oxidation level of both carbon atoms in the functional group) to 0 for a tetrasubstituted alkene (see Figure 2.2), all are of the same functional class. Furthermore, it is evident that, because the lowest possible oxidation level of a single carbon atom in an alkene is -2 while the lowest possible oxidation level of a carbon atom in an alkane is -4, alkenes are oxidized relative to alkanes.

Oxidation States in Other Common Functional Groups

The same process can be carried out to determine the oxidation levels of carbon atoms in several common functional types (Table 2.1). It is clear that by using these procedures, one can assign oxidation levels to carbon atoms in a wide variety of compounds. It is also clear

TABLE 2.1 Oxidation States by Functional Groups

Alkynes	Alcohols	Aldehydes and Ketones	Acids and Derivatives
$H-C{\equiv}C-H$ -1 -1	$CH_3CH_2\text{-}OH$ -1	$\underset{0}{\overset{O}{\underset{H}{\text{C}}}}_{H}$	$\underset{+2}{\overset{O}{\underset{H}{\text{C}}}}_{OH}$ (OR, NH_2, etc)
$R-C{\equiv}C-H$ 0 -1	$(CH_3)_2CH-OH$ 0	$\underset{+1}{\overset{O}{\underset{R}{\text{C}}}}_{H}$	$\underset{+3}{\overset{O}{\underset{R}{\text{C}}}}_{OH}$ (OR, NH_2, etc)
$R-C{\equiv}C-R$ 0 0	$(CH_3)_3C-OH$ $+1$	$\underset{+2}{\overset{O}{\underset{R}{\text{C}}}}_{R}$	

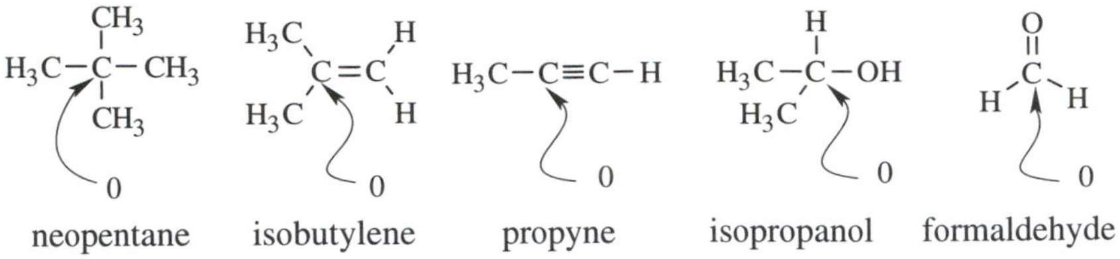

neopentane   isobutylene   propyne   isopropanol   formaldehyde

FIGURE 2.3

that knowing the oxidation level is insufficient for assigning the functional group present. For example, the alkane neopentane, the alkene isobutylene, the alkyne propyne, the alcohol isopropanol, and formaldehyde all have a carbon with an oxidation level of 0, yet all belong to completely different functional classes and have different physical and chemical characteristics (see Figure 2.3). Thus the oxidation level of a given carbon is dependent only on the groups attached to it, not on the functional group to which it belongs.

Oxidation Level Changes During Reactions

Comparing the oxidation levels of various carbon atoms is excellent for illustrating what oxidation state change must occur at a particular carbon in a given reaction of that compound. For example,

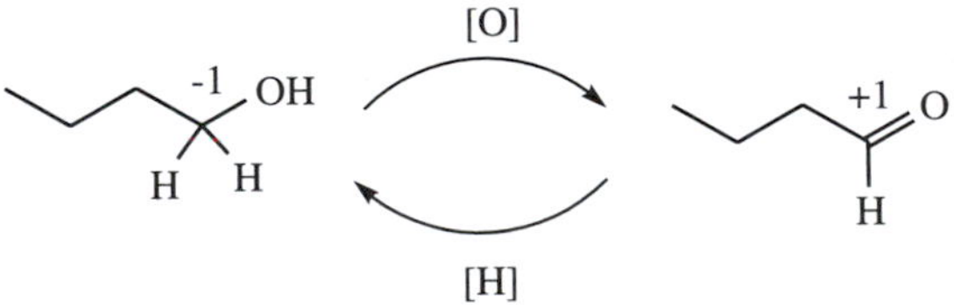

The oxidation level of the primary alcohol (-1) is less than the aldehyde product ($+1$), so this conversion requires an oxidation of the alcohol function to the aldehyde. Any reagent capable of effecting this change must necessarily be an oxidizing agent that will be reduced. The need for an oxidant is noted above the arrow by an [O]. The reverse process, conversion of an aldehyde to a 1° alcohol, is a reduction. Any reagent capable of effecting this change must be a reducing agent. A reduction is commonly indicated by a bracketed [H].

By the same analysis, conversion of an aldehyde to an acetal involves neither oxidation nor reduction. As a consequence no oxidant or reductant is necessary to carry out this reaction:

Similarly, for the process shown below, the oxidation state of the carbon marked by -1 remains the same throughout the sequence; thus the overall sequence involves no change in oxidation level at that carbon, nor does either step.

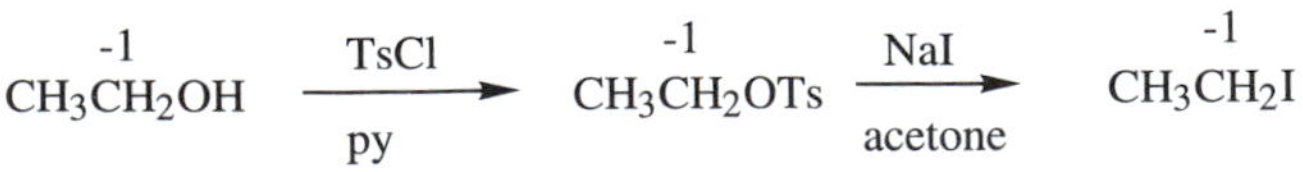

Modifications of substituents or substitution of one electronegative group for another are generally not redox processes.

One often must consider the balanced reaction in order to be certain of any net changes in oxidation state, and similar procedures for determining the oxidation level can be followed for other covalently bound elements. For example, the conversion of methane into ethane is an oxidation of the carbon atoms since the carbons in methane are at the -4 level whereas in ethane they are at the -3 oxidation state.

$$2 \ CH_4 \longrightarrow CH_3CH_3 \ + \ H\text{-}H$$

$$2 \ x \ \text{-}4 = \text{-}8 \qquad\qquad \text{-}3 \ \ \text{-}3 \ + \ \text{-}1 \ \text{-}1 = \text{-}8$$

However, if hydrogen is also considered, the change from being bound to carbon (0) in the reactant to being bound to another hydrogen (-1) in the product means that hydrogen is formally reduced. Thus the sum of oxidation levels in the reactants (-8) is the same as that in the products (-8) and the overall process is neither an oxidation nor reduction. This transformation can be thought of as an internal redox process since part of the reactant (carbon) is oxidized and part (hydrogen) is reduced. Generally such internal redox processes require only a catalyst, not an oxidant or reductant.

On the other hand, if the by-product of the conversion of methane to ethane formation is H^+, the balanced reaction is written as

$$2 \ CH_4 \longrightarrow CH_3CH_3 \ + \ 2 \ H^+$$

$$2 \ x \ \text{-}4 = \text{-}8 \qquad\qquad \text{-}3 \ \ \text{-}3 \ + \ (\ 2 \ x \ \text{+}1) = \text{-}4$$

and a net oxidation is required. An oxidizing agent is thus needed to effect this process. Again, the recognition that the organic reactant (methane) and product (ethane) are both alkanes is not sufficient to determine that an oxidant is necessary.

The Grignard reaction is often one of the first reactions encountered for the preparation of organometallic compounds. As such, it provides a method for the conversion of an alkyl bromide to an alkane.

$$(CH_3)_2CH\text{—}Br \ \xrightarrow[\text{ether}]{Mg} \ (CH_3)_2CH\text{—}Mg\text{-}Br \ \xrightarrow{H_2O} \ (CH_3)_2CH_2 \ + \ HOMgBr$$

$$\phantom{(CH_3)_2CH\text{—}Br \ } 0 \qquad\qquad\qquad \text{-}2 \qquad\qquad\qquad \text{-}2$$

From this example, one can see that the overall change from the organic reactants to the products is from 0 to -2; a reduction has occurred. Magnesium is the reductant and is itself oxidized from 0 to $+2$ oxidation state. The actual reduction takes place in the first step of the process, in which the C—Br bond is converted to a C—Mg—Br bond. The reaction with water is merely a hydrolysis that does not change the oxidation state of carbon.

Reactions of olefins and acetylenes illustrate that the overall change in oxidation level of an organic functional group must be considered when deciding if an overall oxidation level change has occurred in a chemical reaction. For example, addition of hydrogen across an acetylene gives a net reduction of each carbon and thus is a reductive process with respect to the alkyne; the same is true for the hydrogenation of an alkene:

$$R\text{—}C\equiv C\text{—}R \ \xrightarrow[\text{P-2 Ni}]{H_2} \ \underset{R \quad\quad R}{\overset{H \quad\quad H}{C{=}C}} \ \xrightarrow[\text{Pd/C}]{H_2} \ R\text{—}\underset{\underset{H}{|}}{\overset{\overset{H}{|}}{C}}\text{—}\underset{\underset{H}{|}}{\overset{\overset{H}{|}}{C}}\text{—}R$$

$$\phantom{R\text{—}C\equiv C\text{—}R} 0 \ \ 0 \qquad\qquad \text{-}1 \ \ \text{-}1 \qquad\qquad\qquad \text{-}2 \ \ \text{-}2$$

From the point of view of the alkyne and the alkene, the hydrogen can be considered a reducing agent since it undergoes oxidation during the process. The conversion of an alkyne to a *trans*-alkene can be accomplished by heating with LAH or with Li, NH$_3$. Thus H$_2$/P- 2Ni, LAH, and Li, NH$_3$ are reducing agents for alkynes and give alkenes as the reduced products.

In general, any reaction which results in the addition of two hydrogen atoms across a π bond of any type is a reduction. Conversions of aldehydes and ketones to alcohols are reductions; any reagents which are capable of effecting that conversion must therefore function as reducing agents. NaBH$_4$, LAH, and a large variety of other reagents *reduce* aldehydes and ketones to alcohols by the net addition of hydrogen across the C—O π bond. By the same logic, conversion of a primary alcohol to an aldehyde (the reverse process) must be oxidation, and reagents which are capable of effecting this conversion, such as DMSO and acetic anhydride (Swern oxidation) or pyridinium chlorochromate (PCC), are oxidants. Similar considerations hold for other π-bonded functional groups, including acid derivatives and nitriles.

Alkenes also undergo a variety of other addition reactions in which a reagent is added across the double bond. Hydration and hydrohalogenation are classic examples:

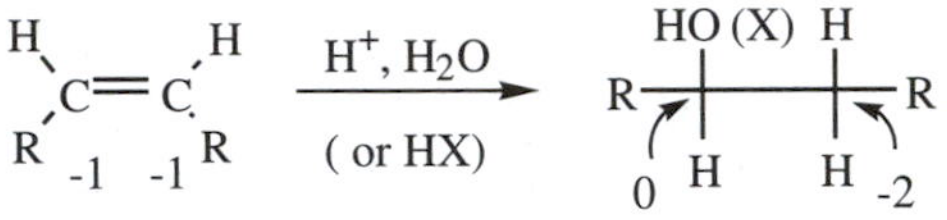

Consideration of the oxidation level reveals that one carbon is reduced (the one to which hydrogen adds) and the other is oxidized (the one to which the oxygen adds). There is no net change in oxidation level of the alkene functional group. Likewise, the reverse processes of these addition reactions, namely, elimination of HX from alkyl halides and dehyration of alcohols to give alkenes, are not redox processes. Addition of water to alkynes is analogous. In this case, the product is a ketone, but the oxidation level of the ketone is the same as the alkyne, so no net change in oxidation level has occurred.

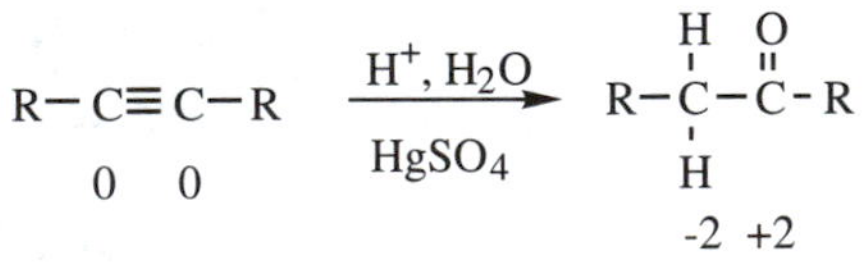

The conversion of alkenes to 1,2-diols by osmium tetroxide is also an olefin addition reaction. In this case a hydroxy group is added to each carbon of the olefin group, and the addition is termed an oxidative addition since the diol product is at a higher oxidation level than the alkene reactant. Oxidation of the carbon atoms of the alkene takes place in the first step, which is reaction with OsO$_4$, to produce the intermediate osmate ester. Zinc serves to further reduce osmium and free the diol product.

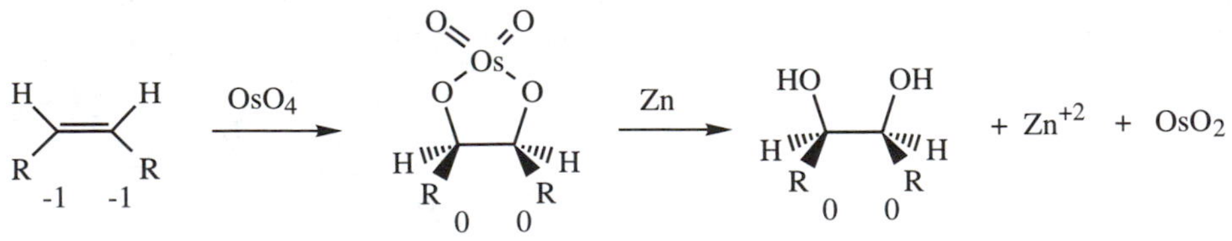

Similar oxidative additions to alkenes occur with bromine, chlorine, IN_3, peracids, and many other electrophiles:

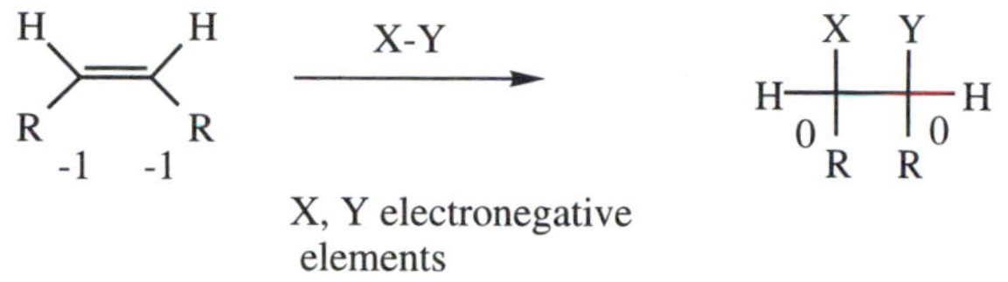

X, Y electronegative
elements

Peracids such as *m*-chloroperbenzoic acid (MCPBA) clearly illustrate the redox nature of oxidative addition. In this reaction the olefin is oxidized and the *meta*-chloroperbenzoic acid is reduced to *meta*-chlorobenzoic acid, which precipitates slowly from solution:

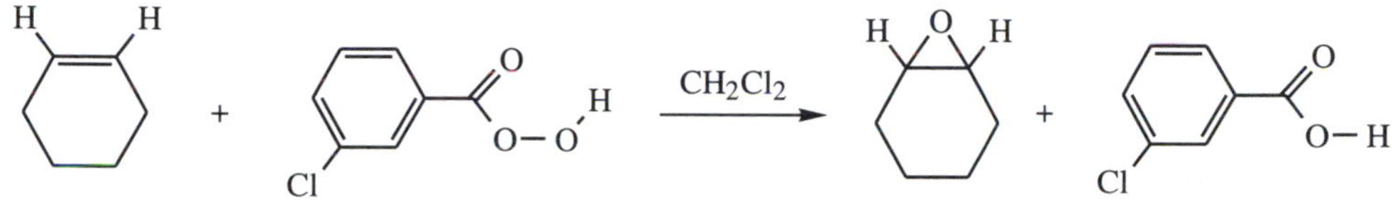

Another common reaction process is one in which one atom or group replaces another atom or group. These are known as *substitution reactions*. When one electronegative group is substituted for another, no change in oxidation level occurs; thus the reagents which carry out such substitutions are neither oxidants nor reductants.

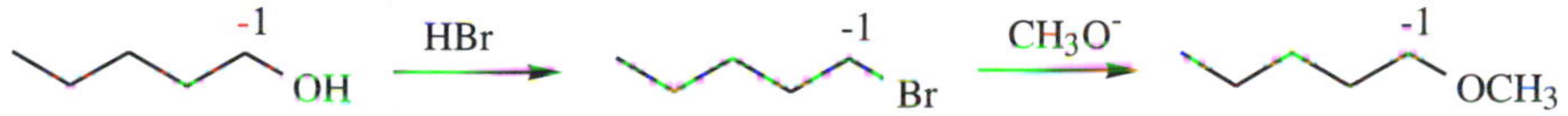

Such substitutions in saturated compounds can be carried out by a variety of strategies involving different nucleophiles and leaving groups, but the oxidation states remain the same. Acyl substitutions are analogous. For this reason carboxylic acid derivatives are treated as a common family of compounds. All have the same oxidation level, and all can be converted from one to another by substitution reactions not requiring oxidation or reduction.

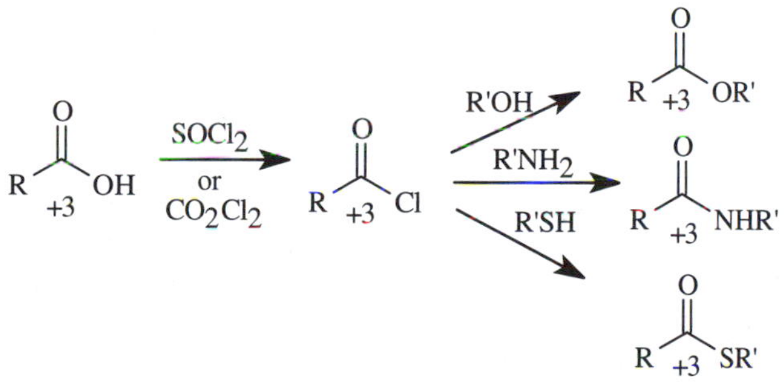

Many useful functional group transformations occur in more than one step, and it is not uncommon to find different redox processes in different steps of the process. From the methods of determining oxidation states, however, it is clear that substitution for an electronegative group by a carbon group or a hydrogen atom is a reduction and requires a reducing agent.

For example, conversion of an acid chloride to a ketone by a lithium organo cuprate reagent involves a reduction of the acid chloride to the ketone oxidation level:

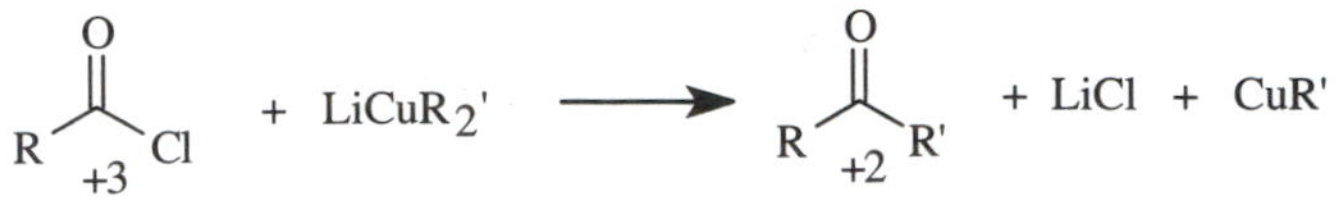

$$
\underset{\underset{+3}{R}}{\overset{O}{\underset{\|}{C}}}\text{Cl} \quad + \text{ LiCuR}_2' \quad \longrightarrow \quad \underset{\underset{+2}{R}}{\overset{O}{\underset{\|}{C}}}\text{R'} \quad + \text{ LiCl } + \text{ CuR'}
$$

Consequently, the copper is oxidized from a cuprate species to an organocopper. By classifying organocuprates as reducing agents toward acid chlorides, one expects that they could act as reducing agents toward other functional groups. It is therefore not surprising that their use as Michael addition reagents can be used to give net reduction of an α,β-unsaturated ketone:

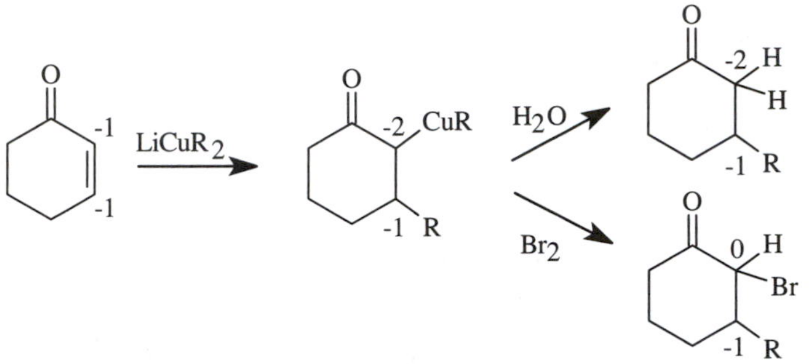

Reaction of the organocuprate intermediate with water gives the fully reduced product. If the organocuprate intermediate is reacted with bromine, the α-brominated product is formed. This product has the equivalent oxidation level as the starting enone but differs in that an additional carbon substituent is present. Functionally, this is equivalent to the addition of HBr to an enone, so no net redox has taken place.

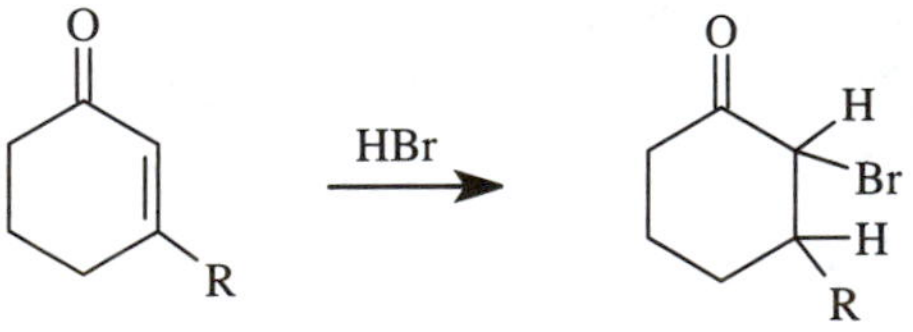

If individual steps are considered, it is clear that the first step (addition of the organo cuprate to the enone) is a reduction and the second step (reaction of the cuprate with bromine) is an oxidation. No net change in the oxidation level has occurred for the overall process, but each step in the sequence can involve an oxidation or reduction. This is an important idea to keep in mind: even though no net change in the oxidation level occurs, individual steps in the sequence may have an oxidation or reduction and thus require oxidants or reductants consistent with the individual step being undertaken.

The realization that many reactions or steps in reactions involve an oxidation or reduction is an important consideration when these reactions are being studied and learned. The change in oxidation level produced is indicative of the transformation and provides an additional organizational category by which reactions can be classified. Reagents can also be classified by their ability to cause oxidation or reduction. For example, from its addition reactions with alkenes and alkynes, bromine can be considered an oxidizing reagent for organic molecules. It is not surprising, therefore, to find that bromine also serves as an oxidant toward other functional groups such as enols, hydrocarbons, aldehydes, and organometallic compounds. Lithium aluminum hydride is well-known as a reductant; thus if it reacts with an organic compound, some functional group is quite likely being reduced by the addition of a hydride. With the concepts of oxidation and reduction, an often neglected, but extremely important view of organic reactions is possible.

Bibliography

For another discussion see M. A. Fox and J. K. Whitesell, *Organic Chemistry,* Jones and Bartlett, 1994, pp. 74–76.

For a different approach to oxidation level see J. McMurry, *Organic Chemistry,* 3d ed., Brooks/Cole, Pacific Grove, CA, 1992, pp. 633–34.

Problems

1. Give the oxidation level of the indicated carbons in the following compounds.

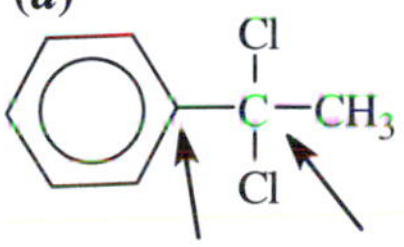

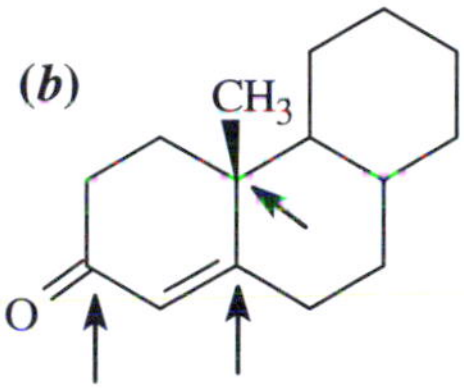

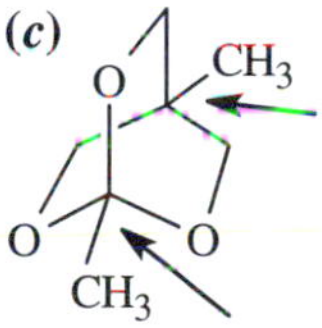

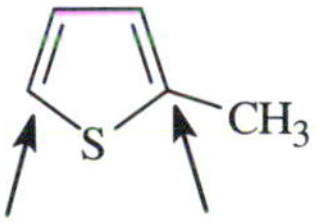

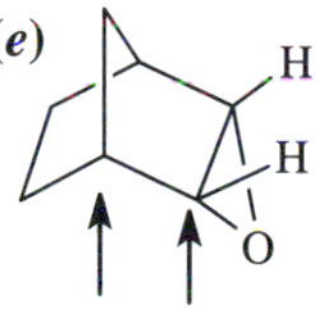

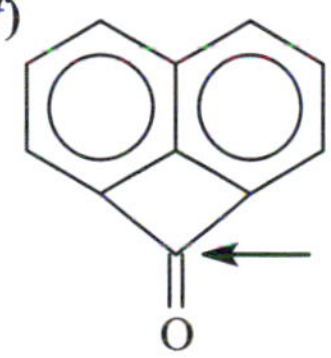

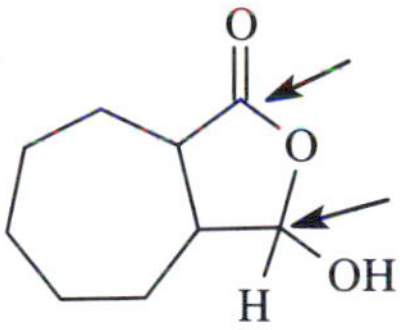

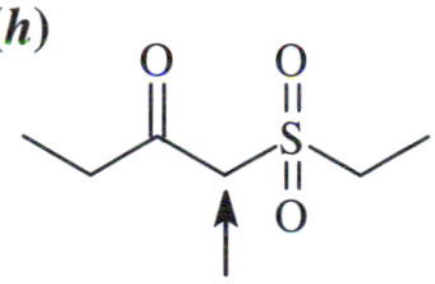

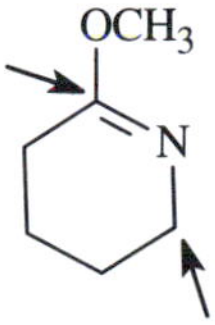

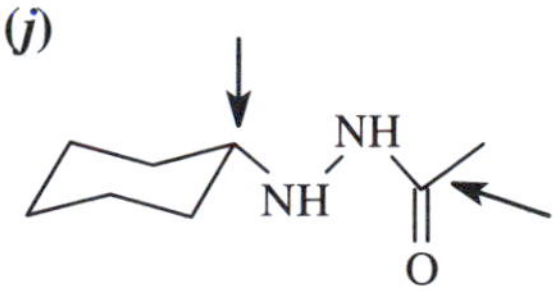

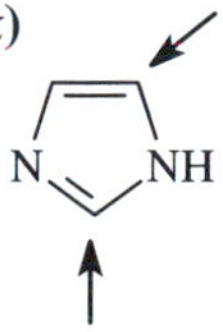

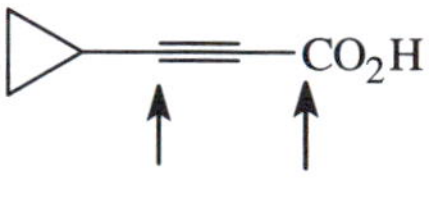

2. For each of the following reactions, (i) write a balanced equation, (ii) determine if an oxidation, reduction, or no change has occurred for the organic substrate(s), (iii) if a redox process has occurred, indicate the oxidized and reduced products, (iv) indicate the reagent responsible for the change in oxidation level of the organic component, and (v) name the functional group in the reactant and indicate what functional group it has been converted to in the product (if possible).

(*a*) $CH_3-\langle\bigcirc\rangle$ + HNO_3 $\xrightarrow{H_2SO_4}$ $CH_3-\langle\bigcirc\rangle-NO_2$

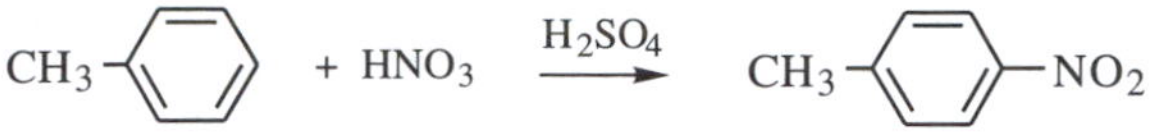

(*b*)

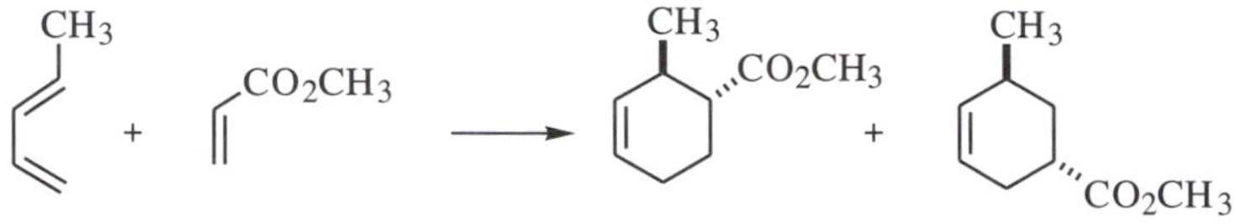

(*c*)

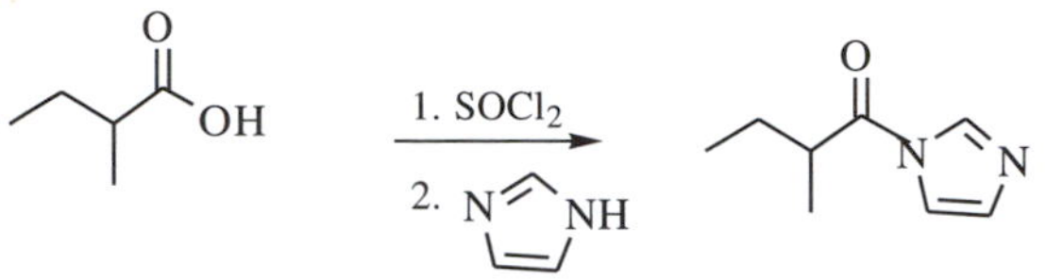

(*d*)

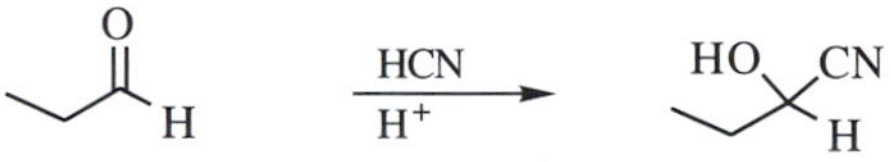

(*e*)

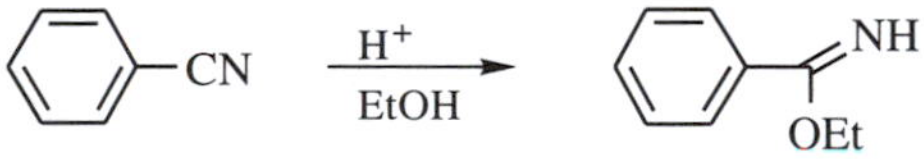

(*f*)

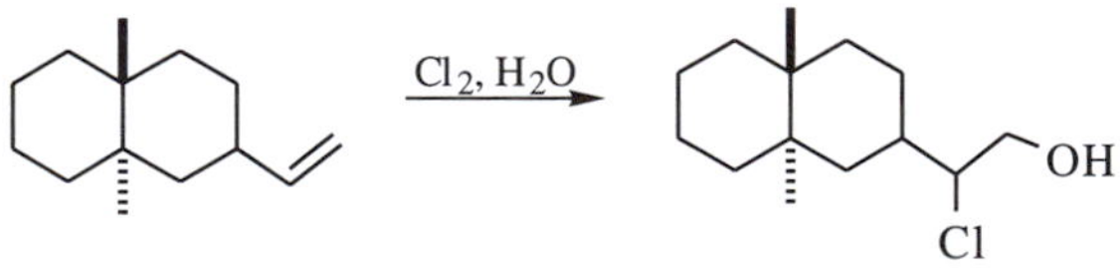

(*g*)

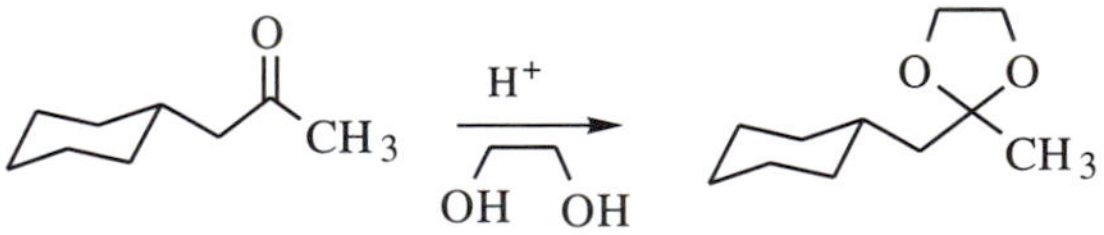

(*h*)

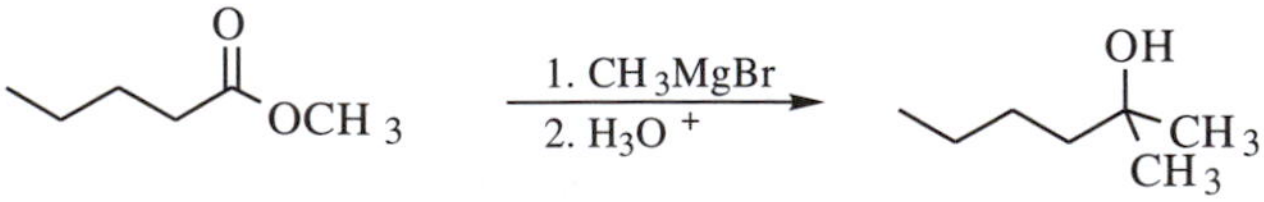

(*i*)

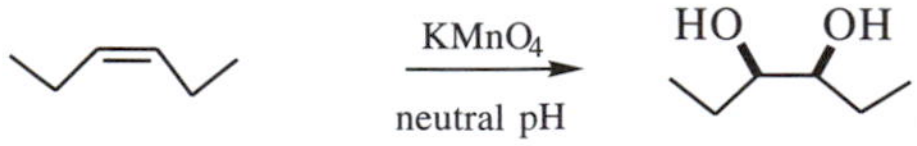

(*j*)

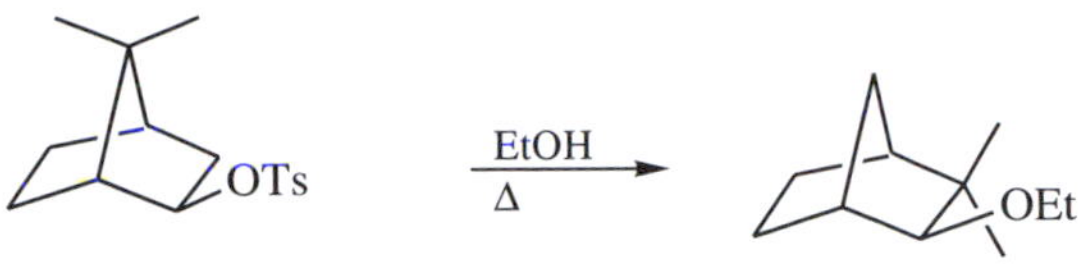
OTs
EtOH
Δ
OEt

(*k*)

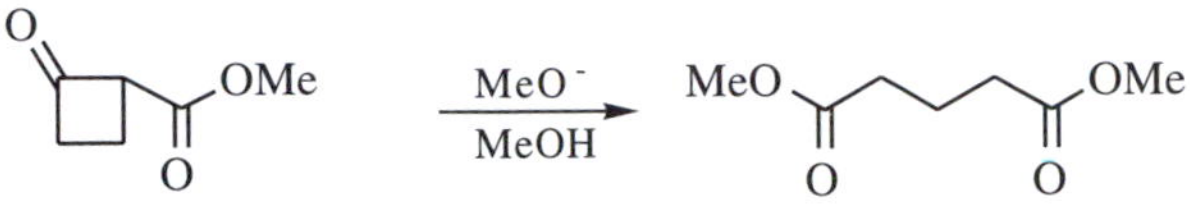
O
OMe
O
MeO⁻
MeOH
MeO
OMe
O
O

(*l*)

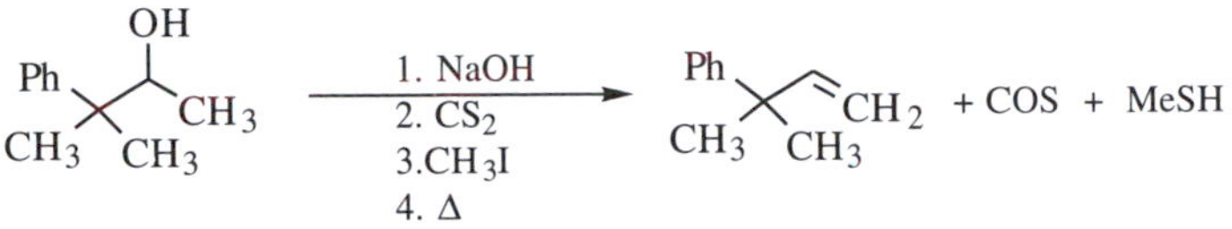
OH
Ph
CH3
CH3 CH3
1. NaOH
2. CS2
3.CH3I
4. Δ
Ph
CH2
CH3 CH3
+ COS + MeSH

(*m*)

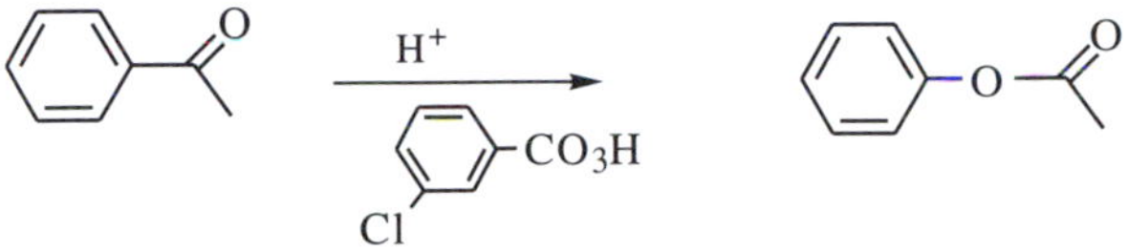
O
H+
CO3H
Cl
O
O

(*n*)

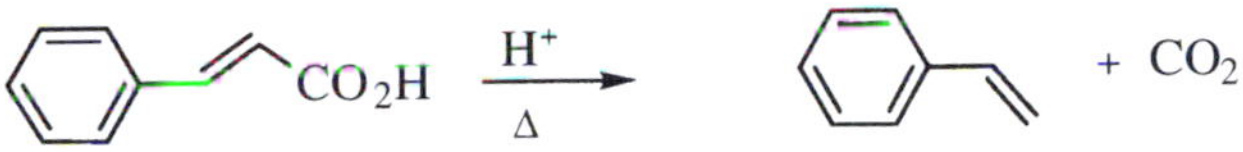
CO2H
H+
Δ
+ CO2

(*o*)

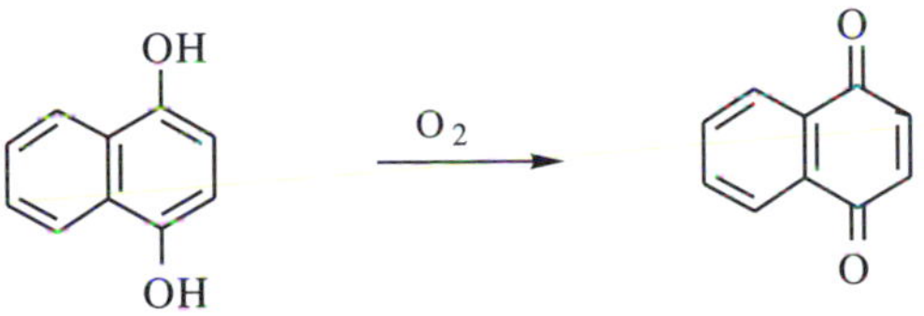
OH
OH
O2
O
O

(*p*)

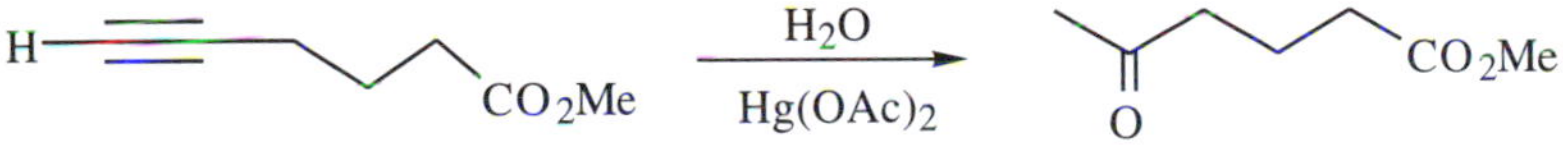
H
CO2Me
H2O
Hg(OAc)2
O
CO2Me

(*q*)

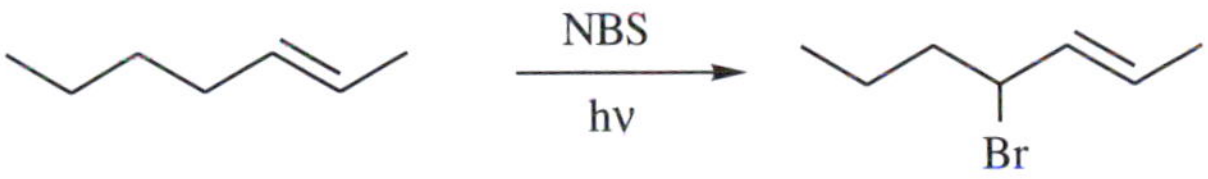
NBS
hν
Br

(*r*)

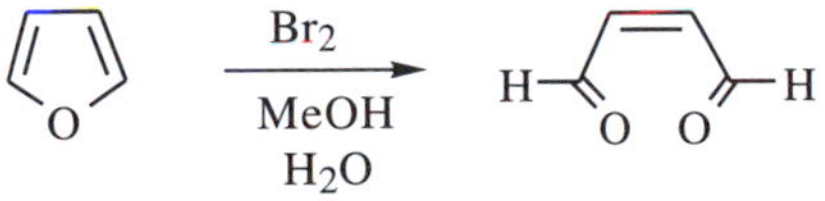
O
Br2
MeOH
H2O
H
O O
H

C H A P T E R

3

ACIDITY AND BASICITY

Bronsted and Lewis Acids and Bases

The notion of acids and bases is one of the first and most important ideas encountered in chemistry, but it is one of the things that often is poorly understood. The concept is actually very simple if a few basic ideas are *always* kept in mind, starting with the definitions of acids and bases and what general structural features make compounds react as acids or bases.

Bronsted acids are defined as proton donors. The dissociation of a Bronsted acid yields a proton *and* the conjugate base (an anion, if the acid is a neutral compound) of the acid.

$$HA \rightarrow A^- + H^+$$

Bronsted bases are proton acceptors. The only way that a Bronsted base can "accept" a proton is to supply an electron pair and form a bond to the proton. Thus Bronsted bases often are compounds with unshared pairs of electrons that can be donated to a proton to form a bond. Upon reaction with a proton, a base is converted to its conjugate acid.

$$B: + H^+ \rightarrow B-H^+$$

So that charge is conserved, anionic bases upon reaction with a proton give neutral conjugate acids; neutral bases upon reaction with a proton give positively charged conjugate acids. Occasionally, shared pairs of electrons can be given up to a proton, such as when olefins react with acids. In such cases, π electrons are the electron pair which bonds to the proton.

The previous two equations describing the behavior of Bronsted acids and bases are not strictly correct because a Bronsted acid does not just dissociate; it donates a proton *to* something which accepts a proton. The proton does not just dissociate and float around in solution but is always attached to something. Furthermore, a Bronsted base does not just find a proton to accept, it accepts a proton from a Bronsted acid. Thus acidity and basicity are paired behaviors—"you can't have one without the other." This is the most common misconception about acids and bases and leads to the greatest amount of difficulty when trying to apply the principles of acidity and basicity to real reactions.

A second description of acidity and basicity is the Lewis definition. A Lewis acid is an electron acceptor and a Lewis base is an electron donor. The definition of Lewis acids includes the proton since the proton can accept electron pairs. Bronsted acids are thus a subset of Lewis acids since all Bronsted acids yield a proton. However, other common Lewis acids (BF_3, $TiCl_4$, $SnCl_4$, $AlCl_3$) are routinely used as organic reagents and all function by accepting unshared or shared electron pairs. In general, the electron pairs that are accepted are readily available (i.e., they are not tightly bound), so they are usually lone pairs or π electrons.

Lewis bases are electron pair donors, and the electrons are given up to Lewis acids (electron acceptors). Unshared electron pairs are the more common type of electron pairs to be donated to Lewis acids although, on occasion, shared pairs can be donated to the Lewis acid. In the case of Lewis acid–base reactions, the product is termed a Lewis acid–base complex.

Given these definitions, it is crucial to emphasize again that no compound is inherently either an acid or a base. A compound functions as an acid (in the Bronsted sense) only if it donates a proton *to* something; that is, there must be a base present (a proton acceptor) to have an acid–base reaction. Thus a compound acts as an acid only in the presence of a base, and a substance can act as a base only in the presence of an acid. A good example of this concept is the fact that HCl in the vapor phase is an undissociated, covalent molecule. This is because there are no molecules present capable of accepting a proton from HCl. As a result, HCl does not function as an acid in the gas phase. If H_2O is added, however, an immediate acid–base reaction takes place in which HCl donates a proton to water, which is capable of accepting a proton from HCl. In this reaction, an unshared pair of electrons on the oxygen atom of water is donated to the proton and an O—H bond is formed. In its reaction with water, HCl can function as an acid only because water can function as a base to accept the proton.

$$HCl \;\xrightleftharpoons{\;\;/\!\!/\;\;}\; H^+ + Cl^- \qquad \text{vapor phase - no reaction!}$$

but

$$HCl + H_2O \;\rightleftharpoons\; H_3O^+ + Cl^-$$

Although the necessity of having *both* an acid and a base present in order to have an acid–base reaction is axiomatic, it is surprising how often this concept is neglected. However, if these conditions are met, a wide variety of organic compounds can donate protons to appropriate bases (they are deprotonated) and a wide variety of compounds can accept protons from appropriate acids (they are protonated).

The same considerations are true for Lewis acids and bases. Pure boron trifluoride does not act as a Lewis acid because there is nothing present capable of donating electrons to it. If diethyl ether is added, boron–trifluoride etherate, a stable Lewis acid–base complex, is produced. Obviously, the electron pairs on the oxygen atom of diethyl ether can be donated to boron trifluoride. When a Lewis base is present, boron trifluoride functions very effectively as a Lewis acid.

$$BF_3 + \overset{..}{:}O(Et)_2 \;\rightleftharpoons\; \overset{\ominus\;\;\oplus}{F_3B-\underset{..}{O}(Et)_2}$$

Most organic compounds do not act as Lewis acids because they are generally closed-shell molecules with filled valence levels and with no unfilled orbitals capable of accepting electrons. By virtue of the unshared pairs of electrons on oxygen and nitrogen atoms, organic compounds which contain these elements can often function as Lewis bases towards many Lewis acids. Lewis acids such as BF_3, $TiCl_4$, and $SnCl_4$ are commonly used to react with oxygen-containing Lewis bases such as carbonyl compounds, alcohols, and ethers.

Acid Strength

Long before taking a chemistry course, we know something of the relative strengths of acids and bases. People would rather spill vinegar than battery acid on their skin. (Even those who have never studied chemistry would make the same choice.) And if asked why, they would say that battery acid is "stronger" than vinegar. The same people would also prefer to ingest a solution of baking soda ($NaHCO_3$) rather than lye (NaOH). Again, they know that lye is stronger than soda, even though they may not know both are bases. Thus the concept of strong or weak acids and bases is known to more people than one might suspect.

Chemists, however, would rather be a little more specific than simply describing acids and bases as either strong or weak. They need to be able to put acid or base strength on a quantitative basis. One way to measure the strength of an acid would be to place your index finger into the acid. The ability of an acid to transfer a proton to your finger, which would act as a base and accept a proton, would result in an audible response—the pain index. From the previous discussion, if battery acid and vinegar were compared by placing the index finger in each, the pain index for battery acid would be much greater than for vinegar because battery acid is "stronger" and can thus transfer a proton to one's finger more effectively than vinegar (see Figure 3.1). Other acids could be tested in a similar fashion, and their acid strength could be ranked by the audible signal. Since the same base, an index finger, is used to test every acid, the pain index is a direct and quantitative measure of the strength of a given acid compared to the other acids that have been tested. Unfortunately, the method is not very reproducible, and the sensor can only be used for a limited number of measurements without degradation.

But the idea is clear: the strength of an acid can be quantitated by knowing how well it transfers a proton to some standard base. Moreover, the extent to which it transfers a proton to the standard base could be compared with the extent to which other acids transfer a proton to the same standard base. Thus by using a single base, one can not only measure the strength of an acid but also compare acids quantitatively.

To make it simple, a molecule is used as the standard base. The acid strength will be a measure of how well the acid transfers a proton to that molecule acting as a base, and we can

$$H_2SO_4 \quad + \quad finger \quad \longrightarrow \quad finger\text{-}H^+ \quad + \quad HSO_4^- \quad \boxed{\text{OUCH}}$$
(battery acid)

$$CH_3CO_2H \quad + \quad finger \quad \longrightarrow \quad finger\text{-}H^+ \quad + \quad CH_3CO_2^- \quad \overline{\text{OUCH}}$$
(vinegar)

Pain Index

FIGURE 3.1

quantitate the acid strength by measuring the amount of the acid that is dissociated in the presence of that base. Thus the ratio of the amount of acid which has transferred a proton to the base, compared to the amount of acid which has not transferred a proton to the base, is a direct measure of the strength of that acid. Water has been chosen as the standard base molecule (in place of a finger) and the *ratio* of ionized acid to undissociated acid is called K_{eq} (the equilibrium constant).

$$HA \ + \ H_2O \ \underset{}{\overset{K_{eq}}{\rightleftharpoons}} \ H_3O^+ \ + \ A^-$$

K_{eq} is expressed as the concentration of products divided by the concentration of reactants present at equilibrium, or $K_{eq} = [H_3O^+][A^-]/[HA][H_2O]$. (Since two product molecules are produced from two reactant molecules, the *ionization ratio* of the acid is expressed as the product of the product concentrations divided by the product of the reactant concentrations.) This numerical ratio can be used to compare the strengths of acids. It is very important to realize just how simple this notion is. The equilibrium constant (in this case the ionization constant of an acid) is merely a *ratio* of the molecules which have donated a proton to water to those which have not.

To further reduce the complexity, since the concentration of water, $[H_2O]$, is a constant in dilute solutions where this treatment is valid, this fraction can be simplified to $K_{eq}[H_2O] = K_a = [H_3O^+][A^-]/[HA]$, which is again essentially a ratio relating the ionized acid molecules to the un-ionized acid molecules. For example, if K_a is $1/10^5$, or 10^{-5}, one of every $10^{2.5}$ acid molecules is ionized. On the other hand, if K_a is $10^6/1$, for every 10^3 molecules ionized, only one is not ionized. The latter acid is a much stronger acid than the former because a much higher fraction is ionized and donates a proton to water.

One can further simplify the comparison by defining $pK_a = -\log K_a$ because this yields small numbers to deal with rather than exponentials. It must be noted, however, that pK_a units are logarithmic units, so one pK_a unit represents a change of 1 in 10 in the ratio of acid molecules which are ionized to those which are not. By the use of pK_a's, the larger the number, the weaker is the acid. For example, if one compares acetic acid ($pK_a = 4.75 = -\log K_a (K_a = 1.8 \times 10^{-5})$) with hydrofluoric acid ($pK_a = 3.2 = -\log K_a (K_a = 6.3 \times 10^{-4})$), it is seen that HF is the stronger acid by about 1.5 orders of magnitude. Likewise, HCl, which has $pK_a = -6.6 (K_a = 3.98 \times 10^6)$, is a stronger acid that acetic acid with $pK_a = 4.75 (K_a = 1.8 \times 10^{-5})$ by more than 10 orders of magnitude (10^{10} stronger). By using a standard, constant molecule as a base, the measure of the fraction of proton transfer to that base becomes a meaningful and quantitative way to compare acid strengths. Table 3.1 is a compilation of pK_a for various acids. It is only a representative group of pK_a values; literally thousands have been precisely measured. It is evident that, although the acid strengths of organic compounds can vary over 50 orders of magnitude, quantitative comparisons can be made between various acids with good accuracy.

Acid–Base Equilibria

Since pK_a values always refer to the ionization of an acid in water under standard conditions (dilute aqueous solution at 25°C), they can be used to predict the positions of acid–base equi-

TABLE 3.1 pK_a Values for Common Organic Acids

Acid	Conjugate Base	pK_a
H_2SO_4	HSO_4^-	-9
HCl	Cl^-	-6
$ArSO_3H$	$ArSO_3^-$	-6 to -7
$R\overset{\overset{+}{O}H}{\underset{}{\|\|}}Y$	$R\overset{O}{\underset{}{\|\|}}Y$	-6 to -7
Y=H, R, OR,		
H_3O^+	H_2O	-1.74
HF	F^-	3.22
$ArNH_3^+$	$ArNH_2$	5
pyridine·H^+	pyridine	5.2
RCO_2H	RCO_2^-	4 to 6
H_2CO_3	HCO_3^-	6.4
$RC(O)CH_2C(O)R$	$RC(O)C^{\ominus}HC(O)R$	9
HCO_3^-	CO_3^-	10.33
ArOH	ArO$^-$	10
RNH_3^+	RNH_2	10
$R_2NH_2^+$	R_2NH	11
R_3NH^+	R_3N	12
$RC(O)CH_2CO_2R$	$RC(O)C^{\ominus}HCO_2R$	11
$RO_2CCH_2CO_2R$	$RO_2CC^{\ominus}HCO_2R$	13
H_2O	OH^-	15.74
RCH_2OH	RCH_2O^-	16
R_2CHOH	R_2CHO^-	17
R_3COH	R_3CO^-	18
$RC(O)NH_2$	$RC(O)NH^-$	18–19
$RC(O)CH_2R$	$RC(O)C^{\ominus}CHR$	19–20
RO_2CCH_2R	$RO_2CC^{\ominus}HR$	24
$RC{\equiv}CH$	$RC{\equiv}C^{\ominus}$	25
NH_3	NH_2^-	33
R_2NH	R_2N^-	35
$R_2C{=}CHCH_3$	$R_2C{=}CHCH_2^-$	43
$R_2C{=}CH_2$	$R_2C{=}CH^-$	44
RH	R^-	50

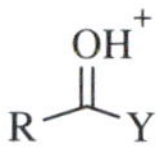

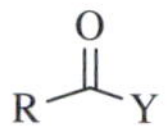

libria. The position of the following equilibrium (pK_{eq}) can be calculated by noting that, in going from left to right, isopropanol is acting as an acid.

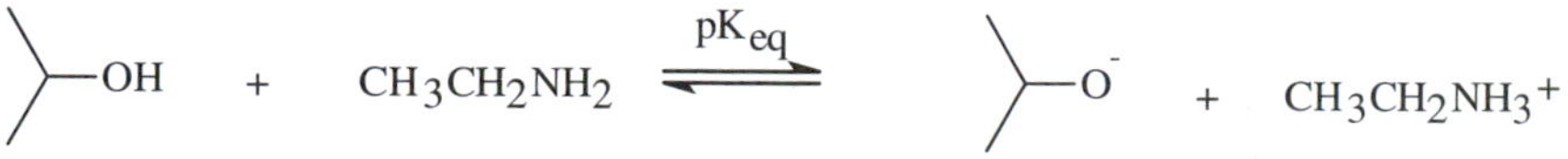

$$\text{(CH}_3)_2\text{CH–OH} + \text{CH}_3\text{CH}_2\text{NH}_2 \underset{}{\overset{pK_{eq}}{\rightleftharpoons}} \text{(CH}_3)_2\text{CH–O}^- + \text{CH}_3\text{CH}_2\text{NH}_3^+$$

In going in the reverse direction, from right to left, the ethyl ammonium ion is acting as an acid. Having identified the compounds acting as acids on either side of the equilibrium, the pK_a values for those acids are found and added to the equation.

$$\text{(isopropyl)}-OH \quad + \quad CH_3CH_2NH_2 \quad \underset{}{\overset{pK_{eq}}{\rightleftharpoons}} \quad \text{(isopropyl)}-O^- \quad + \quad CH_3CH_2NH_3^+$$

$$pK_a = 17 \qquad\qquad\qquad\qquad\qquad\qquad pK_a = 10$$

To evaluate the position of the equilibrium, the pK_{eq} is determined by the equation $pK_{eq} = pK_a$ (reactant acid) $- pK_a$ (product acid). For the preceding example, $pK_{eq} = 17 - 10 = 7$, and based on the definition of pK, $K_{eq} = 10^{-7}$ for this equilibrium. Thus the equilibrium lies far to the reactant side; that is, very little isopropoxide ion or ethyl ammonium ion is present at equilibrium. This technique is applicable for virtually any acid–base equilibrium. The three required steps are to (1) write a balanced equation that describes the equilibrium to be analyzed, (2) identify the species which is acting as an acid on each side of the equilibrium and write down its pK_a, and (3) subtract the pK_a of the product acid from the pK_a of the reactant acid to give the pK_{eq} for the equilibrium in question. It is a requirement that the pK_a of the acids on each side of the equilibrium be known or can be estimated reasonably well. Furthermore, the pK_{eq} that is determined refers to the equilibrium *in the direction it is written.* It is therefore important to write the chemical equilibrium as it will be analyzed.

The reaction of methyl lithium with water is an acid–base reaction. Going from left to right, water donates a proton to CH_3Li so it functions as the acid on the left. Going from right to left, methane donates a proton to LiOH so it functions as the acid on the right. The pK_a's of the acids on either side of the equilibrium reaction are subtracted and $pK_{eq} = -34.3$. Thus the equilibrium constant is $K_{eq} = 10^{34.3}$. This shows that the equilibrium lies very far to the right—so far to the right that, for all practical purposes, the conversion is quantitative.

$$CH_3Li \quad + \quad H_2O \quad \underset{}{\overset{K_{eq}}{\rightleftharpoons}} \quad CH_4 \quad + \quad LiOH \qquad K_{eq} = 10^{34.3}$$

$$pK_a = 15.7 \qquad\qquad\qquad pK_a = 50$$

It is now common experimental practice to react ketones with lithium diisopropyl amide (LDA) in order to generate the enolate of the ketone. This methodology has largely replaced the older approach to enolates, which employed alkoxide bases to remove a proton alpha to the carbonyl group. Comparison of the equilibrium constants for these two acid–base reactions reveals why the LDA method is preferable. The use of the amide base leads to essentially complete conversion of the ketone to its enolate ($K_{eq} \approx 10^{16}$). At equilibrium, there is virtually no unreacted ketone present in solution, only diisopropyl amine and the enolate. These can subsequently react cleanly and controllably with electrophiles which are added to the reaction mixture. In contrast, the use of an alkoxide base results in only partial conversion of the ketone to its enolate ($K_{eq} \approx 10^{-2}$) so that at equilibrium, the enolate, even greater amounts of the un-reacted ketone, and some unreacted alkoxide are present in the reaction mixture. It is therefore difficult to react the enolate with an added electrophile while avoiding competing reactions of both the enolate with the unreacted ketone and the alkoxide with the added electrophile.

Thus control of the reaction is difficult, yields are consequently decreased, and the product mixture is more complex.

$$\overset{\ominus}{\underset{}{(iPr)_2N}}\overset{\oplus}{Li} \quad + \quad \underset{\underset{O}{\overset{\|}{}}}{RC-CH_2R'} \quad \rightleftharpoons \quad (iPr)_2NH \quad + \quad \underset{\overset{|}{OLi}}{RC\!=\!CHR'} \quad K_{eq}=10^{16}$$

$$pK_a = 19 \qquad\qquad pK_a = 35$$

$$\overset{\ominus}{\underset{}{RO}}\overset{\oplus}{Na} \quad + \quad \underset{\underset{O}{\overset{\|}{}}}{RC-CH_2R'} \quad \rightleftharpoons \quad ROH \quad + \quad \underset{\overset{|}{ONa}}{RC\!=\!CHR'} \quad K_{eq}=10^{-2}$$

$$pK_a = 19 \qquad\qquad pK_a \approx 17$$

The ability to make good estimates of acid–base equilibrium constants is an invaluable aid in thinking about organic reactions and processes. Moreover, experimental workup procedures often require pH control that can be easily understood on the basis of pK_a considerations. Thus the concept of acid strength is exceedingly important and should be mastered.

A similar development can be used for the quantitation and comparison of the base strengths of organic bases (the ability to accept protons from acids). To do this, K_b is defined as $K_b = [OH^-][BH^+]/[B]$, the fraction of organic base which removes a proton from the standard acid—water. A scale of K_b was developed with which to compare base strengths quantitatively. The situation can become quite confusing since two sets of constants are defined, K_a and K_b. It turns out that acidities and basicities are inversely related by the ionization constant of water. That is, the stronger an acid in donating a proton to water, the weaker is its conjugate base in removing a proton from water. This seems eminently reasonable since if something gives up a proton easily, it should not take protons back easily. Put in more chemical terms, strong acids have weak conjugate bases, and weak acids have strong conjugate bases. It has become the convention to list only pK_as as a measure of *both* acidity and basicity. The only thing to remember is to assign the appropriate K_a to the reaction in the "acidic" direction.

For example, the base strengths of NH_3 and CH_3O^- minus can be compared in two different ways. First, the reactions of these two bases with water can be written as

$$CH_3O^- + H_2O \overset{K_{eq}}{\rightleftharpoons} CH_3OH + OH^- \qquad pK_{eq} = -1, \; K_{eq} = 10$$
$$ pK_a\,15 \qquad\qquad pK_a\,16$$

$$NH_3 + H_2O \overset{K_{eq}}{\rightleftharpoons} NH_4^+ + OH^- \qquad pK_{eq} = 5.8, \; K_{eq} \approx 10^{-6}$$
$$ pK_a\,15 \qquad\qquad pK_a\,9.2$$

By the preceding analysis, the two equilibrium constants can be estimated by identifying the acids on either side of the two equilibria and subtracting their pK_a. It is seen that the equilibrium for the reaction of methoxide with water lies much farther to the right ($K_{eq} = 10$) than the reaction of ammonia with water ($K_{eq} = 10^{-6}$). Clearly, methoxide is much better at removing a proton from water than is ammonia, by about 10^7. Methoxide is therefore a stronger base than ammonia by about 10^7.

Alternatively, it is noted that the conjugate acids of ammonia and methoxide are the ammonium ion and methanol, respectively, and the equations for their ionization in water are

$$NH_4^+ + H_2O \xrightleftharpoons{K_{eq}} NH_3 + H_3O^+ \qquad pK_{eq} = 11 \,, \; K_{eq} = 10^{-11}$$

$$pK_a \; 9.2 \qquad\qquad\qquad pK_a \; -2$$

$$CH_3OH + H_2O \xrightleftharpoons{K_{eq}} CH_3O^- + H_3O^+ \qquad pK_{eq} = 18 \,, \; K_{eq} = 10^{-18}$$

$$pK_a \; 16 \qquad\qquad\qquad pK_a \; -2$$

Comparing these two equilibria, it is seen that methanol is a much weaker acid than ammonium ion by about 10^7 (i.e., the pK_a of methanol is larger by about seven log units than the pK_a of ammonium); thus its conjugate base methoxide should be a stronger base than ammonia by 10^7. Either method of analyzing the situation is acceptable, and each gives the same relative basicities without the need for two sets of ionization constants, K_a and K_b.

Amphoteric Compounds

A glance at the pK_a values in Table 3.1 reveals that many classes of compounds can act as either acids or bases, depending on the reaction environment. Such materials are termed *amphoteric*. They must have an acidic proton (i.e., a proton attached to an electronegative element or group) and an unshared pair of electrons that can be donated to a proton. For example, water, alcohols, and other hydroxylic compounds, as well as amines and amides, are amphoteric materials. Comparing the pK_a of these materials permits an assessment of the predominant behavior in a given environment. For example, if an amine is dissolved in water, it could function as an acid or a base. To determine which behavior will predominate, the position of the equilibrium can be determined for each process. Comparison of these values will indicate which will be the principal behavior. Thus as an acid the amine would donate a proton to water to give an amide anion and the hydronium ion:

$$RNH_2 + H_2O \xrightleftharpoons{K_{eq}} RNH^- + H_3O^+$$

$$pK_a \quad 33 \qquad\qquad\qquad -2 \qquad K_{eq} = 10^{-35}$$

As a base, the amine would accept a proton from water to give an ammonium ion and hydroxide:

$$RNH_2 + H_2O \xrightleftharpoons{K_{eq}} RNH_3^+ + OH^-$$

$$pK_a \qquad\quad 15 \qquad\qquad 9 \qquad\qquad K_{eq} = 10^{-6}$$

Since the equilibrium constant for the amine acting as an acid is 10^{-35} and that for the amine acting as a base is much larger, at 10^{-6}, reaction as a base will be the main behavior in aqueous solution. The magnitude of the equilibrium constant (10^{-6}) indicates that it is only a weak base.

On the other hand, if an amine such as diisopropyl amine is treated with *n*-butyl lithium in THF, a different behavior is indicated. In this case the equilibrium lies so far to the right as to be virtually irreversible. The amine therefore functions as an acid and donates a proton to the much stronger base butyl lithium. This is the standard method for the preparation of the versatile base LDA.

$$RNH_2 + n\text{-}BuLi \xrightleftharpoons{K_{eq}} RNH^- Li^+ + n\text{-}BuH$$

$$\begin{array}{cccc} pK_a & 33 & & 48 & K_{eq} = 10^{+15} \end{array}$$

Structural Effects on Acidity

Now that a method is in hand to compare acid strengths and predict behavior, a look at Table 3.1 reveals that organic compounds have an enormous range of acidities—from very strong acids, such as arenesulfonic acids ($pK_a = -6.5$) and protonated carbonyl compounds ($pK_a = -7$ to -10), to very weak acids such as alkanes ($pK_a = 50$) and alkenes ($pK_a = 45$). This huge range of acidity, about 10^{60}, is reflective of the huge diversity of structural elements present in organic compounds. Some of the structural features which are major influences on acidity will now be briefly examined.

Again considering the dissociation of an acid in water, it is seen the process has some energy costs and some energy gains. This energy balance (ΔG) determines the equilibrium constant and hence the strength of an acid. The dissociation of an acid in water has an energy cost from the breaking of a bond to hydrogen and the separation of charges produced by the ionization.

$$HA + H_2O \xrightleftharpoons{} A^- + H_3O^+$$

However, there is an energy gain from the formation of the OH bond of the hydronium ion, the solvation of the anion by H-bonding to the solvent, and the solvation of the hydronium by its hydrogen bonding to the solvent. If a number of different acids are compared, it becomes clear that the principal energetic difference in the dissociation of various acids in water is the *stability of the conjugate base* and its interaction with the aqueous solvent:

$$HA_1 + H_2O \xrightleftharpoons{} A_1^- + H_3O^+$$
$$HA_2 + H_2O \xrightleftharpoons{} A_2^- + H_3O^+$$
$$HA_3 + H_2O \xrightleftharpoons{} A_3^- + H_3O^+$$

energy for charge separation similar

bond energies similar

stability & solvation different

bond energies same

This is true because the other energetic factors which influence the equilibria are similar for different acids. Bonds from hydrogen to the first row elements have very similar bond strengths (± 5 kcal/mol), so the energy cost of breaking the bond to hydrogen is relatively constant for most acids of first row elements. This is especially true for acids with the proton bonded to the same element. Moreover, since the solvent is always water, the energy required to separate charges is about the same and, finally, the H—O bond of the hydronium ion is the same, regardless of which acid supplies the proton. The principal differences in the ΔG of ionization for various acids are consequently due to differences in stability of their conjugate bases in the reaction mixture.

This analysis suggests that structural features which stabilize the conjugate base (often an anion) will therefore increase the acidity of an acid. There are exceptions to this general approach (for instance, comparison of the acidities of acids in the second and third rows of the periodic table), but it provides a sound basis for predicting what structural factors can increase or decrease the acidity of organic acids.

There are three principal factors that lead to increased stability of anions: (1) the electronegativity of the atom carrying the negative charge, (2) inductive effects which can stabilize negative charge, and (3) resonance effects which delocalize the negative charge over several atoms and hence stabilize the anion.

Electronegativity

Increased electronegativity of an atom allows it to carry negative charge more readily, and increases the stability of the anion. It is for this reason that the order of acidity of first row hydrides is CH < NH < OH < FH. Transfer of a proton from these substances to water yields a series of anions whose stabilities are ordered according to the electronegativity of the negatively charged atom.

$$-\overset{\displaystyle |}{\underset{\displaystyle |}{C}}-^{-} \qquad \overset{\diagdown}{\underset{\diagup}{:N}}^{-} \qquad -\ddot{\overset{..}{O}}{:}^{-} \qquad F^{-}$$

⟶ Increasing electronegativity ⟶

⟶ Increasing stability ⟶

Thus

$$-\overset{\displaystyle |}{\underset{\displaystyle |}{C}}\text{-H} \qquad \overset{\diagdown}{\underset{\diagup}{:N}}\text{-H} \qquad -\ddot{\overset{..}{O}}\text{-H} \qquad \text{F-H}$$

⟶ Increasing acidity ⟶

Such ordering is valid only for elements in the same row in the periodic table. Comparisons among acids in which the proton is lost from elements from different rows are not valid because the bond strengths to the acidic hydrogen change from row to row. Because bond strength can contribute significantly to acid strength, it cannot be neglected when significant bond strength differences among acids are present.

The effective electronegativity of the atom carrying the charge is also dependent on the hybridization of that atom. As the s character of an orbital increases, electrons in that orbital are more stable, due to greater attraction to the nucleus. Thus the effective electronegativity of the atom increases. This effect is clearly seen in the relative acidities of hydrocarbons. Removal of a proton from alkanes, alkenes, and alkynes produces conjugate bases with electron pairs in sp^3, sp^2, and sp orbitals, respectively. As the amount of s character increases from 25 to 33 to 50 percent in this series, the stability of the conjugate base increases and accounts for the marked increase in acidity in the series. Based on these data, it is expected that cyclopropane,

TABLE 3.2

Acid	Conjugate base	s Character Percentage	pK_a
$R-CH_3$	$R-CH_2^-$	25	≈ 50
$RCH=CH_2$	$RCH=CH^-$	33	44
$R-C\equiv CH$	$R-C\equiv C^-$	50	25

which because of ring strain has the hydrogens bonded to carbons that are hybridized at about an $sp^{2.5}$ level (29 percent s character), should have a pK_a between that of an alkane and an alkene. In fact, the pK_a of cyclopropane is 46 as predicted (see Table 3.2).

The increase in acidity by 25 orders of magnitude between sp^3- and sp-hybridized carbon acids is similar to that found for the difference in acidity between an ammonium ion (sp^3 hybridization) and a protonated nitrile (sp hybridization). It is clear that hybridization can play a major role in stabilizing electron pairs and thus in influencing effective electronegativity of an atom.

$$R-C\equiv \overset{+}{N}\text{-H} + H_2O \rightleftharpoons R-C\equiv N\colon + H_3O^+ \qquad pK_a = -10$$

$$R\text{-}NH_3^+ + H_2O \rightleftharpoons R\text{-}\overset{\cdot\cdot}{N}H_2 + H_3O^+ \qquad pK_a = 10$$

Inductive Effects

The inductive effect is the ability of a substituent or *group* near the acidic proton to alter the electron distribution at the reaction center by through-bond displacement of electrons. The result is that substituents which withdraw electrons from the reaction center by the inductive effect stabilize anions and thus increase the acidity of the conjugate acids of those anions. Conversely, groups which donate electrons make the reaction center more electron-rich and thus make the formation of the anion at that center more difficult. The conjugate acid is thus a weaker acid.

This is easily demonstrated by considering a group of substituted acetic acids (Table 3.3). Compared to acetic acid (X = H), replacement of a hydrogen by more electron-withdrawing groups (Cl, F, $(CH_3)_3N^+$) leads to an increase in the acidity. Replacement of hydrogen with an electron-donating *t*-butyl group decreases the acidity. One can understand these changes in terms of the inductive effect. Comparing the conjugate bases of acetic acid and chloroacetic acids, it is seen that the carbon chlorine bond has a dipole moment associated with it. This bond dipole induces smaller dipole moments in adjacent bonds, which in turn induce ever smaller dipole moments in adjacent bonds.

$$\overset{\longleftarrow \quad \longleftarrow}{Cl-CH_2-\overset{O}{\overset{\|}{C}}-O^-} \qquad \overset{\longrightarrow \quad \longrightarrow}{(CH_3)_3C-CH_2-\overset{O}{\overset{\|}{C}}-O^-}$$

The results of this inductive effect are that the electron density on the carboxylate anion is reduced, the negative charge is distributed over more atoms, and the chloroacetate anion is

TABLE 3.3 Acidities of Substituted
Acetic Acids $X-CH_2CO_2H$

X	pK_a
t-Bu	5.05
H	4.75
Cl	2.86
F	2.66
$(CH_3)_3N^+-$	1.83

stabilized relative to acetate. Because the chloroacetate anion is more stable than the acetate ion, its conjugate acid, chloroacetic acid, is a stronger acid than the conjugate acid of the acetate ion acetic acid (Table 3.1).

As expected, groups with higher electronegativity ($X = F$) or electron deficiency result in greater inductive electron withdrawal; hence the anion is more stable and the acidity increased. Conversely, a group such as *t*-butyl is electron donating relative to hydrogen. Its inductive effect serves to increase the electron density on the carboxylate group, destabilizing the anion and thus decreasing the acidity of its conjugate acid.

As mentioned, inductive effects operate through bonds by successive bond polarizations. As such, they diminish rapidly with distance, so very little effect results if an inductive effect must be transferred through more than four bonds. As shown in Table 3.4, placement of a chlorine substituent next to the carboxyl group causes a hundredfold increase in acidity. Moving it to the β position reduces the effect significantly, whereas a γ-chloro substituent causes almost no acidity increase.

Inductive effects serve to alter the electron distributions in molecules and they are consequently very important influences on many types of reactions, not just on acidity and basicity. To the extent that electronic changes occur during the conversion of reactants to products, inductive effects can facilitate or impede those electronic changes and thus change the rates of conversion. It is important to keep them in mind when other examples of reactivity changes are discussed.

TABLE 3.4 Acidities of Chlorobutanoic Acids as
the Position of Chlorine Attachment Is Varied

Acid	pK_a
$CH_3CH_2CH_2CO_2H$	4.88
$CH_3CH_2CHClCO_2H$	2.80
$CH_3CHClCH_2CO_2H$	4.06
$CH_2ClCH_2CH_2CO_2H$	4.52

Resonance Effects

A final structural effect which influences acidity is the delocalization of electrons via resonance. In terms of acid–base behavior, resonance delocalization can stabilize the conjugate base of an acid, thus making the acid a stronger acid. For example, alcohols have pK_a of about 16, whereas carboxylic acids have $pK_a \approx 5$. In each case, the acidic proton is lost from oxygen. The bond strength to the proton and the electronegativity of the atom carrying the charge (oxygen) are identical; thus these factors cannot account for the large difference in acidity. On the other hand, the alkoxide ion is a localized anion with the oxygen atom carrying a full negative charge whereas the carboxylate ion is resonance delocalized. In the carboxylate ion, the electron pair (and negative charge) is distributed between both oxygens, so each oxygen carries only a partial negative charge (actually about $-1/2$) and the anion is greatly stabilized. Thus, carboxylic acids are more acidic than alcohols by approximately 10^{11}.

$$ROH + H_2O \xrightleftharpoons{\quad pK_a = 16 \quad} RO^- + H_3O^+$$

$$R\overset{O}{\underset{OH}{\diagdown}} + H_2O \xrightleftharpoons{\quad pK_a = 5 \quad} H_3O^+ + R\overset{O}{\underset{O^-}{\diagdown}} \longleftrightarrow R\overset{O^-}{\underset{O}{\diagdown}}$$

The groups of compounds in Figure 3.2 illustrate the profound effect that resonance de-localization has on the stability of anions and hence on the acidity of the conjugate acids. To compare the acidities of these acids, the conjugate bases can be ranked according to their resonance stabilization and that ranking of anion stabilization is predictive of the acidity orders.

While resonance stabilization is greatest for those compounds which have more electron density distributed to more electronegative elements (compare **1**, **2**, and **3**), delocalization of charge over any elements results in significant anion stabilization and a corresponding increase in acidity of the conjugate acid of that anion (e.g., **6**, **7**, **8**). Moreover, electronegativity effects can be considered in addition to resonance effects, where applicable. Both amides **5** and methyl ketones **8** have resonance stabilization, but in amide anions, the negative charge is shared between nitrogen and oxygen whereas in ketone enolates, the negative charge is shared between carbon and oxygen. Due to the greater electronegativity of nitrogen over carbon, the amide anion is more stable and hence amides ($pK_a \approx 18–19$) are somewhat more acidic than ketones ($pK_a \approx 20–21$)

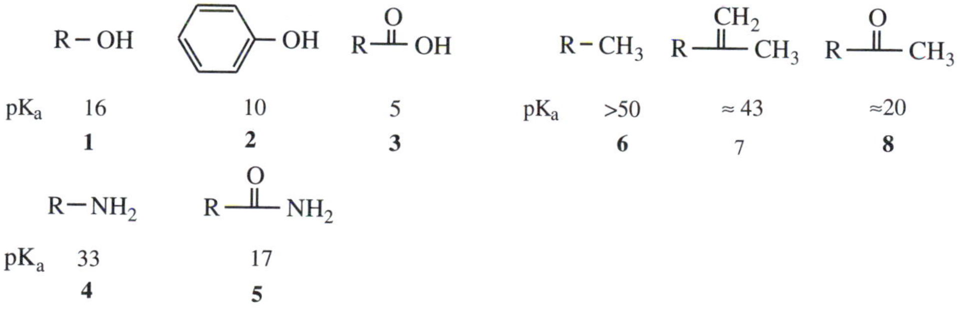

R – OH	⬡– OH	R–C(=O)–OH	R – CH₃	R–C(=CH₂)–CH₃	R–C(=O)–CH₃
pK_a 16	10	5	pK_a >50	≈ 43	≈20
1	2	3	6	7	8

R–NH₂	R–C(=O)–NH₂
pK_a 33	17
4	5

FIGURE 3.2

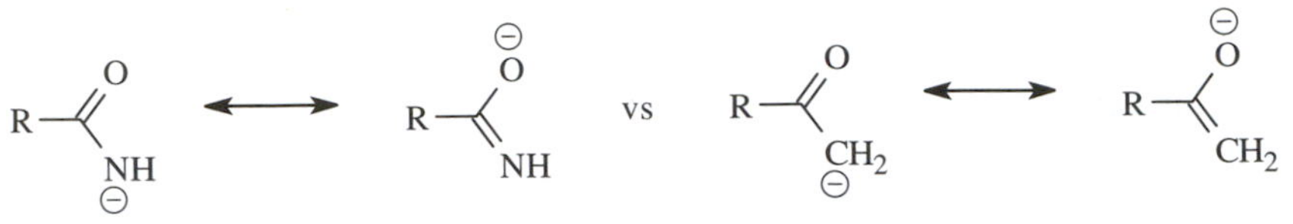

A particularly strong type of resonance stabilization is found for those compounds which form an aromatic ring upon removal of a proton. The enhanced aromatic stability of the conjugate base translates into a large increase in acidity of the acid. Whereas the doubly allylic proton of 1,4-pentadiene is predicted to have a $pK_a \approx 40$ due to resonance stabilization of the anion, the doubly allylic proton in cyclopentadiene has a $pK_a = 16$ because the resulting anion produces an aromatic π system.

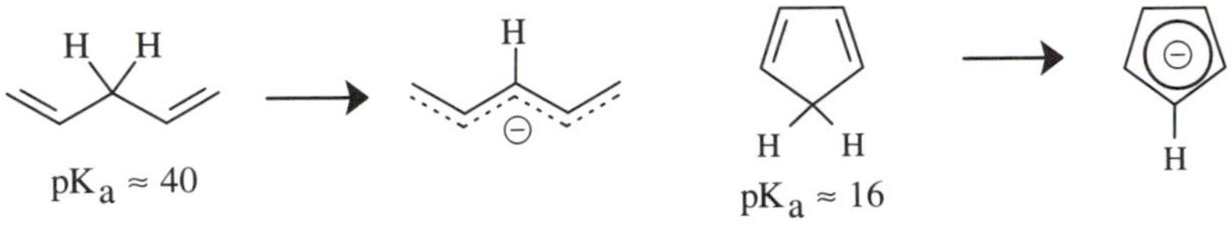

Aromaticity also explains why tropolone ($pK_a \approx -5$) is slightly more basic than a normal ketone ($pK_a \approx -7$). The conjugate acid is stabilized upon protonation by the formation of an aromatic tropylium ion.

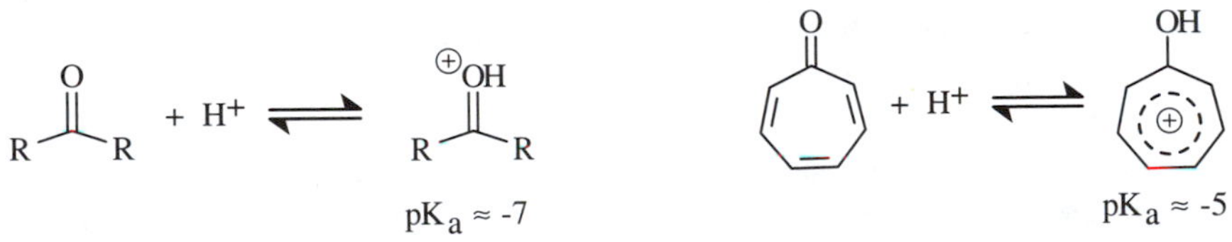

These inductive and resonance effects can significantly alter the electron distributions in molecules and can influence not only acidity but many other reactions as well. A general understanding of these effects will be important in many different transformations.

Bibliography

For an excellent discussion see M. A. Fox and J. K. Whitesell, *Organic Chemistry,* Jones & Bartlett, Boston, 1994, pp. 208–16.

T. H. Lowry and K. S. Richardson, *Mechanism and Theory in Organic Chemistry,* 3d ed., Harper & Row, New York, 1987, pp. 293–322, is a very good discussion of acid strengths.

Another very good discussion is J. A. March, *Advanced Organic Chemistry, Reactions, Mechanisms, and Structure,* 4th ed., Wiley, New York, 1992, pp. 248–72.

Problems

1. Using the pK_a data from Table 3.1, predict the position of the following equilibria and give an estimate of the equilibrium constant.

(a) $NH_3 + OH^- \rightleftharpoons NH_2^- + H_2O$

(b) $CH_3CO_2H + CO_3^- \rightleftharpoons CH_3CO_2^- + HCO_3^-$

(c) $RCO_2H + R_2NH \rightleftharpoons RCO_2^- + R_2NH_2^+$

(d) $R_3COH + RC(O)\overset{\ominus}{C}HR' \rightleftharpoons R_3CO^- + RC(O)CH_2R'$

(e) $NaF + HCl \rightleftharpoons NaCl + HF$

(f) $ArOH + RNH_2 \rightleftharpoons ArO^- + RNH_3^+$

(g) $RC(O)OR' + H_2SO_4 \rightleftharpoons RC(OH^+)OR' + HSO_4^-$

(h) $RC(O)CH_2C(O)R' + R_3N \rightleftharpoons RC(O)\overset{\ominus}{C}HC(O)R' + R_3NH^+$

(i) $RC \equiv C^- + RCH_2OH \rightleftharpoons RC \equiv CH + RCH_2O^-$

(j) $CH_3Li + (CH_3)_2C=CH_2 \rightleftharpoons CH_4 + (CH_3)_2C=CHLi$

2. Predict the positions of the following equilibria (left, right, or center) and give an approximate pK value for the equilibrium. It may be necessary to look up or estimate pK_a values using structural effects on the pK_a values of known compounds. If the pK_a's are estimated, rationalize the method of estimation.

(a)

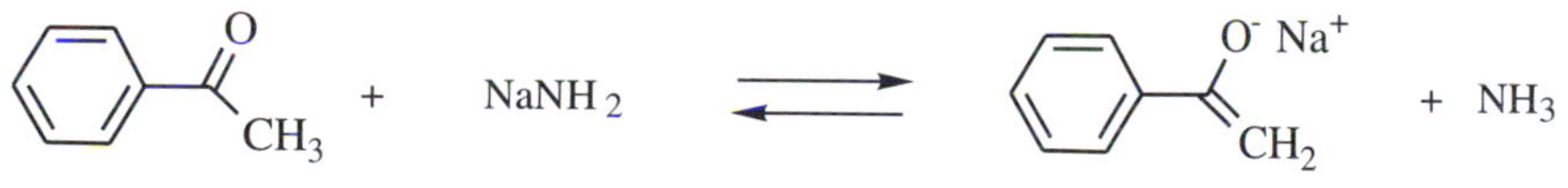

(b) $(CH_3)_2CH_2NH_3^+ + H_2O \rightleftharpoons (CH_3)_2CH_2NH_2 + H_3O^+$

(c)

(d) $CH_3NO_2 + CH_2\underset{H}{\overset{O}{\underset{\|}{C}}} \rightleftharpoons {}^-CH_2NO_2 + CH_3\underset{H}{\overset{O}{\underset{\|}{C}}}$

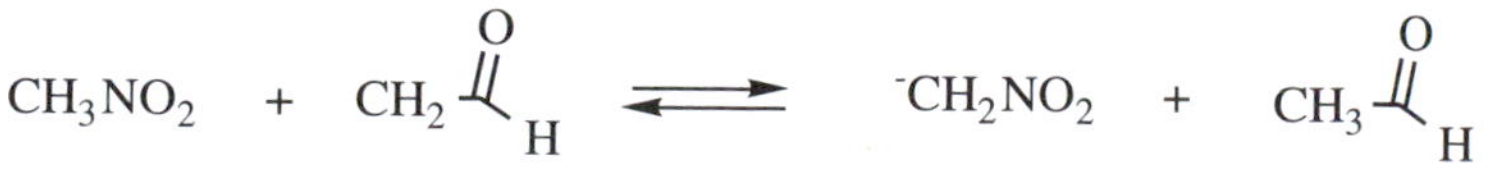

(e)　　F-CH$_2$CO$_2^-$　+　Cl-CH$_2$CO$_2$H　$\rightleftharpoons$　F-CH$_2$CO$_2$H　+　Cl-CH$_2$CO$_2^-$

(f)　 + HCO$_3^-$ $\rightleftharpoons$ + H$_2$CO$_3$

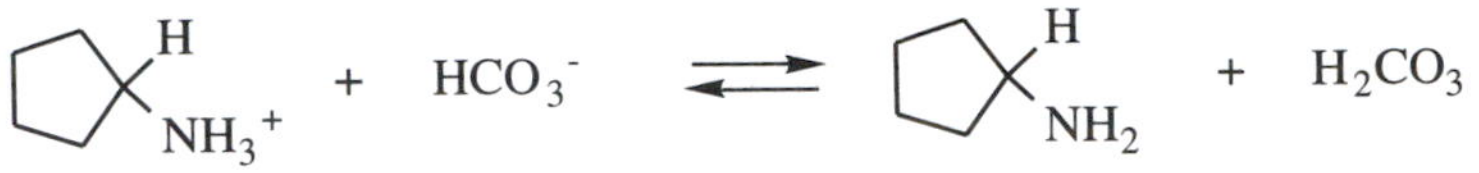

(g)　 + CH$_3$OH $\rightleftharpoons$ + CH$_3$OH$_2^+$

(h)　 + (C$_6$H$_5$)$_3$C$^-$ $\rightleftharpoons$ + (C$_6$H$_5$)$_3$CH

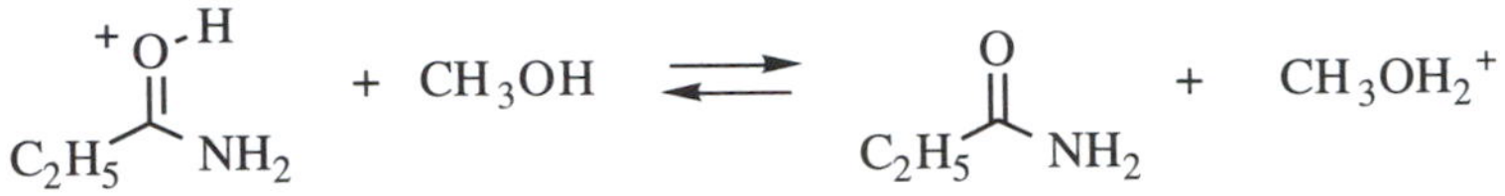

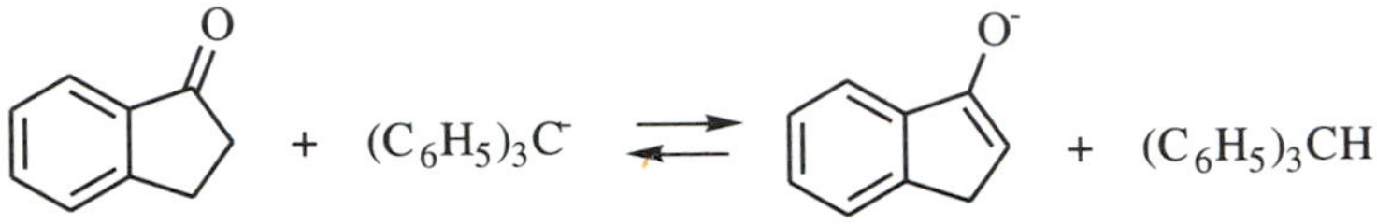

3. List (by number) in order of increasing acidity of the underlined protons.

(a) (1) CH$_3$CH$_2$—$\underline{H}$
　　(2) CH$_3$O—$\underline{H}$
　　(3) $\underline{H}$—F
　　(4) CH$_3$CH$_2$NH—$\underline{H}$

(b) (1) $\underline{H}$—Cl
　　(2) $\underline{H}$—F
　　(3) $\underline{H}$—Br
　　(4) $\underline{H}$—I

(c)　1.　CH$_3$CHCH$_3$　2.　Cl—O—$\underline{H}$　3.　H$_3$C—C—O$\underline{H}$ 4.　H$_3$C—C—$\underline{H}$
　　　　　|　　　　　　　　　　　　　　　　　　‖　　　　　　　　　‖
　　　　　O$\underline{H}$　　　　　　　　　　　　　　　　O　　　　　　　O

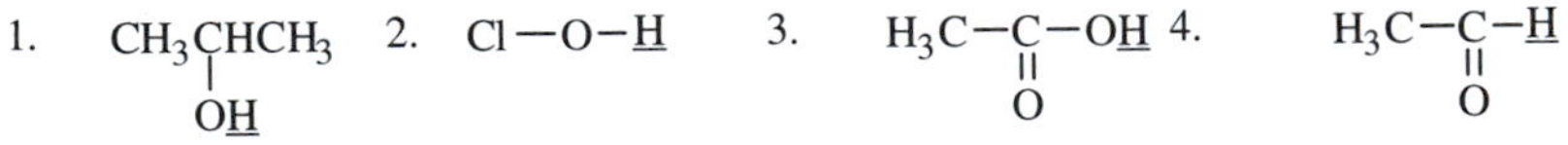

(d)　1.　　O　　2.　　　O　　3.　　　O　4.
　　　　　　‖　　　　　　‖　　　　　　‖
　　　CH$_3$CH$_2$C—O$\underline{H}$　　CH$_3$CH$_2$C—CH$_2$$\underline{H}$　　CH$_3$CH$_2$C—N$\underline{H}$　　CH$_3$CH$_2$OCH$_2$—$\underline{H}$
　　　　　　　　　　　　　　　　　　　　　　　　　　H

(e)　1.　　　　　　　2.　F　　　3.　Cl　4.　Cl
　　　　　　　　　　　　　|　　　　　　|　　　　|
　　　CH$_3$CH$_2$C—O$\underline{H}$　　CH$_3$CHC—O$\underline{H}$　　CH$_3$CHC—O$\underline{H}$　　CH$_2$CH$_2$C—O$\underline{H}$
　　　　　　‖　　　　　　　‖　　　　　　‖　　　　　‖
　　　　　　O　　　　　　　O　　　　　　O　　　　　O

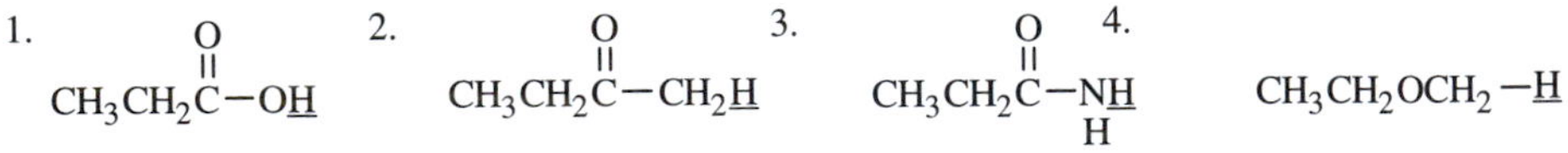

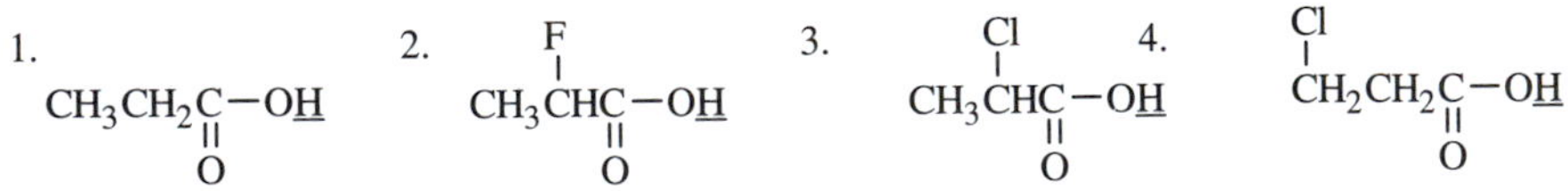

(f)　1.　　　　　　2.　　　　　　3.　　　　　4.
　　　CH$_3$O—$\underline{H}$　　(CH$_3$)$_2$CHO—$\underline{H}$　　CH$_3$CH$_2$O$\underline{H}$　　(CH$_3$)$_3$CO—$\underline{H}$

(g)　1. 2. 3, 4.

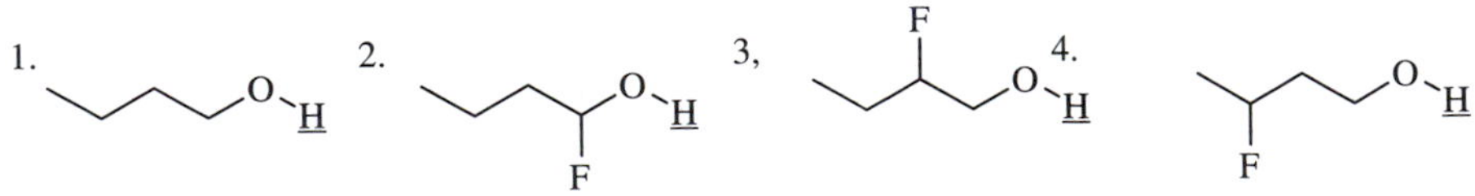

(h)

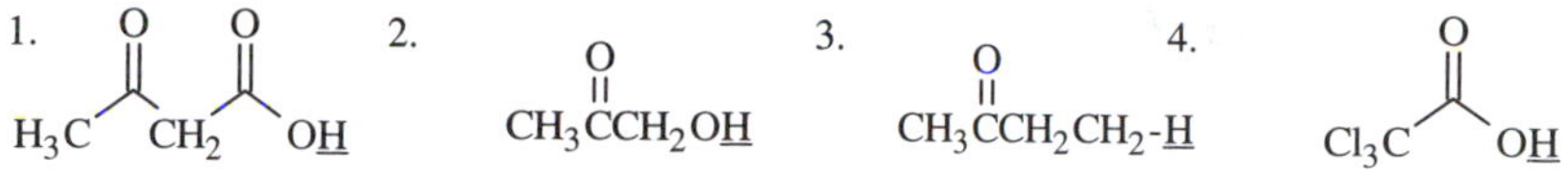

1.

$$H_3C \overset{O}{\underset{}{\parallel}} \overset{O}{\underset{}{\parallel}} OH$$
(H₃C–CO–CH₂–CO–OH)

2.

$CH_3\overset{O}{\overset{\parallel}{C}}CH_2\underline{OH}$

3.

$CH_3\overset{O}{\overset{\parallel}{C}}CH_2CH_2\text{-}H$

4.

$Cl_3C \overset{O}{\underset{}{\parallel}} OH$

(i)

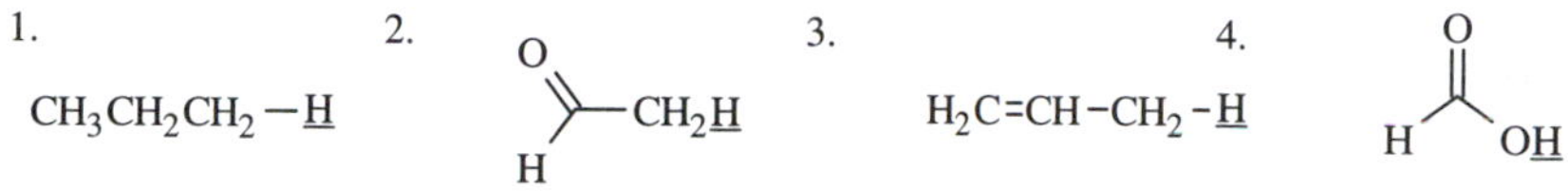

1.

$CH_3CH_2CH_2-H$

2.

$\underset{H}{\overset{O}{\diagup}}-CH_2\underline{H}$

3.

$H_2C{=}CH-CH_2-\underline{H}$

4.

$H \overset{O}{\underset{}{\diagup}} O\underline{H}$

(j)

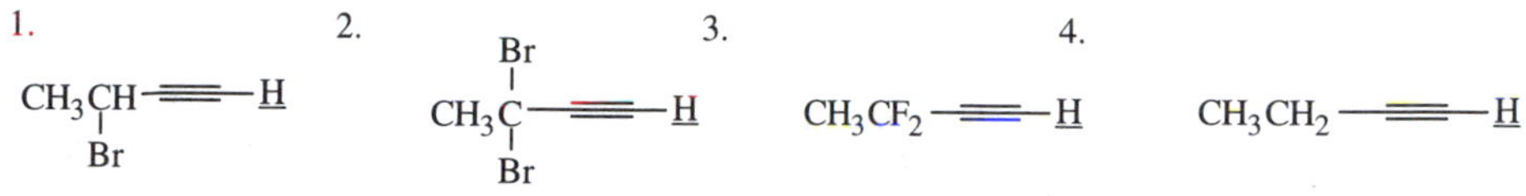

1.

$CH_3\underset{Br}{\overset{}{\underset{}{C}}}H\!\!=\!\!=\!\!=\!\!H$

2.

$CH_3\overset{Br}{\underset{Br}{C}}\!\!=\!\!=\!\!=\!\!-H$

3.

$CH_3CF_2\!\!=\!\!=\!\!=\!\!-H$

4.

$CH_3CH_2\!\!=\!\!=\!\!=\!\!-H$

(k)

1.

2.

3.

4.

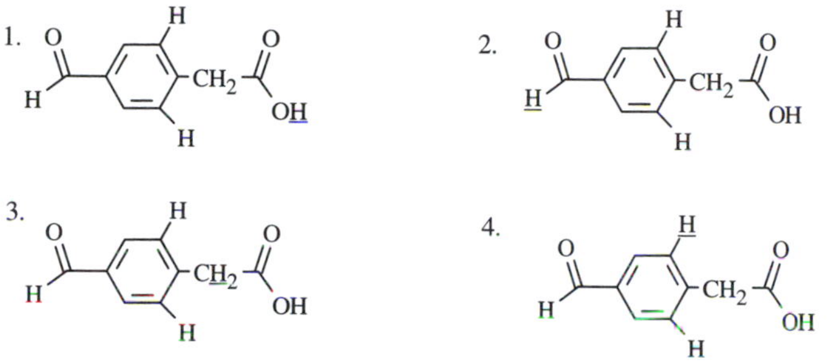

4. Give explanations for the following observations.

(**a**) Amides are protonated on oxygen rather than on nitrogen.

(**b**) Ethers are better Lewis bases than ketones.

(**c**) Tetramethyl guanidine (p$K_a \approx 12$) is a much stronger base than N,N-dimethylacetamide (p$K_a \approx -0.5$).

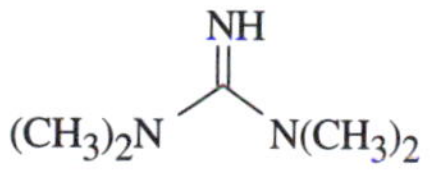

$$(CH_3)_2N \overset{NH}{\underset{}{\diagdown \diagup}} N(CH_3)_2$$

tetramethylguanidine

(**d**) Boron trifluoride BF_3 is a stronger Lewis acid than trimethyl borate $(CH_3O)_3B$.

(**e**) Piperidine is a much stronger Lewis base than pyridine.

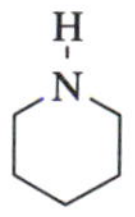

piperidine 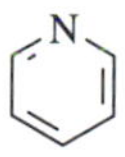   pyridine

(**f**) *p*-Nitrophenol has a pK_a ≈ 8 whereas phenol has a pK_a ≈ 10.

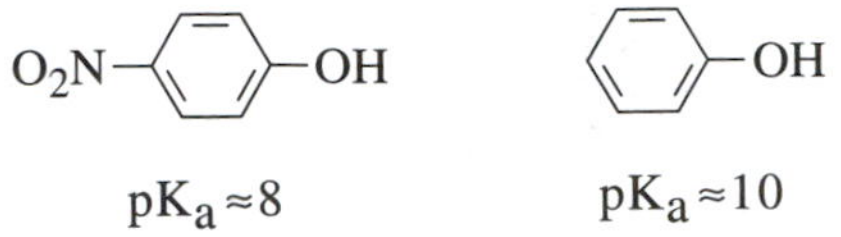

$$O_2N-\langle\ \rangle-OH \qquad \langle\ \rangle-OH$$

$$\text{p}K_a \approx 8 \qquad\qquad \text{p}K_a \approx 10$$

(**g**) *o*-Chloroaniline is a weaker base than *p*-chloroaniline.

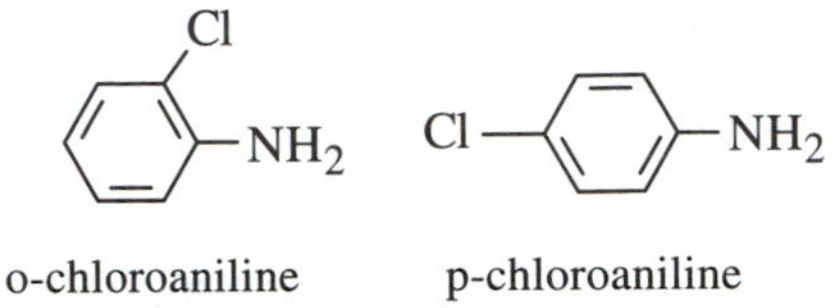

o-chloroaniline   p-chloroaniline

(**h**) Sodium borohydride in alcohol does not reduce imines effectively. If BF_3 is added to the mixture, however, the reduction proceeds much more rapidly and efficiently.

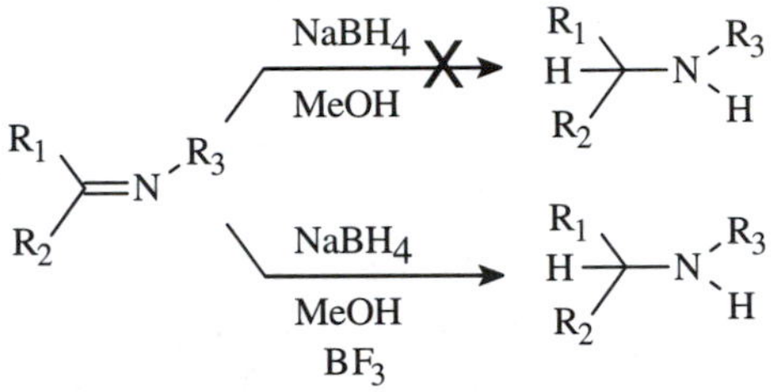

4

CURVED-ARROW NOTATION

Electron Movement

In a very simple sense, most organic reactions are accomplished merely by the movements of electrons. Molecules are composed of atoms held together by bonds, and covalent bonds are merely shared pairs of electrons. Therefore, the conversion of one molecule to another by changes in chemical bonds between reactants and products can be described simply as a change in electron pairs that are shared between the various nuclei in the reactants. While this is a gross oversimplification, it nevertheless provides a very important tool with which to keep track of bonding changes that occur during the transformation of one molecule into another.

One can keep track of electron pairs by noting changes in their locations in molecules by means of curved arrows. The curved arrow, *from* the tail of the arrow *to* the head of the arrow, depicts movement of an electron pair. For example, in the reaction of a proton with water to produce the hydronium ion, a new bond is formed from oxygen to hydrogen. The electrons in that bond start out as a lone pair on oxygen and are donated to, and shared with, the proton. One can easily show this electron movement with a curved arrow from the lone pair on oxygen to the proton.

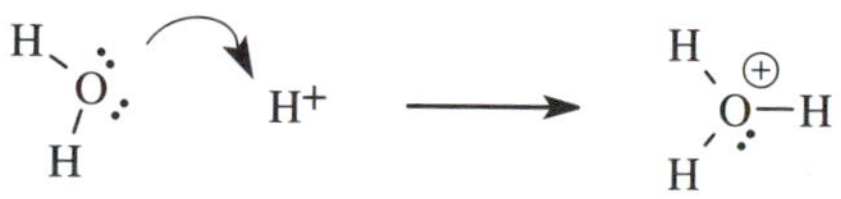

$$\underset{H}{\overset{H}{\diagdown}}\!\!\ddot{O}\!\!: \qquad H^{+} \longrightarrow \underset{H}{\overset{H}{\diagdown}}\!\!\overset{\oplus}{O}\!\!-\!H$$

In a similar sense, the Lewis acid–base reaction between ammonia and boron trifluoride can be depicted by a curved arrow from the lone pair of electrons on nitrogen to the boron atom. This electron movement creates a new bond between nitrogen and boron, but the curved-arrow notation clearly illustrates that the electron pair of that bond is supplied by the nitrogen. An additional consequence is that the nitrogen atom gains a formal positive charge and the boron atom gains a formal negative charge.

$$H_3N: \quad BF_3 \quad \longrightarrow \quad \overset{\oplus}{H_3N} - \overset{\ominus}{BF_3}$$

Before further applications of electron movement using curved arrow notation are presented, it is important to recognize just why it is such a useful tool for describing reactions. First, it permits one to keep track of valence shell electrons during a chemical reaction and thus serves as a method for electronic bookkeeping. Second, it shows how *changes* in bonding result from changes in electron distribution. Third, it can show likely mechanisms for chemical reactions in terms of the breaking and making of chemical bonds.

In order for curved-arrow notation to be used correctly, however, the structural and bonding principles which have already been presented must be followed. Thus, donor–acceptor properties, oxidation states, hybridization and octet structure of atoms, normal valences, formal charges, and reactive intermediates must all be taken into account as curved-arrow notation is used to track changes in electron distribution that occur during chemical reactions. Within the framework of these principles, however, curved-arrow notation can be a powerful and effective tool for depicting bonding changes during chemical reactions. To do this, it is necessary to first understand the general processes of bond formation and bond cleavage that are commonly encountered.

Heterolytic Bond Cleavages

Simple bond cleavages can proceed with the shared pair ultimately residing on either of the previously joined elements. In either case, one atom has a sextet electronic structure and a positive charge. The other has a valence octet with at least one lone pair (the pair that was formerly shared) and a negative charge. Such bond cleavage, in which the previously shared pair goes with one of the bonded atoms, is termed *heterolytic cleavage* and necessarily results in the formation of charged species.

$$-\overset{|}{\underset{|}{A}} - \ddot{B}: \quad \longrightarrow \quad -\overset{|}{\underset{|}{A}} \oplus \quad + \quad :\overset{..}{\underset{..}{B}}: \ominus$$

$$-\overset{|}{\underset{|}{A}} - \ddot{B}: \quad \longrightarrow \quad -\overset{|}{\underset{|}{A}} : \ominus \quad + \quad \overset{..}{B}: \oplus$$

The movement of electrons during heterolytic cleavage generally follows the direction of bond polarity. In a polar covalent bond, the shared pair is displaced toward the more electronegative element. Upon cleavage, the pair of bonded electrons is transferred completely to the more electronegative element, which becomes negatively charged, and the more electropositive element loses the bonded electron pair and becomes positively charged.

If one of the bonded elements is positively charged to begin with, it can gain the bonded pair upon bond cleavage and become neutral. Note, however, that net charge is always conserved in any reaction. Moreover, bond cleavages are depicted the same for either σ or π bonds.

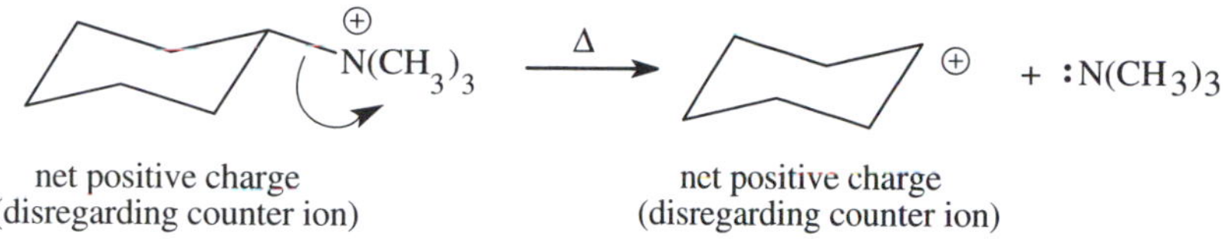

net positive charge
(disregarding counter ion)

nct positive charge
(disregarding counter ion)

Heterolytic Bond Formation

The formation of a bond between two atoms can proceed by one of the atoms donating an electron pair and the other atom accepting the electron pair. As before, charge must be conserved, and the loss and gain in electrons by the donor and acceptor, respectively, must be accompanied by a corresponding change in formal charge.

The atom donating the electron pair must obviously have an electron pair that is not tightly bound and is thus available for donation. Commonly, lone pairs and π-bonded electron pairs can be most easily donated, but electron pairs in σ bonds can occasionally be donated if the σ bond is weak or electron-rich.

The atom accepting the electron pair must have an unfilled orbital available which the donated electron pair can populate. This can be an unfilled valence shell orbital, as is the case if the acceptor atom has a valence sextet, or it can be an accessible antibonding orbital, either $\sigma*$ or $\pi*$.

Thus the reaction of acetone with BF_3 is a Lewis acid–base reaction in which a lone pair of the ketone oxygen atom is donated to an unfilled valence orbital of BF_3. Bond formation is accompanied by the development of formal charges on both oxygen and boron.

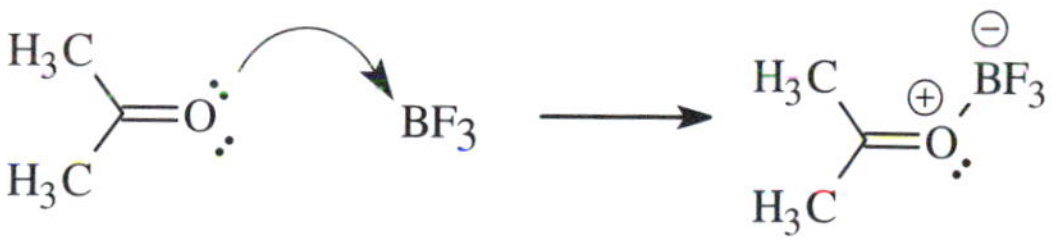

The capture of carbocations by alcohols involves a similar donation of a lone pair of electrons on oxygen to the vacant 2p atomic orbital of the sp^2-hybridized, sextet carbocation. Note that charge must be conserved, so the first formed product is a positively charged oxonium ion.

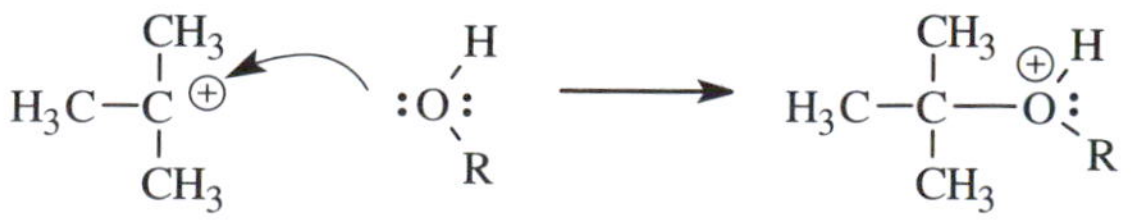

Reaction of an alkene with a nitronium ion involves donation of the π electron pair of the alkene into a $\pi*$ orbital of the nitronium ion. Donation of a bonded electron pair necessarily means that the bond from which it comes is broken. Likewise, population of an antibonding level by electron donation generally results in breaking the bond to which the antibonding orbital corresponds. In this case, electron donation of the olefinic π electron pair results in the rupture of the olefinic π bond and acceptance into the N—O $\pi*$ orbital also results in breakage of the N—O π bond. Note that a new bond is formed and net charge is conserved.

net charge = +1   net charge = +1

Addition of an alkyl lithium reagent (which has a very electron-rich σ bond between carbon and lithium) to a ketone involves electron donation of the σ electrons to the $\pi*$ orbital of the ketone to give a new carbon–carbon bond. The lithium counterion also plays a role in the addition by complexing with the lone pairs of the carbonyl oxygen, thus making the carbonyl group more electron deficient. This effectively lowers the energy of the $\pi*$ orbital, so its energy better matches the energy of the electron donor, and donation of electrons into that orbital is facilitated.

It is important to remember that the requirements for any two-electron bond-making process are that, first, there must be an available pair of electrons to be donated and, second, there must be an unoccupied orbital of suitable energy available into which the electrons can be donated.

Homolytic Bond Breaking and Bond Making

Heterolytic processes make up a large proportion of organic transformations because most bonds are somewhat polarized. Heterolytic cleavage is merely an increase of this polarity to the point at which point there is no bond remaining; that is, electron movement follows in the direction established by the bond polarity to give a cation–anion pair.

If a bond is particularly weak or nonpolar, bond cleavage can occur by a nonpolar, or *homolytic,* process. One electron of the shared pair goes with each of the two bonded atoms. Bond breaking then is the movement of single electrons rather than electron pairs, and is indicated in curved-arrow notation as "half-headed" arrows. Homolytic cleavage of a bond does not result in the formation of charge but does result in the formation of unpaired electron intermediates called free radicals. Free radicals normally have seven electrons in the valence shell and are consequently very reactive intermediates. Common examples of

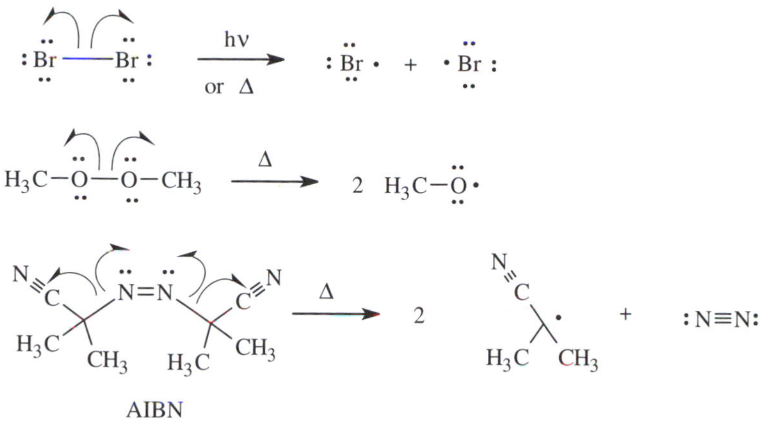

FIGURE 4.1

compounds which undergo homolytic bond cleavages include halogens (Br_2, Cl_2, F_2), peroxides ($R-O-O-R$), and azocompounds ($R-N=N-R$), seen in Figure 4.1. All of these free radical precursors are characterized by relatively weak, nonpolar bonds which, upon heating, break to give free radical intermediates. Free radicals are very reactive and proceed to products by a variety of one-electron, or homolytic, reactions.

Homolytic bond formation can occur when two free radical species contact each other. Each has an available unpaired electron and, if these two electrons are shared, a new bond will result.

This is simply the reverse of the homolytic cleavage. It is a very exothermic process (by an amount equal to the energy of the bond being formed) and it occurs at a *very* fast rate.

Homolytic bond formation can also occur by the reaction of a free radical with a bonded pair of electrons. Two common examples of this behavior are hydrogen (or other atom) abstraction reactions and free radical addition to double bonds (Figure 4.2). Atom abstraction reactions take place by the interaction of a free radical with a σ-bonded atom. One electron of the σ bond pairs with the unpaired electron of the free radical to produce a new bond. The remaining electron of the σ bond remains on the fragment from which the atom has been abstracted and produces a new free radical species. This process is energetically driven by bond strengths; that is, atom abstraction occurs only if the bond that is formed is stronger than the one that is broken. In the example of hydrogen abstraction that follows, a phenyl radical readily abstracts a benzylic hydrogen from toluene to give benzene plus the benzyl free radical, because the aromatic C—H bond (103 kcal/mol) that is formed is appreciably stronger than the benzylic C—H bond (85 kcal/mol) that is broken.

Addition to π bonds is a second very common reaction of free radicals. Interaction of the free radical with the π electron pair causes one of the π electrons to pair up with the unpaired electron of the free radical to produce a new bond to one of the π-bonded atoms. The remaining π electron is now unpaired and thus forms a new free radical species. The process is often very

Hydrogen Abstraction

π-Addition

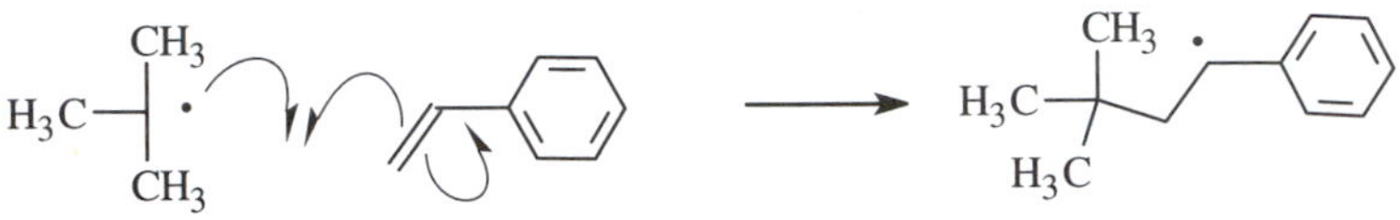

FIGURE 4.2

favorable since the new σ bond (70–90 kcal/mol) formed in the addition process is normally much stronger than the π bond (60 kcal/mol) which is broken in the reaction. In the previous example, a new carbon–carbon σ bond is formed by free radical addition to produce a new carbon-centered free radical; however, a wide variety of other free radical species add readily to olefins.

Resonance Structures

Curved-arrow notation is also a very useful device with which to generate resonance structures. In this application it is truly a bookkeeping system. Since individual canonical forms do not exist but are only thought of as contributors to the description of a real molecule, the use of curved arrow notation to convert one canonical form to another is without physical significance. Nevertheless, it provides a useful tool for keeping track of electrons and bonds in canonical structures. For example, the structures of carboxylate resonance contributors can be interconverted as follows:

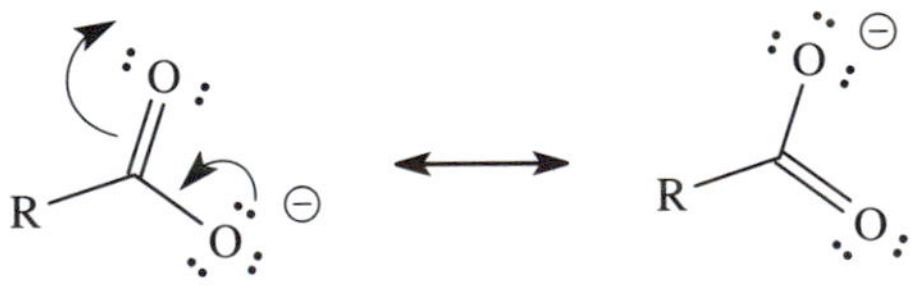

Likewise, the Kekulé forms of benzene can be shown as

Allyl cations, for example, can be shown nicely while keeping track of charges, electrons, and bonds:

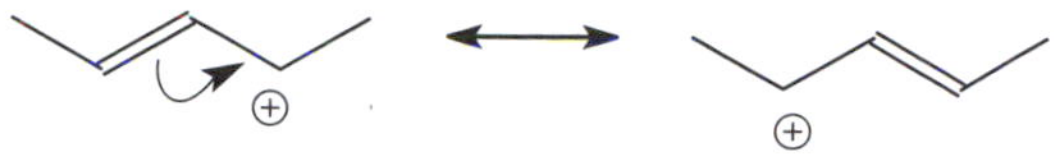

Similar considerations can be used for a variety of intermediates and structures. In using curved-arrow notation to generate contributing resonance structures, the same rules of valence, charge, bonding, and so on must be applied. Given these criteria, however, it is a straightforward exercise to generate complete sets of resonance structures which can then be evaluated.

Depiction of Mechanism

Use of curved-arrow notation to depict the mechanisms of organic reactions requires that appropriate mechanistic principles be superimposed on the correct use of curved arrows to denote movement of electrons. The mechanism of a reaction is the stepwise process by which reactants are converted to products, and generally each step involves bond making or bond breaking that can readily be depicted by curved-arrow notation.

Simple substitution reactions are shown in curved-arrow notation as

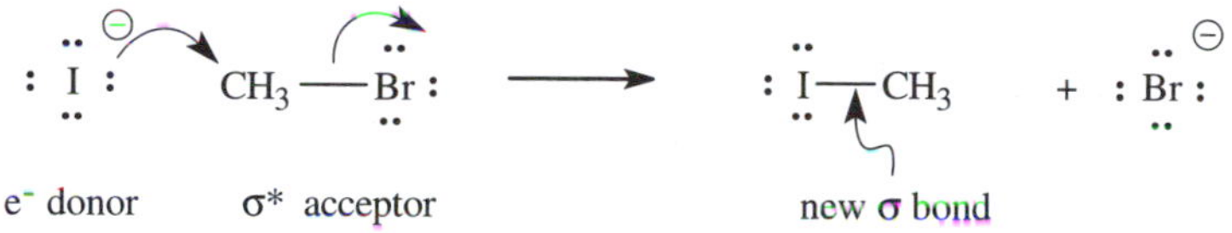

In this example, a nucleophile donates electrons to the electrophile, in this case a carbon with a leaving group attached, to produce a new σ bond. As the new σ bond is formed, the bond to the leaving group breaks and the substitution of one group for another is completed. The iodide nucleophile has unshared pairs of electrons which are donated. The $\sigma*$ orbital of the C—Br bond is the acceptor orbital. In any donor–acceptor interaction which leads to a chemical change, identification of both the donor and acceptor is crucial in predicting what change will occur and determining how the change might occur.

Addition of Grignard reagents to carbonyl groups involves donation of the electrons of an electron-rich carbon–metal σ bond to the $\pi*$ orbital of the carbonyl group. As shown, carbon–carbon bond formation is accompanied by carbon–oxygen π-bond cleavage, oxygen–metal bond formation, and a corresponding change in geometry from trigonal to tetrahedral:

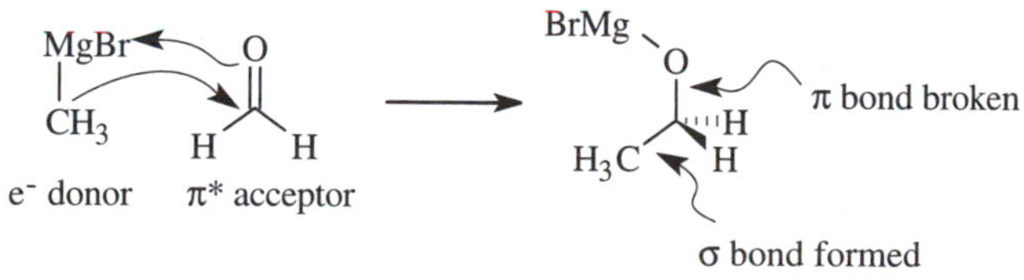

Ring opening of epoxides by alkoxides is used to emphasize that charge must be conserved during each mechanistic step. Because the reactants as written (neglecting spectator ions) have a net negative charge, the products *must* have a net negative charge.

σ* acceptor   e⁻ donor

The conservation of charge is a fundamental law for all processes, such as the additions of nucleophiles to π systems or acid–base reactions. The first step of the basic hydrolysis of nitriles has the hydroxide ion adding to the π bond of the nitrile. For purposes of mechanistic discussion, the hydroxide is shown without its counter ion, and the net charge on the reactant side of the equation is a negative one. Consequently, the product of this first step (and each subsequent step) must also have a net negative charge.

In aqueous solution, proton transfer to the first-formed intermediate is *very* rapid. However, again for illustrating the stepwise changes that must occur on the way from reactants to products using curved-arrow notation, these steps are shown independently.

Similarly, the production of enolates from carbonyl compounds involves base removal of a proton from the α position. The enolate is negatively charged and has delocalized electrons.

Although this process can be written to give a single canonical form **A**, it must be realized that the enolate is a delocalized species and resonance forms **A** and **B** can be generated, as discussed previously, using curved-arrow notation. This is *not* a mechanistic step since the delocalized product is a resonance hybrid of **A** and **B**. That is, **A** is not converted to **B**, but rather the curved arrows merely indicate the changes in electron distribution that must be used to describe the canonical form, **B**.

These previous examples are reactions in which the electron donor (nucleophile) supplies the "electronic push" to accomplish bond breaking. Many nucleophiles, either neutral or anionic, have lone pairs of electrons that are easily donated. They can be donated to even weak electron acceptors (electrophiles).

There are other reactions, also easily describable by electron movement (curved arrows), in which π electrons are donated. In such reactions, the π electrons are bonded electrons, and hence the π donor nucleophile is a weak electron donor. Consequently, a much stronger electron acceptor (stronger electrophile) is required for the "electronic pull" for electron donation to occur successfully. However, such descriptions are simply a matter of semantics, because

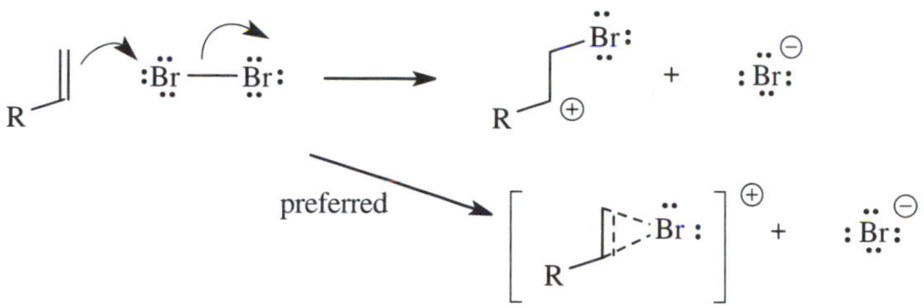

FIGURE 4.3

curved-arrow notation only shows changes in electrons, and does not indicate driving force. For a donor–acceptor interaction to occur productively, there has to be an energetic driving force for the process, and the energy levels of the donor and acceptor must be matched so that electron movement from the donor to the acceptor can occur.

For example, the protonation of a double bond has a proton as the electrophile and the π bond as the electron donor.

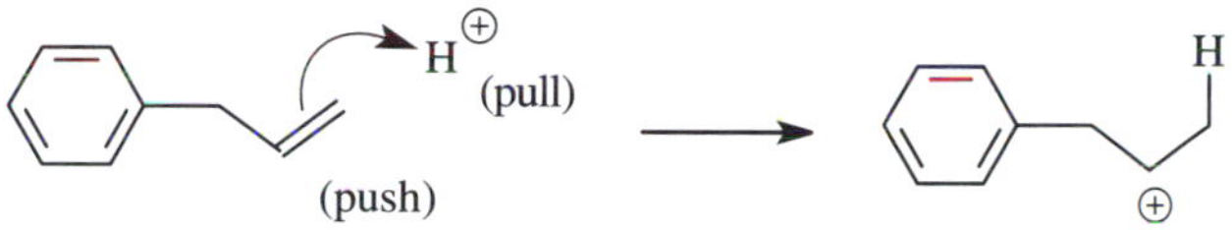

When bonded electrons are donated, it is important to remember that one of the bonded atoms which shared that pair is now left without a valence octet. That is, removal of a bonded pair must result in a sextet atom. Such is the case after protonation of a double bond, as previously seen. Because of the instability of a sextet electronic configuration, several strategies are available to stabilize it.

Neighboring atoms with unshared pairs of electrons can undergo bridging interaction with the cationic center to give structures in which all atoms have valence octets. An archetypical example is the bromination of olefins (Figure 4.3). Electrophilic addition of bromine to the double bond is predicted to give an α-bromocarbocation. However, formation of the bridged bromonium ion prevents a sextet configuration of carbon and is thus formed preferentially. Bridging interactions occur when a carbocation is generated vicinal to substituents such as $-OR, -Cl, -F, -SR,$ and $-NR_2$, all of which have lone pairs capable of bridging interactions.

Resonance stabilization can also make π-electron donation much more effective by preventing the formation of a sextet carbocation. Lone pair donation from the oxygen of enol derivatives is very important to the good donor ability of these compounds. The resulting oxonium ion has all valence octets (although positively charged) and is thus stabilized over sextet canonical forms.

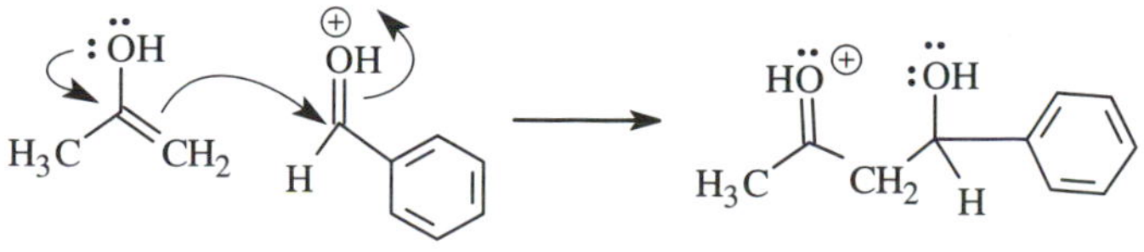

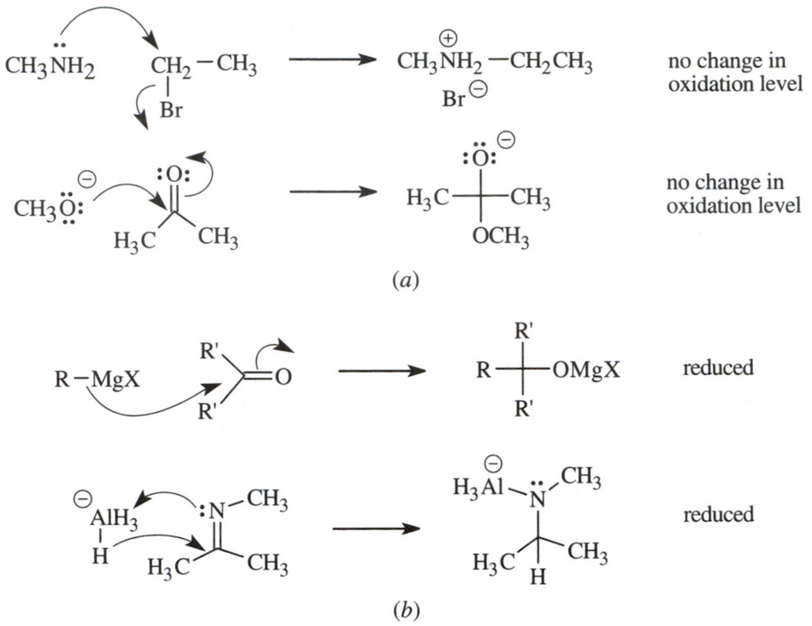

(a)

(b)

FIGURE 4.4

Resonance stabilization is also important in electrophilic aromatic substitution. Although each of the canonical forms of the Wheland intermediate has a sextet carbon atom, the charge is distributed over five atoms by resonance and is thus greatly stabilized.

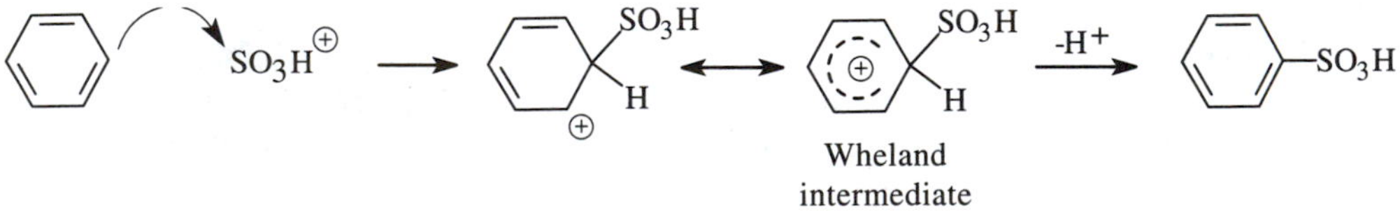

Wheland
intermediate

   The reactions of nucleophiles with electrophiles also relate to the overall oxidative change of a reaction. As expected, nucleophilic atoms which are more electronegative than carbon are not reductants and usually give no change in oxidation state; conversely, nucleophiles which are carbanion or hydride equivalents are reductants. (See Figure 4.4a and b, respectively.)

   Carbocation or proton electrophiles give no change in oxidation level, whereas electrophiles which are electronegative elements (Br$_2$, Cl$_2$, NBS, peracids, etc.) are oxidants (see Figure 4.5). Besides intermolecular reactions, curved-arrow notation is also useful in indicating bonding changes in intramolecular reactions and rearrangement. For example, Cope-type rearrangements are seen to involve changes in three pairs of bonded electrons.

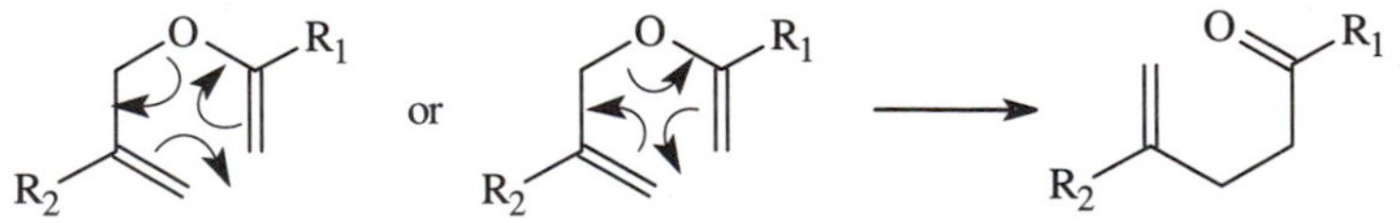

The arrows can be written in either directional sense, since these reactions are concerted

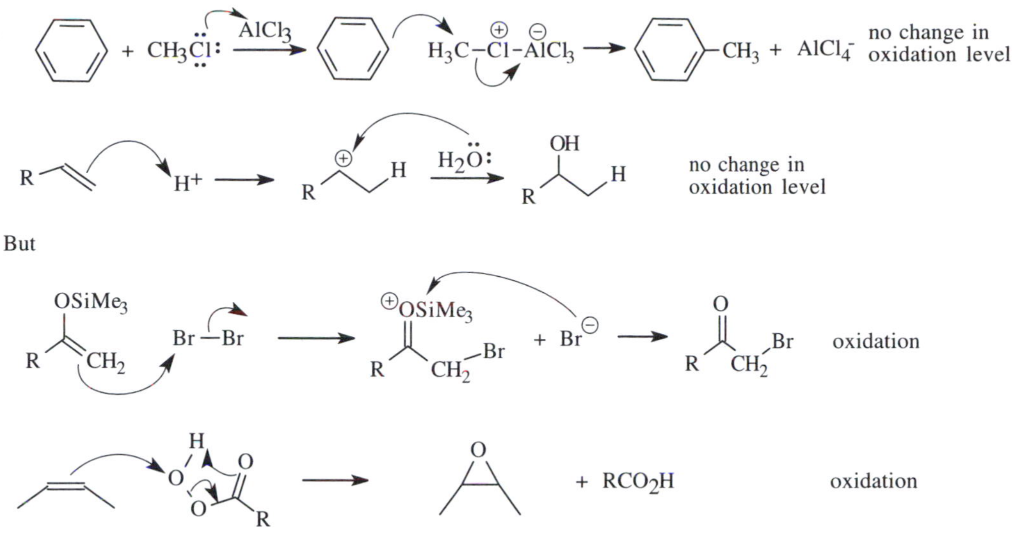

FIGURE 4.5

rearrangements with all bond making and bond breaking taking place at the same time. This example emphasizes the fact that curved-arrow notation is merely an electron bookkeeping method.

Cationic rearrangements are also handled easily by keeping track of where electron pairs come from and where they go. For example, the neophyl-type rearrangement which follows leads to skeletally rearranged products:

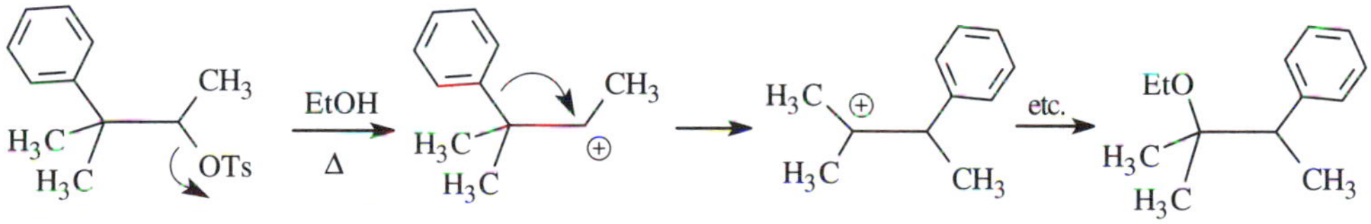

The curved-arrow notation clearly shows the electron flow needed to effect the rearrangement. What curved-arrow notation does not show is the timing of these events; that is, whether loss of leaving group precedes or is concerted with 1,2-phenyl migration, or if a bridged ion is an intermediate. Such considerations, if known, can be included in more detailed mechanistic sequences.

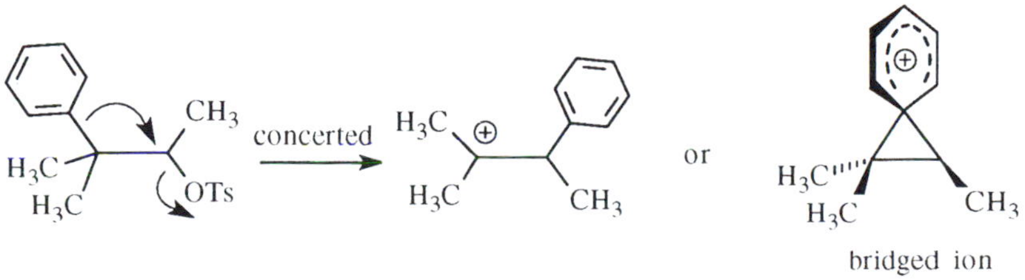

With these considerations, the steps one goes through to use electron movement to generate a possible reaction mechanism include the following:

1. Write a balanced equation for the reaction. While spectator ions may be neglected, it is imperative to write correct Lewis structures for reactants and products. This step is very important but often neglected.

2. Note the connectivity changes that occur, the changes in oxidation level that occur, and the reagents or reactant types necessary for the conversion.

3. Write a stepwise process for the reaction using curved arrows to account for bonding changes. The use of curved arrows for electron movement should be guided by bond polarities, donor–acceptor properties, electronegativities, and structural factors, and should result in a reasonable series of bonding changes from reactant to product.

4. Evaluate intermediates for stability and valence. If they fit normal chemical expectations, the mechanism is potentially correct. There may be other mechanisms operating, or the timing of individual steps (synchronous, concerted, etc.) may be different, but the previously stated process can be used to generate them as well.

Thus, used properly, curved-arrow notation for electron movement is indispensable to the organic chemist as a way to depict chemical change in complex molecules. Furthermore, it can be extended to include a method for showing mechanism, if the ground rules are understood and followed carefully.

Bibliography

Most undergraduate texts have a short section on curved-arrow notation, or electron movement, and these discussions are tied in with the development of reaction mechanisms. For example,

M. A. Fox and J. K. Whitesell, *Organic Chemistry,* Jones & Bartlett, Boston, 1994, pp. 233–36, and L. G. Wade, *Organic Chemistry,* 2d ed., Prentice Hall, Englewood Cliffs, NJ, 1991, pp. 28–30.

For a very nice discussion, see P. H. Scudder, *Electron Flow in Organic Chemistry,* Wiley, New York, 1992, pp. 7–10.

An excellent discussion with many examples is found in F. M. Menger and L. Mandell, *Electronic Interpretation of Organic Chemistry,* Plenum, New York, 1980, pp. 47–116.

Problems

1. For the following reactions, show the complete structures of the reactants and products of the step which is shown, point out the bonds which have been made or broken, identify the electron donors and acceptors, and use curved-arrow notation to indicate electron flow.

(*a*) $CH_3\text{-}S^{\ominus}$ + $\underset{\displaystyle OTs}{CH_2CH_3}$ $\longrightarrow$

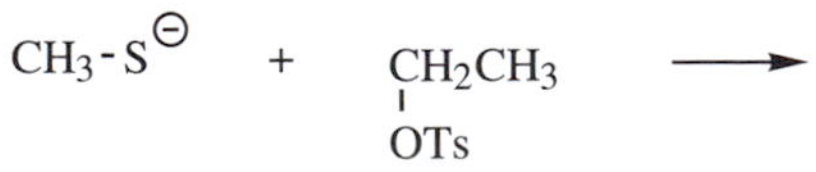

(*b*) $\underset{\displaystyle CH_3 \quad\; OCH_3}{\overset{\displaystyle O}{\|}}$ + $^{\ominus}OH$ $\longrightarrow$

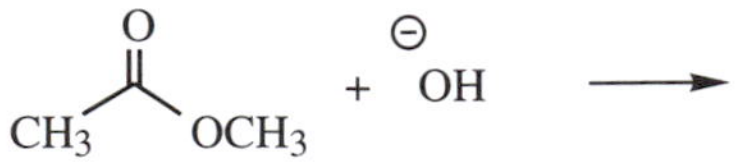

(*c*) $NaBH_4$ + C_6H_5CHO $\longrightarrow$

(d) H_3C-⬡ $+$ NO_2^+ ⟶

(e) $H_3C-C(CH_3)_2-CHBr-CH_3$ with CH3 Br groups, $\xrightarrow[\Delta]{\text{EtOH}}$

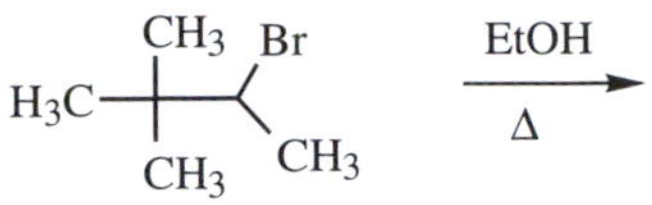

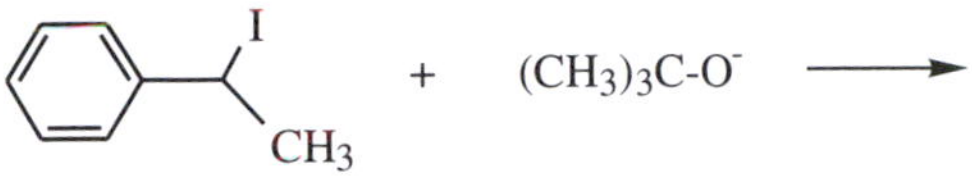

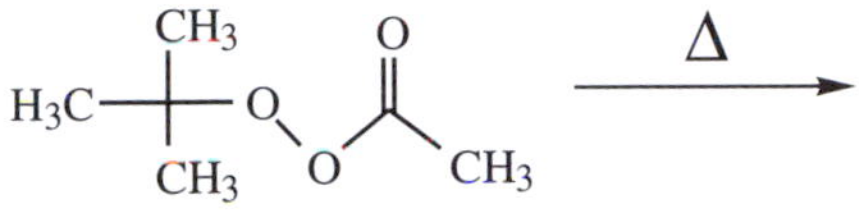

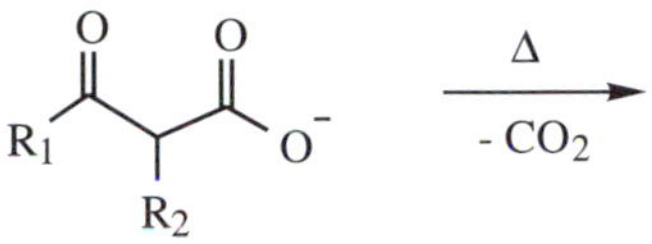

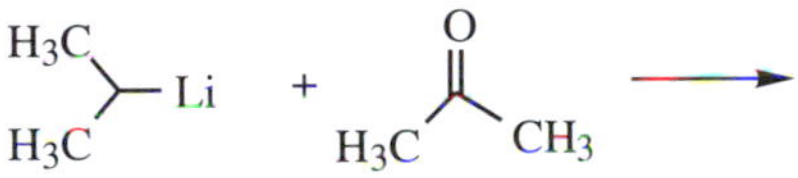

(f) ⬡—CHI(CH3) $+$ $(CH_3)_3C-O^-$ ⟶

(g) $H_3C-C(CH_3)_2-O-O-C(=O)-CH_3$ $\xrightarrow{\Delta}$

(h) $R_1-C(=O)-CH(R_2)-C(=O)-O^-$ $\xrightarrow[-CO_2]{\Delta}$

(i) $(H_3C)_2CH-Li$ $+$ $H_3C-C(=O)-CH_3$ ⟶

(j) $CH_3O-CH_2-CH=CH_2$ (allyl) with $CH_2=CH-CO_2Me$ ⟶

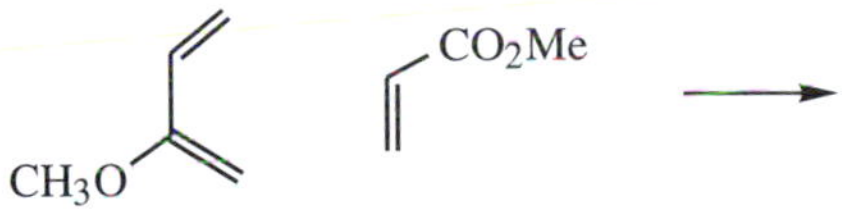

2. For the following transformations, give an acceptable mechanism, as indicated by electron movement.

(a) ⬡—CH(H)(H)—ONO_2 $\xrightarrow{\text{NaOCH}_3}$ ⬡—CHO

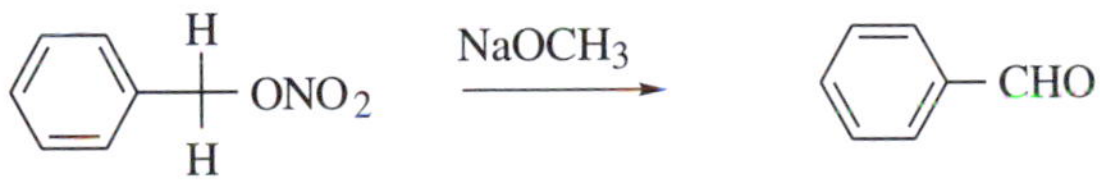

(b) $CH_3CH_2-CH(OSO_2CF_3)-CO_2CH_3$ $\xrightarrow[\text{CH}_3\text{CN}]{\text{NaN}_3}$ $CH_3CH_2-CH(N_3)-CO_2CH_3$

(c) $CH_3CH_2CH_2CH_2-CHBr-C(=O)-CH_3$ $\xrightarrow{\text{NaOEt}}$ $CH_3CH_2CH_2CH_2-CH(-O-)(-OEt)$ (epoxide with OEt)

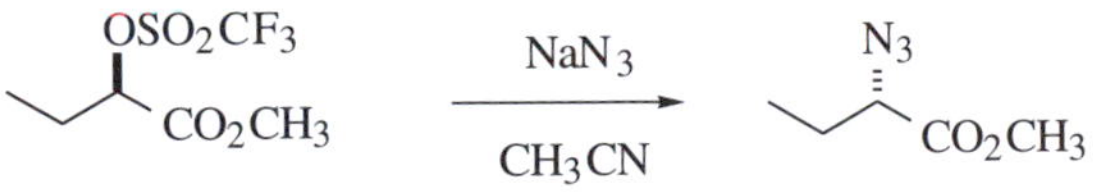

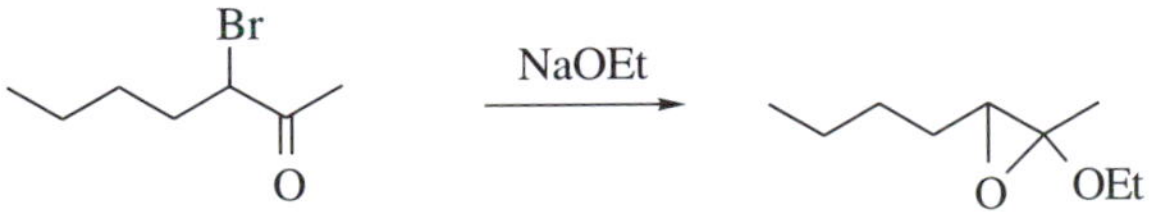

(*d*)
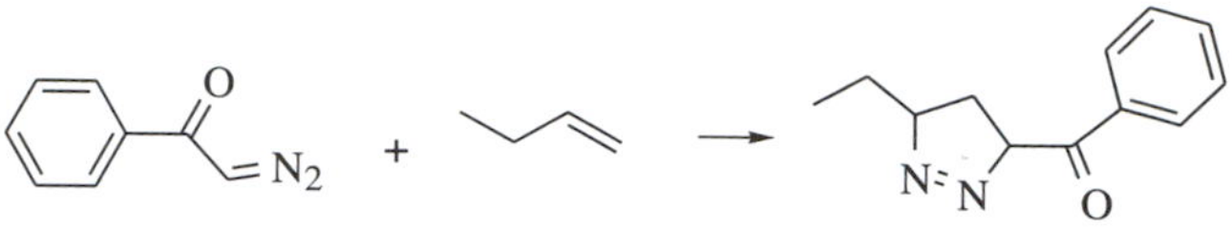
O
N2
+
N=N
O

(*e*)
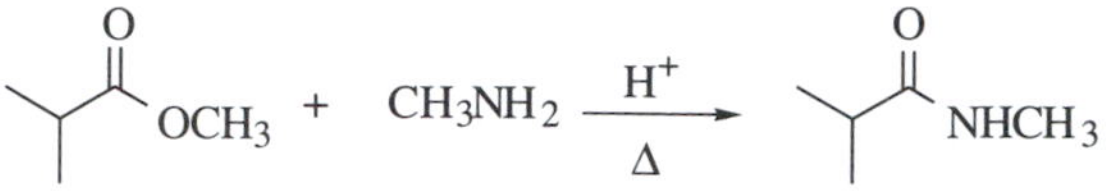
O
OCH3
+
CH3NH2
H+
Δ
O
NHCH3

(*f*)
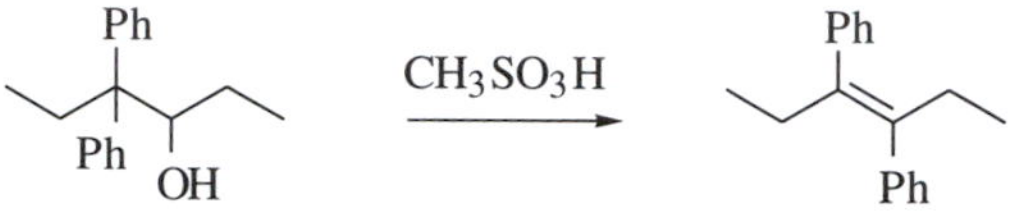
Ph
Ph
OH
CH3SO3H
Ph
Ph

(*g*)
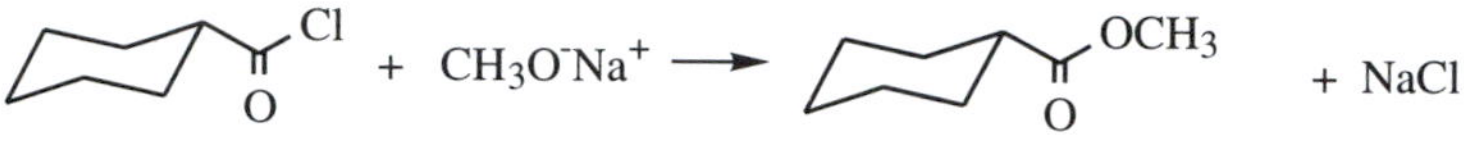
Cl
O
+ CH3O-Na+
OCH3
O
+ NaCl

(*h*)
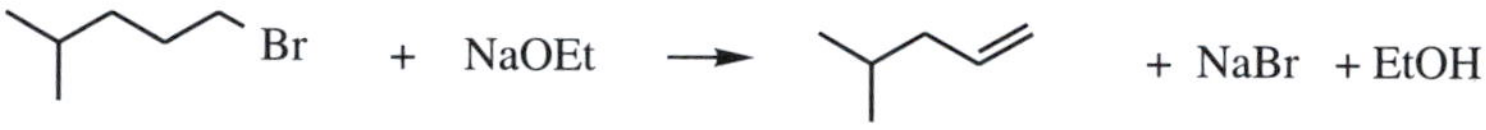
Br
+ NaOEt
+ NaBr + EtOH

(*i*)
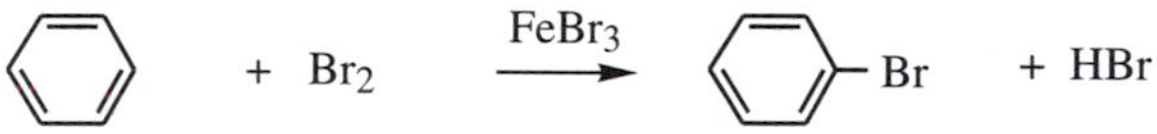
+ Br2
FeBr3
Br
+ HBr

(*j*)
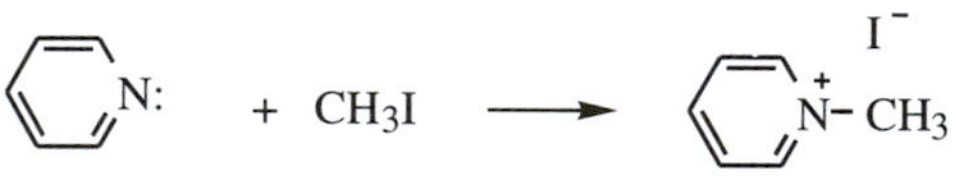
N:
+ CH3I
I-
N+ CH3

(*k*)
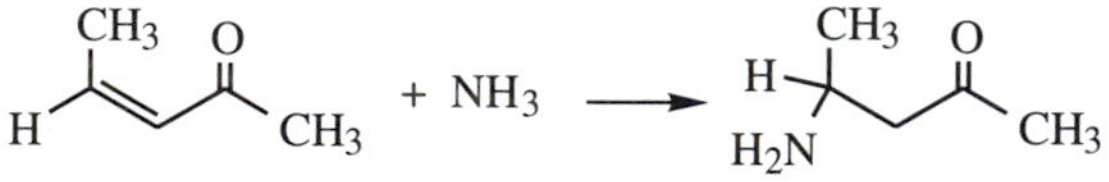
CH3
O
H
CH3
+ NH3
CH3
O
H
H2N
CH3

(*l*)
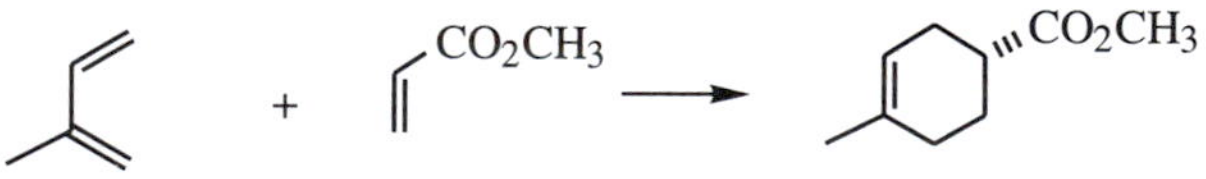
+
CO2CH3
CO2CH3

3. Using curved arrow notation, show how to derive three principal resonance structures of each structure below.

(a)

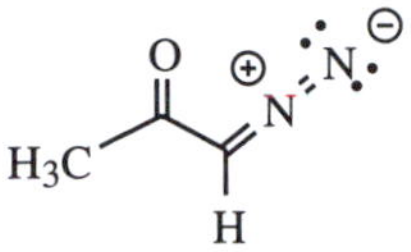

(b)

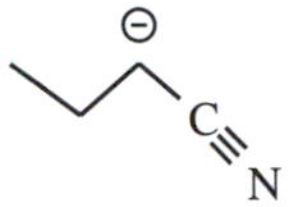

(c) CH₃O— —CH₂

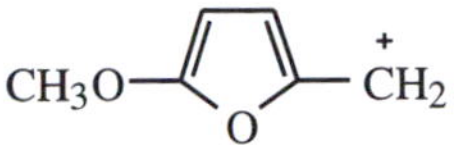

(d)

(e) H₂C

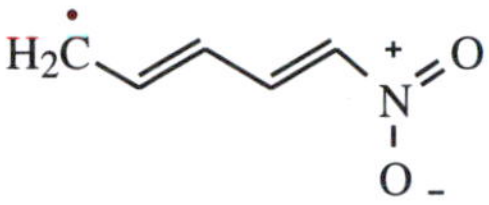

(f)

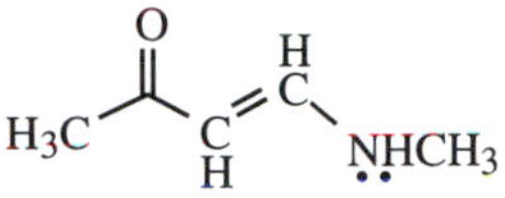

(g)

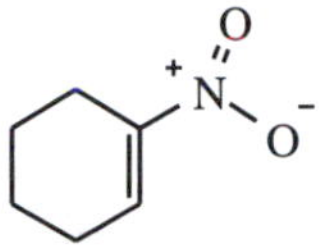

(h) R—N=N=N

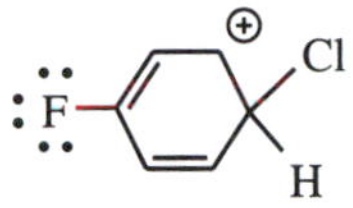

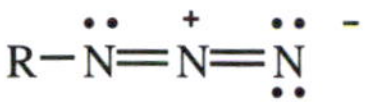

(i)

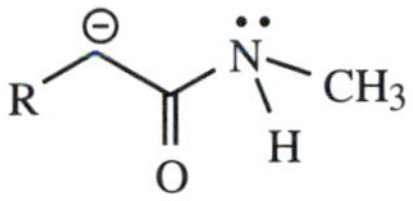

C H A P T E R **5**

STEREOCHEMISTRY AND

CONFORMATION

Stereochemical Structures

An important aspect of molecular structure is the stereochemistry, or spatial arrangement of groups, of the molecule. Furthermore, in many types of functional group manipulations and chemical reactions, the stereochemical outcome of the reaction may be just as important as the chemical outcome. Therefore, it is necessary to learn how to deal with stereochemical issues in molecules and how to control the stereochemistry of a chemical process.

In order to discuss spatial relationships of groups in molecules, we first have to be able to draw structures in such a way that the stereochemical features will be represented unambiguously. Thus a system is needed to depict in two dimensions the spatial relationships between groups in molecules that occur in three dimensions.

An early method to picture the three-dimensional properties of molecules is the use of Fischer projections. In Fischer projections, bonds are drawn either vertically or horizontally. Bonds which are vertical project into the space behind the plane of the paper (or blackboard, or computer screen). Bonds which are horizontal project into the space in front of the plane of the paper (or blackboard or computer screen).

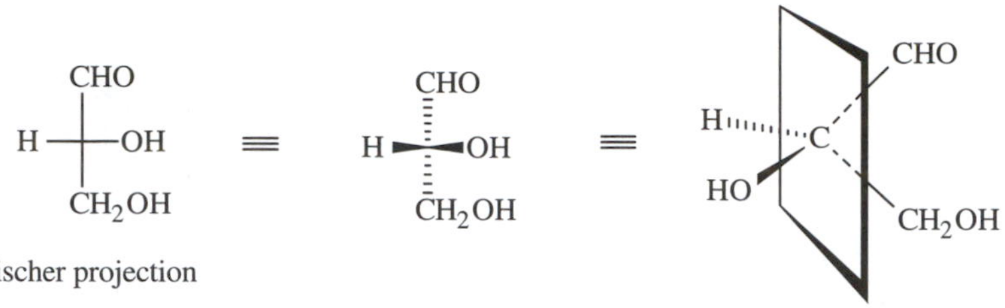

Fischer projection

This provides a perfectly good way to denote the stereochemical structure of a molecule if a few rules are followed.

Although it is a planar figure, the figure can be rotated only in increments of 180° *in the plane* of the paper, and it may not be taken out of the plane of the paper and flipped over. As seen in Figure 5.1, rotating the molecule 180° gives the identical molecule, whereas rotating the molecule 90° gives a nonsuperimposable isomer (the enantiomer). Within these constraints, however, Fischer projections are quite valid for showing the stereochemistry of a molecule.

A more recent approach to the three-dimensional depiction of molecules is the use of wedged and dashed lines. Simple lines are used to denote bonds in the plane of the paper (or blackboard, or computer screen). Wedged lines denote bonds projecting into the space in front of the plane, and dashed lines denote bonds projecting into the space behind the plane. These bonds are not restricted to any particular orientation, but are general. Bearing in mind the tetrahedral geometry of saturated carbon atoms, these figures can be turned and rotated at will as long as you can keep track of what happens to each of the four valences as the molecule is tumbled.

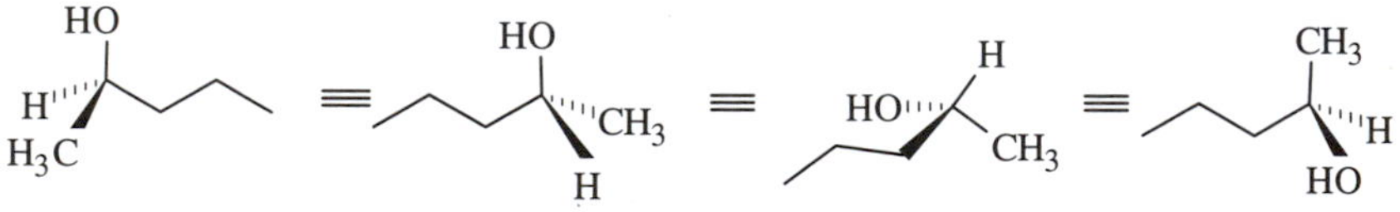

It may help to make models and practice manipulating these structures. An added benefit of depicting molecules using wedge and dash figures is that other types of molecules can be depicted using this convention. For example, the planar geometry of olefins and the twisted geometry of allenes are easily pictured using wedge and dashed bonds.

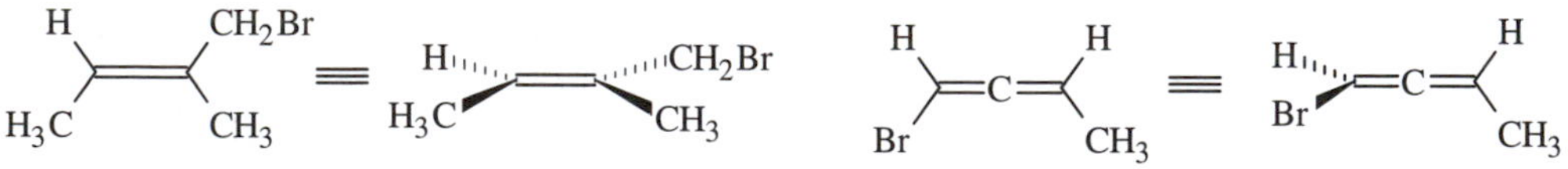

A further simplification of stereochemical notation for saturated carbon centers is to stretch out the carbon skeleton in the plane of the paper. Valences of atoms or groups other than hydrogen are indicated by a bold line if they project into the space in front of the plane, and with a dashed line if they project into the space behind the paper (or blackboard, or computer screen).

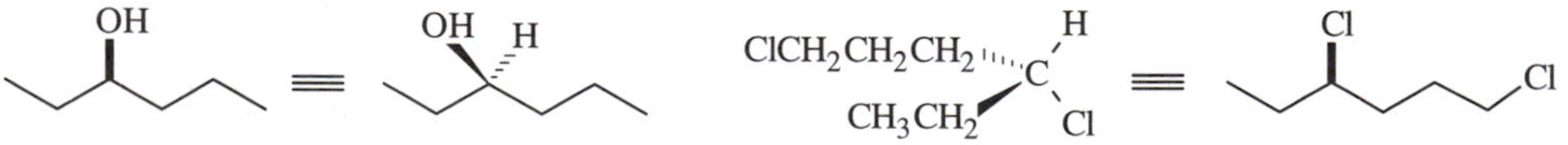

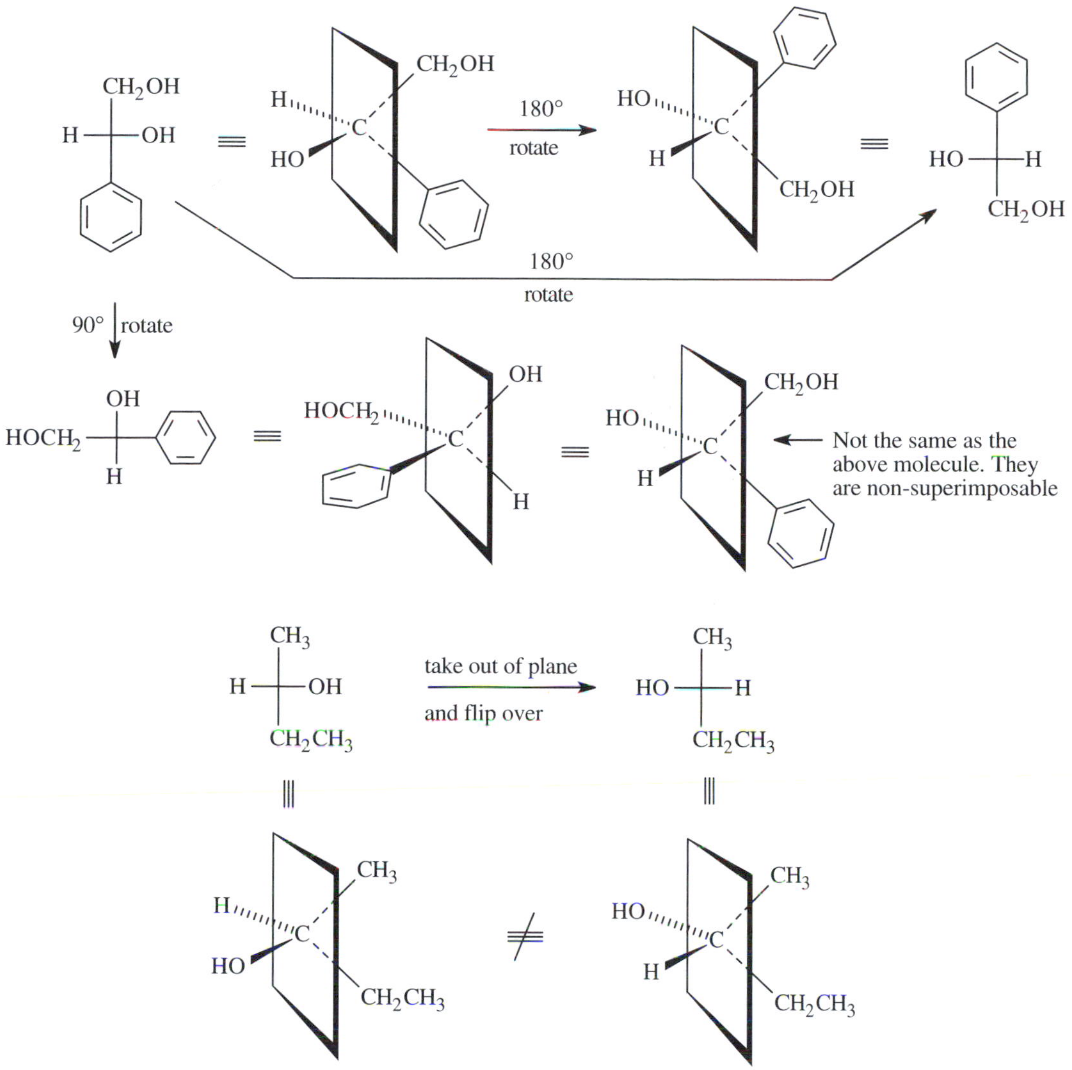

FIGURE 5.1

Other common methods for representing the three-dimensional structures of molecules include Newman projections, for showing conformational relationships, and sawhorse figures. Newman projections look down a carbon–carbon bond so that the front carbon, designated by a circle, obscures the carbon directly behind it. Valences (bonds) to the front carbon extend to the center of the circle, whereas bonds to the rear carbon stop at the circle. Sawhorse projections have the carbon–carbon bond at oblique angles and attempt to represent a perspective drawing of the molecule. Thus for 2-chloro butane, if one chooses to examine the 2,3—bond, the sawhorse and Newman projections would be

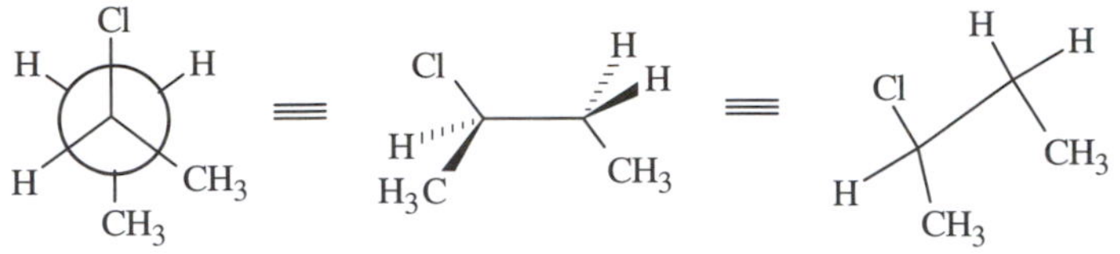

Newman projection   Sawhorse projection

Keeping in mind the three-dimensional properties of molecules, Newman projections can be converted to wedge–dashed structures or Fischer projections as desired. It is important to develop facility for manipulating structures and visualizing the three-dimensional properties of molecules from various stereostructures. The use of molecular models in conjunction with two-dimensional structures is often helpful in making the visual connections.

Chirality

Because of the tetrahedral geometry of saturated carbon and the associated three-dimensional properties, molecules can have chirality as one stereochemical feature. Any object is *chiral* if it is different from its mirror image (non-superimposable). Likewise, a molecule is chiral if it is non-superimposable on its mirror image. This requirement does not consider conformational changes (rotations about single bonds) as valid conditions for non-superimposability. Thus, for the molecules in Figure 5.2, the first is achiral because it is superimposable on its mirror image and the second is chiral because it is not superimposable on its mirror image.

When a molecule is chiral, it will have two isomeric forms called *enantiomers,* each of which is the mirror image of the other but which are non-superimposable. Enantiomers are distinct stereoisomers because they are compounds that have the same molecular formula and sequence of bonded elements, but differ in the spatial arrangement of groups in the molecule. In general, a molecule is chiral, and thus has two enantiomers, if it does not have planes or axes of symmetry.

There are several bonding patterns which can lead to asymmetry (lack of symmetry elements) in a molecule, but the most common is a tetrahedral carbon atom which is bonded to four different groups (Figure 5.3). A tetrahedral carbon with four different groups attached is

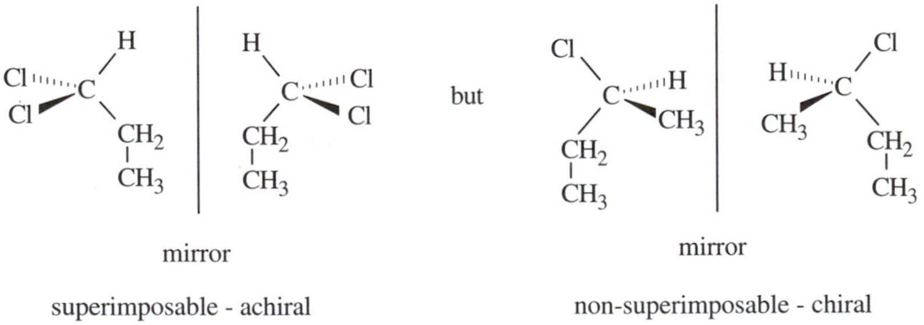

mirror   mirror

superimposable - achiral   non-superimposable - chiral

FIGURE 5.2

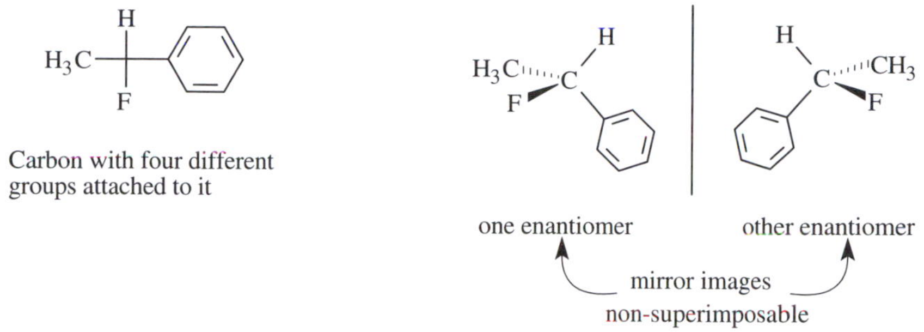

Carbon with four different groups attached to it

one enantiomer   other enantiomer

mirror images
non-superimposable

FIGURE 5.3

described variously as a chiral center, a chiral carbon, a stereogenic center, or an asymmetric center because that carbon lacks symmetry elements. On the other hand, a tetrahedral carbon with two or more of the same groups attached automatically has a plane or axis of symmetry associated with it, so it is achiral (not chiral). When a molecule contains a chiral center, that molecule lacks symmetry elements and is thus chiral, and it can have two enantiomers which are non-superimposable mirror images of each other.

The four different groups attached to a chiral carbon can be different elements, isotopes, or functional groups, and chiral centers can be present in either open-chain molecules or cyclic compounds. The recognition of chirality and chiral centers in molecules is an important step in determining the numbers of stereoisomers that are possible for a given compound.

Configuration of Chiral Centers

As noted, when a single chiral center is identified in a molecule, there will be two stereoisomers (enantiomers) of that molecule (Figure 5.4). Enantiomers differ only by the spatial arrangements of groups around the chiral center; however, they are distinct isomers. In order to discuss enantiomers, it is necessary to have some way to name, or denote, which one is which; that is, one must convert the configuration (spatial arrangement) of groups around the chiral

chiral center   chiral center   achiral

enantiomers   enantiomers   identical

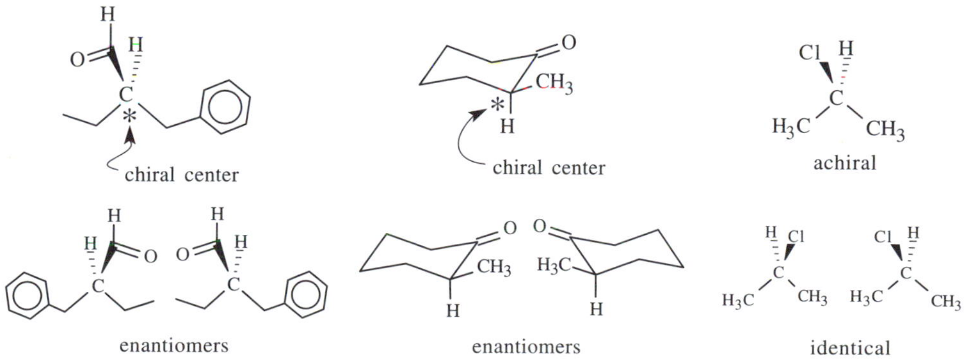

FIGURE 5.4

center into a name or designation. An early method was to designate enantiomers as either D or L, based on the relationship of their chiral center to the chiral center in D- or L-glyceraldehyde. This system of nomenclature was difficult to apply to larger molecules and particularly ones with more than one chiral center.

In response to this nomenclature dilemma, the Cahn–Ingold–Prelog (IUPAC) system of nomenclature was developed and is now the standard method to specify the relative configuration of chiral centers in molecules. Each chiral center will have two possible mirror image configurations which are designated as either R or S.

The strategy for determining whether the chiral center has an R or S configuration is based on the symmetry properties of a tetrahedral carbon. First, one assigns priorities to the four groups attached to the chiral center. Next, one orients the molecule so as to sight directly down the bond from the chiral carbon to the group of lowest priority. The remaining three bonds will form a trigonal array (as in Newman projections).

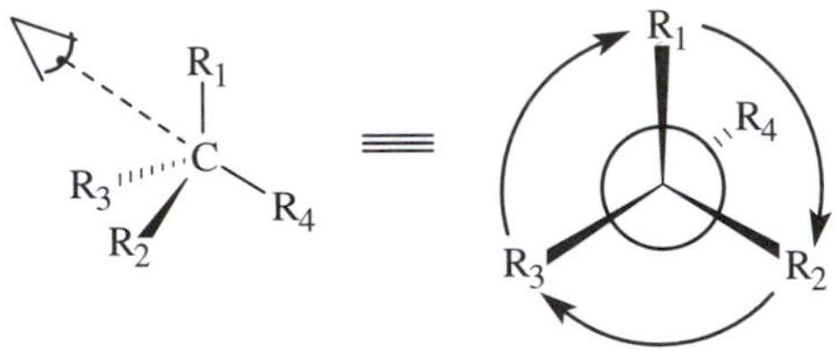

clockwise ∴ R configuration

(If this is not apparent from stereostructures, build a model and demonstrate that it is so.)

If one follows these three groups around a circle, from the group of highest priority to the group of lowest priority, one must proceed either clockwise (right), which is designated as the R configuration, or counterclockwise (left), which is designated as the S configuration. This provides a general method for assigning the configuration to any chiral center:

1. Assign priorities to the groups attached to the chiral center.

2. Orient the molecule so the group of lowest priority points directly away from one's eye.

3. Follow the direction of the remaining groups from the highest to lowest priority. If the procession is clockwise, the configuration is designated R; if the procession is counterclockwise, the configuration is designated S.

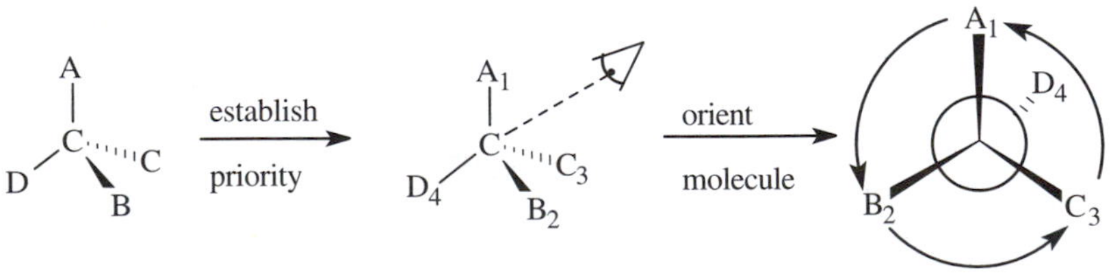

highest to lowest priority
is counterclockwise ∴ S

The other enantiomer will obviously have the R configuration.

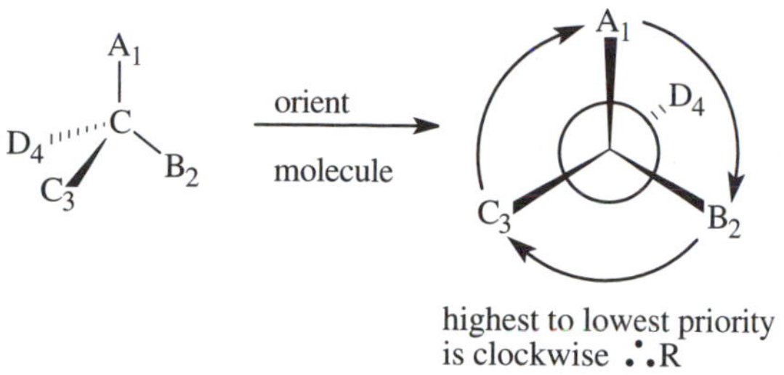

highest to lowest priority
is clockwise ∴R

The only remaining task is to assign priorities to the groups attached to the chiral center. This is done by the following rules:

1. Priority is first assigned on the basis of the atomic number of the atoms attached directly to the chiral center. Atoms of higher atomic number are given higher priorities. Thus for 1-bromo-1-fluoroethane, the ordering of priorities is $Br > F > C > H$, on the basis of their respective atomic numbers of $35 > 9 > 6 > 1$.

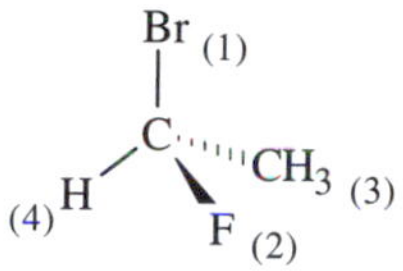

2. When assignment of priority cannot be made on the basis of atoms attached directly to the chiral center, proceed away from the chiral center and examine the next sets of atoms for differences in atomic numbers of attached atoms. Thus for 2-chlorobutane, two carbon atoms are attached to the stereogenic center. In order to establish which carbon group takes priority, note that the next atoms are H, H, H for the methyl group and H, H, C for the ethyl group. Thus the latter has higher priority because of the greater atomic number of carbon.

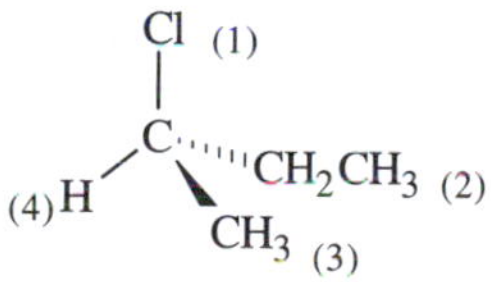

3. Groups containing multiple bonds are assigned priority as if both atoms were doubled or tripled. Thus a vinyl group is equivalent to a 2-butyl group by so-called phantom atoms.

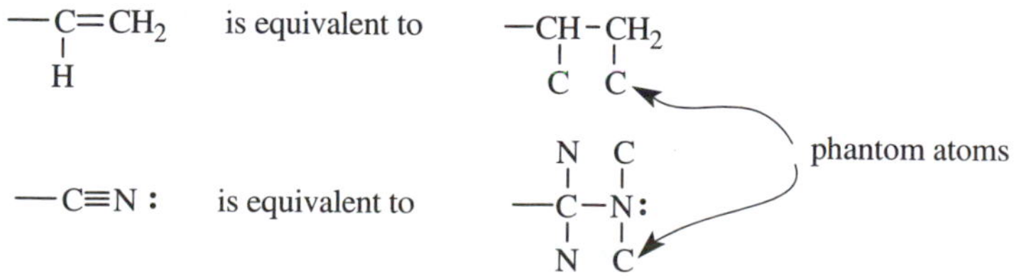

These rules and applications are summarized completely in most introductory organic texts. It is important to be able to assign R and S configurations to stereogenic centers in molecules and to construct chiral molecules, given the R or S configuration. In this way, it is very easy to determine the stereochemical relationships between stereoisomers.

Multiple Stereocenters

When there is more than one stereogenic center in a molecule, the number of possible stereoisomers increases. Since each stereogenic center can have either the R or S configuration, for a molecule of n chiral centers, there will be 2^n possible stereoisomers. Thus 3-phenyl-2-butanol has two stereogenic centers and four possible stereoisomers. These are shown with the configuration of each chiral center designated:

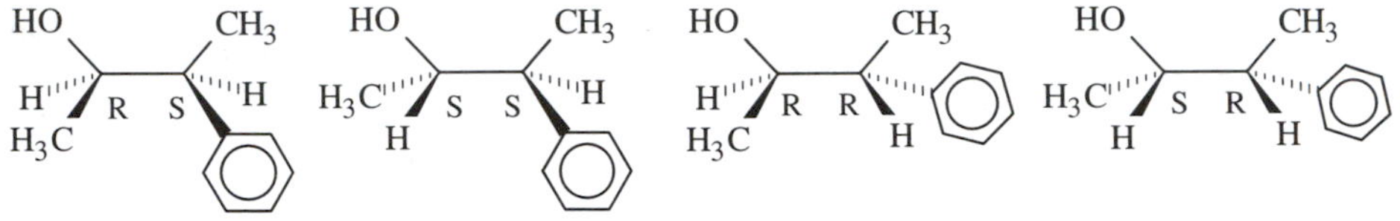

The configuration of each stereogenic center can be determined in the usual way, but there is a much faster way to draw the four stereoisomers. First draw a stereostructure of the molecule, such as **A**, then go through the process of determining the R or S configurations. Since each stereogenic center can be only R or S, and since the configuration of a particular stereogenic center can be changed merely by exchanging any two valences, the remaining stereoisomers of **A** can now be generated easily by switching valences of one stereo center (**B**), then the other (**C**), then both (**D**). Furthermore, if the relative configurations (R, S) are known for **A**, the configurations of the other stereoisomers are immediately known, since exchanging any two valences inverts the configuration (R goes to S). For 3-phenyl-2-butanol, structure **A** has the configurations (2R,3S)-3-phenyl-2-butanol. (The stereochemical information is denoted by the number of the carbon and its configuration enclosed in parentheses before the name of the compound.) By exchanging the hydrogen and methyl groups (C-2) of **A**, the configuration of C-2 is inverted and isomer **B** is produced, the 2S,3S isomer. Exchange of the hydrogen and phenyl groups at C-3 of **A** inverts the configuration of C-3, and isomer **C** is produced, the 2R,3R isomer. Exchanging the hydrogen and methyl group at C-2 and the hydrogen and phenyl group at C-3 of **A** gives isomer **D**, the 2S,3R isomer. The same configurational changes at C-2 of **A** could be accomplished by exchanging any two valences, such as hydrogen and hydroxyl or hydroxyl and methyl. The same is true for configurational changes at C-3.

Thus given the configurations for the three chiral centers of an aldopentose such as **D**, one can rapidly write down the configurations of the chiral centers of **E**, **F**, and **G** as in Figure 5.5. Since these are stereoisomers of **D**, one need not apply the Cahn–Ingold–Prelog rules for each isomer. Rather, one must simply note which substituents are switched relative to those in **D**. If the configuration is switched from that in **D**, the designation (R,S) will also be switched from that in **D**. Isomers **D–G** are only four stereoisomers out of eight possible stereoisomers for an aldopentose which has three stereocenters ($2^3 = 8$).

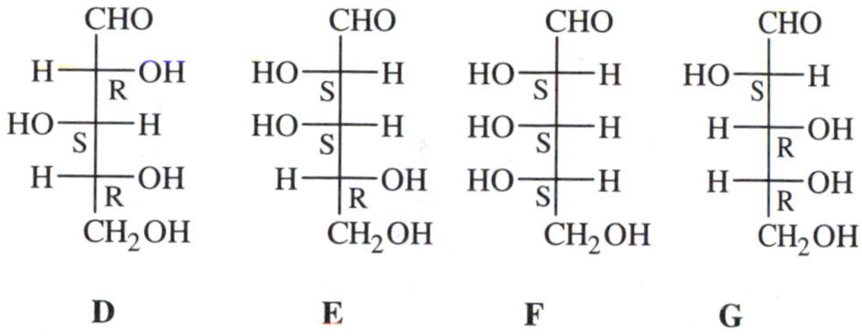

FIGURE 5.5

Now that one can generate the stereoisomers of compounds with more than one chiral center, it is appropriate to ask what the relationships are between these isomers. Thus 2-bromo-3-acetoxy butane has two chiral centers and four stereoisomers:

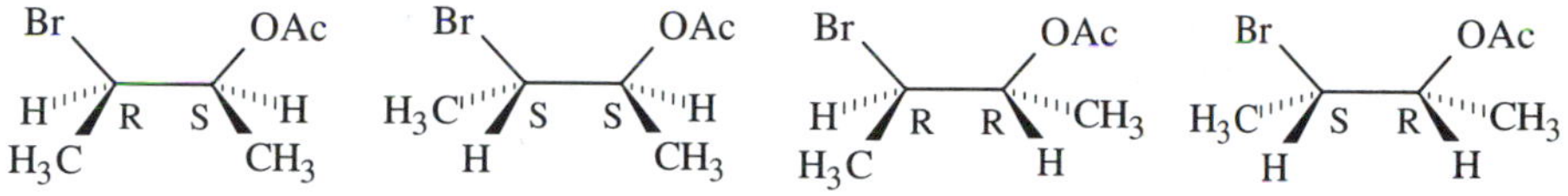

Since each carbon of **1** is the mirror image configuration of the carbons in **4** (i.e., C-2 of **1** is R, C-2 of **4** is S, C-3 of **1** is S, and C-3 of **4** is R), the molecules themselves are mirror images, but they are non-superimposable. They are thus enantiomers. This relationship can also be shown by reorienting the molecules to see that they are mirror images (Figure 5.6*a*) but non-superimposable, and are therefore enantiomers (Figure 5.6*b*) A similar analysis reveals that **2** and **3** are also enantiomers. Comparison of any other pairs of stereoisomers, **1** and **2**, for example, shows that they are not mirror images: the C-2 of **1** is R and the C-2 of **2** is S, *but the*

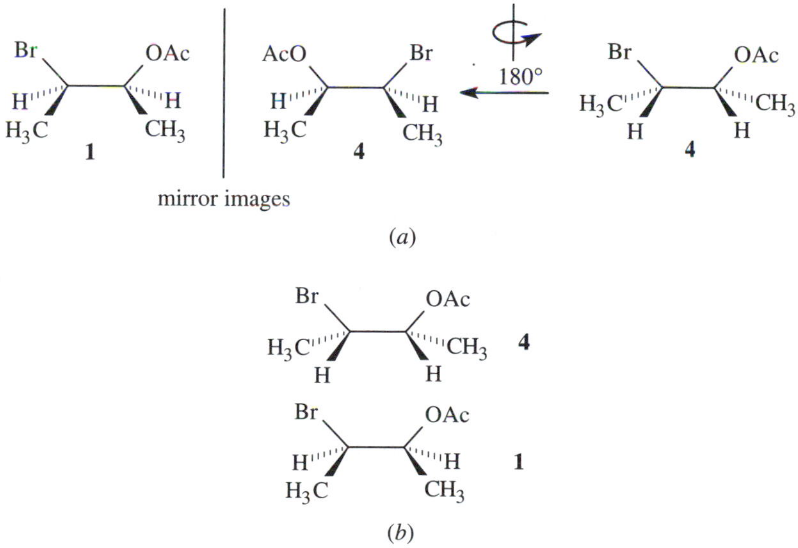

mirror images

(*a*)

(*b*)

FIGURE 5.6

C-3 of both **1** and **2** is S. Isomers **1** and **2** are also not superimposable. So **1** and **2** are a second type of stereoisomer and are non-superimposable, non-mirror images which are called *diastereomers*. Diastereomers have the same molecular formula and sequence of bonded elements but different spatial arrangements and are non-superimposable, non-mirror images.

A third type of stereoisomer occurs when a molecule with several stereogenic centers contains an internal plane of symmetry. This usually happens when two of the stereogenic centers are attached to the same four different valences. For example, 2,4-dibromopentane has two stereogenic centers and thus four stereoisomers (**5-8**).

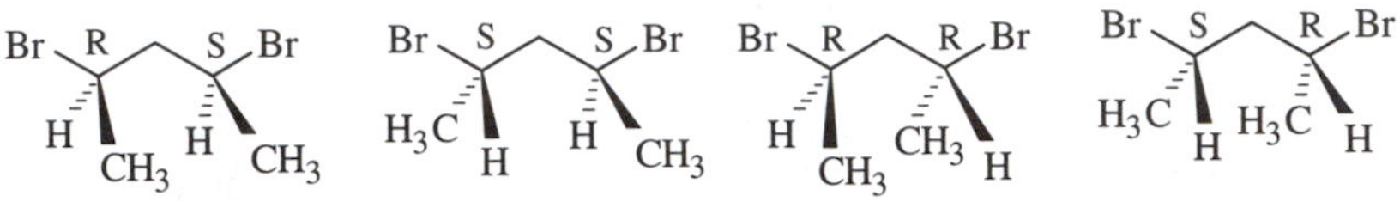

It is easily seen that **6** and **7** are enantiomers, **5** and **6** are diastereomers, and so on. However, **5** and **8** are identical. Although there are two chiral centers in **5** (and **8**), the molecule itself is achiral because it contains an internal mirror plane. Thus it has a plane of symmetry. Structure **8** is superimposable on **5** by a 180° rotation and is thus the same compound. This molecule is called a *meso* isomer, which is a compound that contains chiral centers but which itself has a plane of symmetry. Even though 2,4-dibromopentane has two stereogenic centers, there are really only three stereoisomers, a pair of enantiomers and a *meso* compound, which is diastereomeric with the enantiomeric pair.

It is clear from the previous examples that the presence of stereogenic centers in molecules leads to stereoisomers. There is another type of molecule which itself is chiral, but which has no chiral center. The molecular chirality arises from the presence of a screw axis in the molecule. Allenes and biphenyls are examples of such compounds, and because they are achiral, they exist as enantiomers.

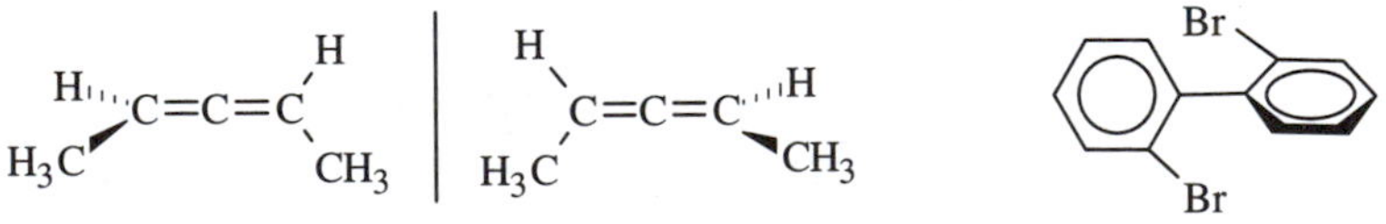

It has been shown that when more than one stereocenter is present in a molecule, both enantiomers and diastereomers are possible. Distinguishing among enantiomers requires the relative configurations of each stereogenic center to be specified. However, to distinguish diastereomers, only the relative spatial orientation of groups needs to be specified. For example, aldotetroses have two stereocenters and four stereoisomers, as seen in Figure 5.7. The enantiomeric

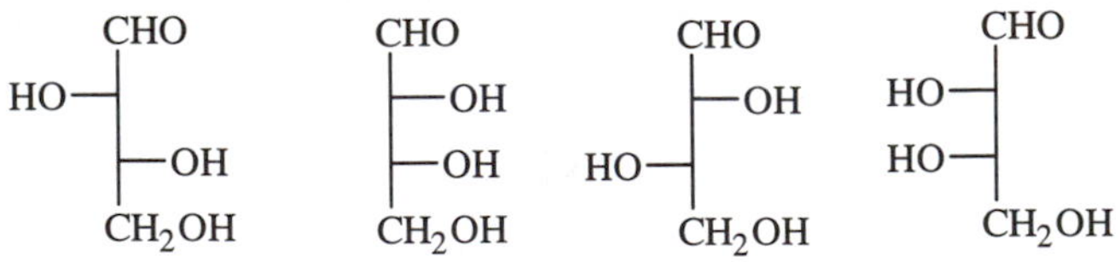

FIGURE 5.7

erythro isomer   threo isomer

FIGURE 5.8

relationship between D-threose and L-threose is specified by the 2S,3R and 2R,3S configurations (each stereocenter is the mirror image of the other). Moreover, the enantiomeric relationship between D-erythrose and L-erythrose is clear from the 2R,3R and 2S,3S configurations.
However, threose and erythrose are diastereomers. The different spatial orientation of the —OH
groups extending from the chain in the Fischer projections makes the diastereomeric relationship obvious without specifying the configuration; that is, they are clearly non-superimposable
and non-mirror images.

By extension, other diastereomeric pairs of molecules which contain two adjacent stereogenic centers can be designated as threo or erythro, depending on whether substituents extend
to opposite (threo) or the same (erythro) sides of the Fischer projection of the molecule. For
example, see Figure 5.8. The threo and erythro designations denote a diastereomeric relationship of the isomers. Each threo and erythro isomer will also have enantiomers which will also
have a threo–erythro diastereomeric relationship to each other.

More recently, a new method for designating the stereochemical relationship of diastereomers has been developed. In this method, the carbon backbone is extended in the plane of
the paper, blackboard, or computer screen in the horizontal direction. Groups will extend from
this backbone either in front of the plane or behind it and are designated by bold or dashed
bonds, respectively. If two substituents extend in the same direction, their spatial relationship
is designated *syn*; if they extend in opposite directions, their spatial relationship is designated
anti.

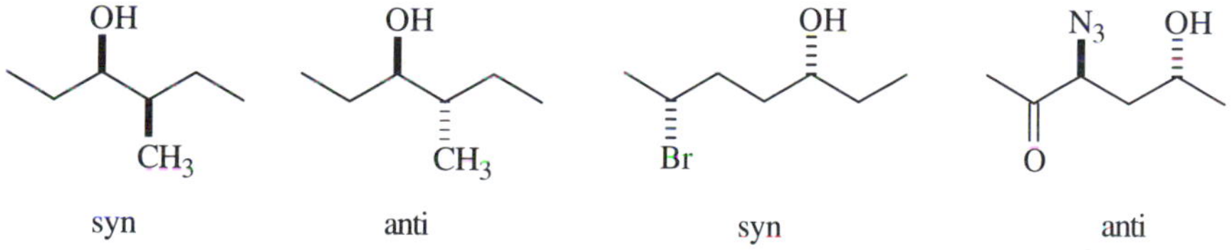

syn   anti   syn   anti

Molecules which are *syn–anti* isomers of each other are diastereomers, and there will be two *syn*
enantiomers and two *anti* enantiomers. The *syn–anti* designation is not restricted to substituents
on vicinal carbon atoms as is the threo–erythro designation, and is thus more versatile.

Optical Activity

We have seen how stereochemical relationships can be designated and distinguished. Now let
us see how the stereochemistry influences the chemical or physical properties of molecules.

Individual pure enantiomers are identical to each other in most respects in that they have the same physical properties, melting point, boiling point, refractive index, polarity, and solubility. The only difference between individual enantiomers is that they behave differently in chiral environments. For example, each enantiomer of an enantiomeric pair produces a rotation of the plane of plane polarized light to an equal but opposite extent, because plane polarized light is itself chiral and each enantiomer interacts differently with the light (Figure 5.9).

If a compound rotates plane polarized light, it is termed optically active. To be optically active a compound must be chiral, and one enantiomer of the compound must be present in excess over its mirror image. The enantiomer which produces clockwise rotation of plane polarized light is designated the (+) enantiomer, and the enantiomer which produces counterclockwise rotation of plane polarized light is designated the (−) enantiomer. At a given wavelength under standard conditions of concentration (1 g/mL) and path length (10 cm), a pure enantiomer will give the maximum rotation in degrees. Its pure mirror image will give an equal but opposite rotation.

The rotation of plane polarized light by a pure enantiomer is an inherent property of that enantiomer. However, the amount of rotation actually measured is dependent on the concentration of molecules in the light beam, the pathlength, and the wavelength of the light used. To account for these variables, the observed rotation is converted to the specific rotation $[\alpha]$, which is defined as the rotation observed for a solution of 1 g/mL concentration in a 10-cm pathlength cell. Furthermore, a wavelength of 589 μm, the D-line of sodium, is normally the standard wavelength for measuring the specific rotation, and 25°C is the standard temperature of the measurement. The specific rotation, $[\alpha]_D^{25}$, in which the superscript indicates the temperature of the measurement and the subscript D (or number) indicates the wavelength of light used to measure the optical rotation, is calculated from

$$[\alpha]_D^{25} = \frac{\alpha_{obs}}{l(cm)/10 \times C(g/ml)}$$

where α_{obs} is the observed rotation, l is the pathlength of the cell used for the measurement, and C is the concentration of the sample in grams per milliliter.

When both enantiomers are present in solution, the observed rotation will reflect the enantiomeric composition of the mixture. If equal amounts of enantiomers are present, the solution will not exhibit optical activity, because for each molecule that rotates light in one direction, there will be another molecule that rotates light in the opposite direction and the net rotation is zero. Such a mixture is called a *racemic mixture* and is indicated by (±). Thus (±)-2-butanol is an equal mixture of the R and S enantiomers of 2-butanol. In the liquid state, racemic mixtures have the same physical properties as the individual enantiomers.

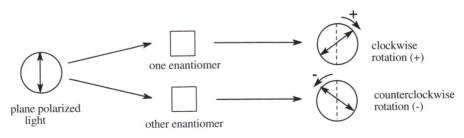

FIGURE 5.9

If one enantiomer is present in excess over the other, the solution will have a net rotation corresponding in sign (+ or −) to that of the more abundant enantiomer. The composition of the mixture is denoted by the optical purity or the percent enantiomeric excess (ee%). The enantiomeric excess is defined as ee = % major enantiomer −% minor enantiomer, and is a measure of the optical purity of the sample. Values range from 100% (pure enantiomer, ee = 100% − 0%) to 0% (racemic mixture or ee = 50% − 50%). A sample which has an optical purity of 92% is thus a mixture of 96% of one enantiomer and 4% of the other enantiomer.

Absolute Configuration

The observation of optical activity and the measurement of optical rotation distinguishes one enantiomer (+) from the other (−). The sign of rotation (+ or −) is thus an experimental way to differentiate the relative configurations of enantiomers. A second way to designate these stereoisomers is to assign the configurations as R or S based on stereostructures. The R and S configurations are relative configurations based on structures which are drawn on paper. However, it is not possible to predict a priori whether the R enantiomer (for example) will be dextrorotatory (+) or levorotatory (−).

The absolute configuration of an enantiomer is determined only when the optical rotation (+) or (−) and the relative configuration (R) or (S) can be matched. For example, the absolute configuration of lactic acid has been found to be R-(−) in that the R enantiomer is levorotatory. Conversion to the methyl ester does not change the configuration of the stereocenter, which remains R. However, the rotation is found to be positive, so the absolute configuration is R-(+).

$$CH_3$$
$$HO_2C \quad \overset{|}{\underset{OH}{\diagdown}} {}^{''''}H \qquad\qquad MeO_2C \quad \overset{|}{\underset{OH}{\diagdown}} {}^{''''}H$$

R-(-)-lactic acid   R-(+)-methyl lactate

This illustrates that, whereas the relative configuration (R, S) can be used to show the structures of stereocenters, the absolute configuration is required to show the changes in configuration that occur during a chemical sequence.

Physical Properties of Enantiomers

If both enantiomers are present in a solid sample, the melting point and the solubility of the solid mixture are often found to be different from those of the pure enantiomers. This is due to the fact that the solid state interaction of two R enantiomers or two S enantiomers is often different from the solid state interaction of an R and an S enantiomer. (In fact, these interactions are diastereomeric.) The result is that three different scenarios are possible when a racemic mixture is crystallized from solution:

1. If an enantiomer has a greater affinity for molecules of like configuration, two sets of crystals will be formed, those of the (+) form and the (−) form. This racemic mixture

is called a conglomerate and behaves as a typical mixture — the melting point is lower than the pure enantiomeric components and the solubility is higher. This is a relatively rare situation.

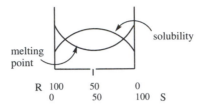

2. If an enantiomer has a greater affinity for molecules of opposite configuration, crystals are produced which contain equal numbers of the (+) and (−) forms. The solid compound, which has properties different from either pure enantiomer and exists only in the solid state, is called a *racemate* or a racemic compound or a racemic mixture. A racemate is often higher melting and less soluble than a pure enantiomer and behaves as a mixture in the presence of either pure enantiomer.

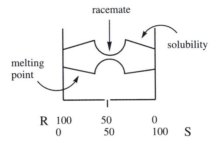

This is the most common situation and allows an unequal mixture of enantiomers to be purified. Upon crystallization, the racemate will precipitate first, leaving behind the enantiomer in excess.

3. If one enantiomer has similar affinity for molecules of either configuration, the enantiomers are randomly distributed in the crystal and the solid is a *racemic solid solution*, or mixed crystal. Such solids are identical with either enantiomer.

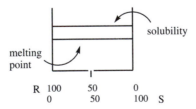

Resolution of Enantiomers

It has been shown that individual enantiomers have identical physical properties, and can be distinguished only in a chiral environment. Plane polarized light is such a chiral environment, and one enantiomer is dextrorotatory and one is levorotatory. Another way to distinguish enantiomers is to allow them to react (or interact) with other chiral molecules. The interaction of a mixture of enantiomers with a single enantiomer of a chiral molecule produces a mixture of diastereomers:

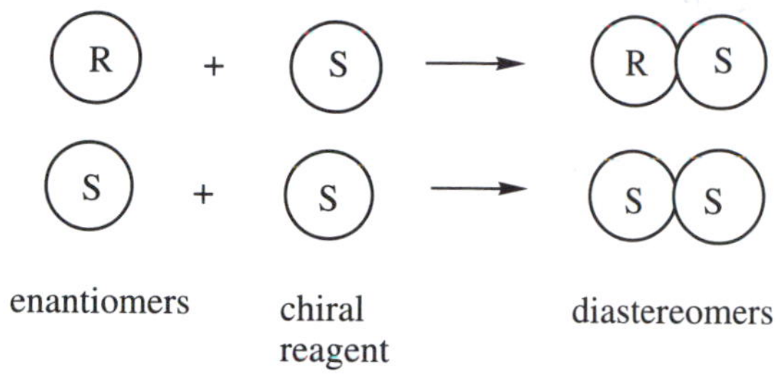

enantiomers   chiral reagent   diastereomers

Since diastereomers have different physical properties, they can then be separated on the basis of those physical properties. After separation of the diastereomers, the individual enantiomers are reclaimed, and in this way the two enantiomers will have been separated. Such interactions form the basis for all separations of racemic mixtures into pure enantiomers, termed *resolution* of enantiomers. There are several experimental techniques used to resolve enantiomers, but all utilize a chiral reagent of some type to furnish the chiral environment needed to distinguish the enantiomers.

Crystallization has been a traditional method for separating the diastereomers produced from a racemic mixture and a chiral resolving agent. For example, racemic carboxylic acids can be treated with an optically active alkaloid (which is basic) and the resulting diastereomeric salts are separated by crystallization. The individual enantiomeric acids are then regenerated from the salts. A variety of alkaloids have been used as resolving agents for racemic acids. They include brucine strychnine, ephedrine, quinine, morphine, and α-phenyl ethylamine among others.

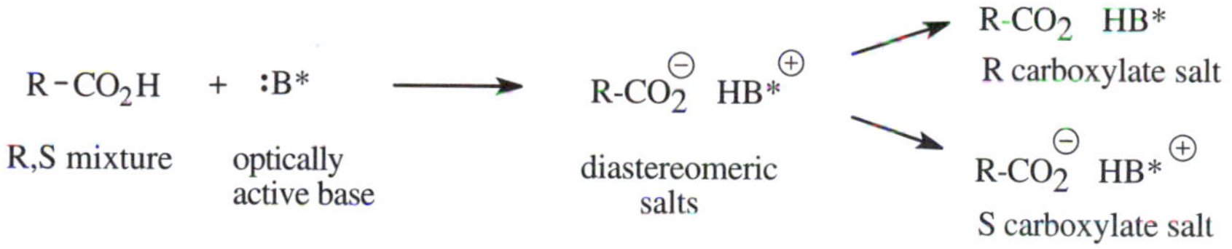

Racemic bases can be resolved by treating them with an optically active acid and separating the resulting diastereomeric salts by fractional crystallization. The individual enantiomeric bases are regenerated from the salts. Common acid-resolving agents include camphorsulfonic acid and derivatives of it, tartaric acid, malic acid, and pyroglutamic acid among others.

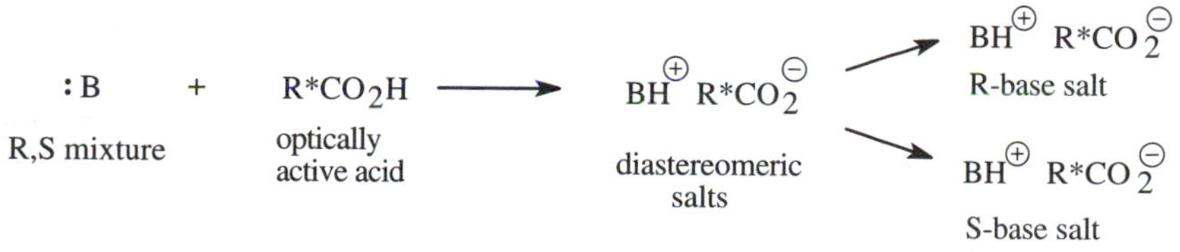

Alcohols are often resolved by conversion to half-esters of phthalic acid or succinic acid which are then resolved as typical acids. The alcohol is then regenerated from the resolved half-ester by either hydrolysis or reductive cleavage with LAH. A second method for resolving alcohols is to convert them to esters of optically active acids. This gives a mixture of diastereomeric esters which is separated by fractional crystallization and the alcohol is recovered by either hydrolysis or reductive cleavage.

In recent times, chromatography has become a major technique for separations and has increasingly supplanted fractional crystallization as a way to separate diastereomeric compounds. Not only do diastereomers have different solubilities; they also interact with surfaces such as silica gel or alumina differently. The mixture of diastereomeric esters obtained by coupling a racemic alcohol to optically active acid can often be separated by HPLC, radial chromatography, or flash chromatography. Chromatography is often much faster and more efficient than crystallization. The individual alcohols can be regenerated in the usual fashion.

The preceding methods for the resolution of enantiomers rely on the formation of strongly bound diastereomers (ionic or covalent) which are then separated. It has become more and more common to use weak interactions as a means of resolving enantiomers. Chiral chromatography columns are useful for the separation of a variety of compounds, including amino acids. A chiral substance is permanently attached to the column surface. If a mixture of enantiomers is passed over the surface, the individual enantiomers will interact with the chiral surface differently and thus will elute along the column at different rates. (The enantiomer which interacts with the surface more strongly will elute more slowly.) They can thus be collected individually. A variety of chiral stationary phases are available to separate an ever-increasing number of examples.

The use of enzymes to resolve enantiomers has become an extremely popular method only recently. Enzymes are chiral catalysts which often exhibit very high selectivity for one enantiomer of a racemic mixture. Since enzymes are soluble in aqueous solution, it is often impossible to get sufficiently high concentrations of organic substrates in the aqueous medium to achieve conversion at any reasonable rate. The finding that a variety of esterases (lipases) can function very well in organic solvents has removed this major stumbling block to the practical utilization of biochemical transformations for resolution. In addition, a variety of enzymes are available commercially.

Thus it is very easy to acetylate a racemic alcohol and treat the racemic mixture of acetates with a lipase. One enantiomer is hydrolyzed to the alcohol and the other remains as the ester. These are separated chromatographically and each component is obtained with high optical purity. This technique is becoming more important and could be the most general technique for resolution in the near future.

$$
\text{ROH} \xrightarrow{\text{Ac}_2\text{O}} \text{ROAc} \xrightarrow{\text{lipase}} \text{R*OH} \quad + \quad \text{R*OAc}
$$

R,S mixture	R,S mixture	resolved alcohol R or S	resolved ester S or R

The use of kinetic resolution to obtain a single enantiomer from a mixture of enantiomers is often useful for particular functional groups. Since individual enantiomers react at different rates with chiral reagents, treatment of a racemic mixture with a limited amount (0.5 equivalent) of a chiral reagent will convert one of the enantiomers to product in preference to the other. After workup, one enantiomer will be recovered unchanged while the other will have been converted to a new product. The efficiency of the kinetic resolution will depend on the relative rates of reaction of the two enantiomers. If rates of reaction (selectivity factor) vary by more that 100, the recovered enantiomer will be greater than 99 percent optically pure. Lower selectivity factors will lead to less pure enantiomers.

Optically active diisopinocamphenylborane can be used to resolve racemic olefins. The reagent adds to one enantiomer, and the other is unchanged. Optical purities on the order of about 37–65 percent are possible. Chiral allylic alcohols can be resolved with chiral epoxidizing agents derived from tartrate complexes of titanium. One enantiomer is epoxidized and the other is not, and is thus separable. Use of the other tartrate isomer reverses the stereoselectivity. Selectivities on the order of > 100 are possible with this method. As in any kinetic resolution, however, only one enantiomer can be recovered. The other is converted to a different chiral product.

Even if the separation of enantiomers by any of the preceding methods is not completely successful, it is often possible to further raise the enantiomeric excess by crystallization or chromatography. As a result, many pure enantiomers are now available.

Stereoselective Reactions

Chiral compounds are very important substances. Many natural products, medicinal compounds, and biomolecules exist as single, optically active stereoisomers. Furthermore, the opposite enantiomer or diastereomer may not have any physiological activity and may, in fact, have a detrimental physiological effect. There is therefore great interest in reactions in which only one stereoisomeric form of a compound is produced by a particular synthetic sequence.

The simplest approach is to use a starting material of known absolute configuration and manipulate it to the final product, using reactions whose stereochemical outcome at the chiral center is known or predictable. For example, the synthesis of unusual amino acids as single enantiomers can begin with a "normal" amino acid from the chiral pool. (The *chiral pool* is a large group of molecules which are readily available and whose absolute configurations are known with certainty. Often these molecules are, or are derived from, natural products which can be isolated from natural sources in enantiomerically pure form.)

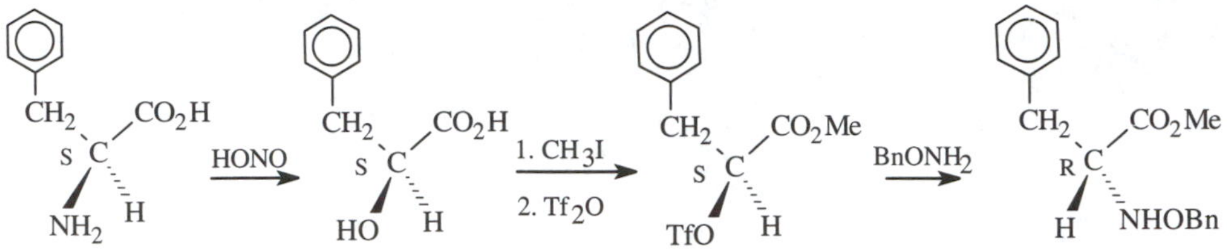

Since the stereochemical changes in each reaction of the sequence are known, a particular amino acid starting material (R or S) will give a particular configuration in the product. In this strategy of asymmetric synthesis, all or part of the final molecular skeleton is derived from the chiral precursor. While simple, this strategy is limited by the size of the chiral pool and by the types of reactions which occur stereospecifically at tetrahedral centers.

It is much more common for reactions to produce new chiral centers from achiral starting materials. Consequently, if one is to use the whole arsenal of synthetic methods available and at the same time produce single stereoisomers, one must be able to control (or at least understand) the stereochemistry of reactions occurring at achiral centers.

Formation of Enantiomers

Since chiral centers are generally tetrahedral, the conversion of trigonal centers to tetrahedral centers by some type of addition process is the most common way in which new chiral centers are created. The reaction of carbonyl groups with nucleophiles is a classic example. If substituents on the carbonyl group and the nucleophile are all different, a new chiral center is produced, as in the reaction of acetophenone with sodium borohydride to produce 1-phenylethanol:

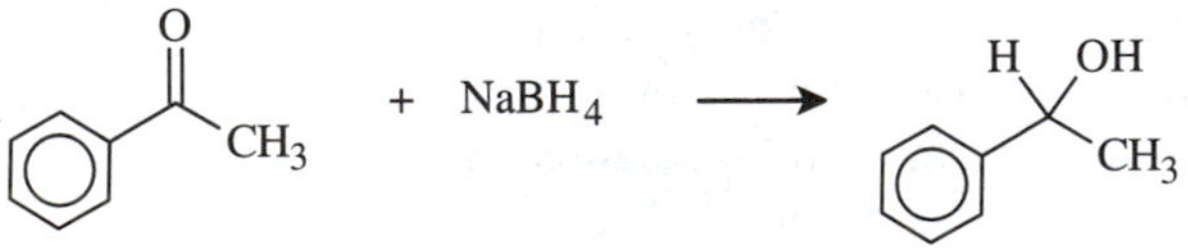

The carbonyl group is trigonal and planar and can be thought of as having two faces. Addition of hydride to one face gives one enantiomer, and addition to the opposite face gives the opposite enantiomer. As rewritten, attack from above gives the R enantiomer, and attack from below gives the S enantiomer. The faces are stereochemically nonequivalent since different stereoisomers are produced.

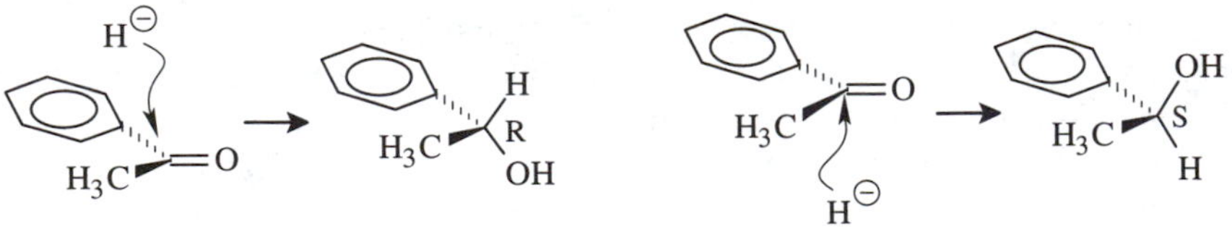

In order to differentiate the faces of a carbonyl group, the Re–Si nomenclature has been developed. The groups around the carbonyl carbon are given priorities by the same rules used in

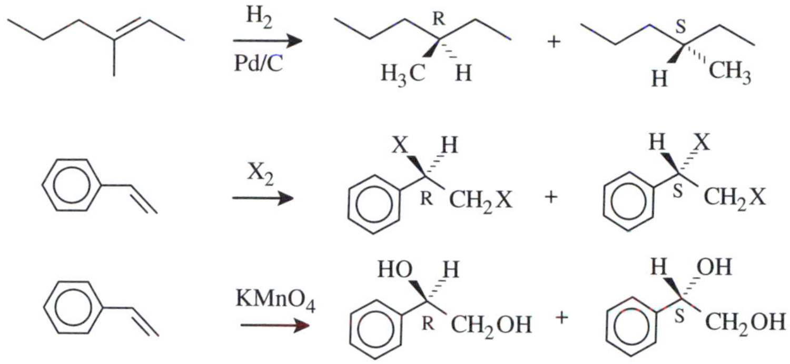

FIGURE 5.10

the Cahn–Ingold–Prelog system for R,S nomenclature. Then, going from the group of highest priority to the group of lowest priority around the face of a carbonyl group, proceeding in the clockwise direction defines the Re face and proceeding in the counterclockwise direction defines the Si face.

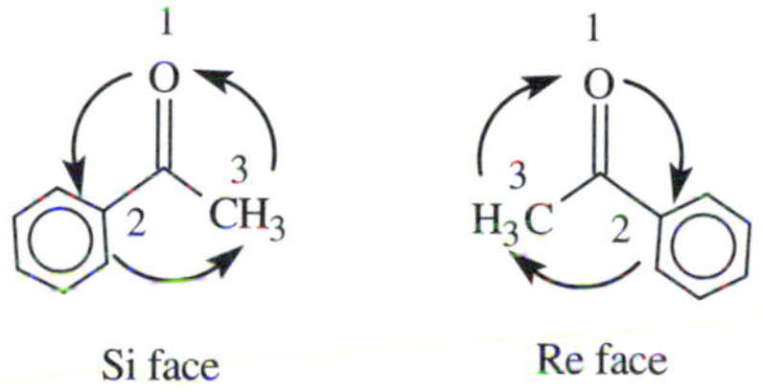

Si face   Re face

The Re–Si nomenclature enables the faces of a carbonyl group to be differentiated stereochemically; however, the carbonyl group itself is achiral. Moreover, the Re–Si designation is not indicative of the stereochemistry of the chiral center produced by addition. In the preceding example, hydride addition to the Si face gives the R enantiomer, whereas hydride addition to the Re face gives the S enantiomer. If ethyl lithium were added, the stereochemistry would be reversed; that is, Si → S and Re → R.

New chiral centers are also produced by addition reactions to other trigonal centers. Hydrogenation of 3-methyl-3-hexene gives 3-methylhexane. Clearly, the addition of hydrogen to one face of the planar olefinic system gives one enantiomer and addition to the opposite face gives the opposite enantiomer. Likewise, reaction of styrene with chlorine or bromine (X_2) or potassium permanganate produces products with a new chiral center. Formation of the two possible enantiomers results from addition to either face of the olefin (see Figure 5.10).

Reactive intermediates which are planar can also produce enantiomers. The acid-catalyzed addition of water to 1-pentene proceeds via a secondary carbocation. Because the carbocation is a planar intermediate, water can add to either face to give the R or S enantiomers.

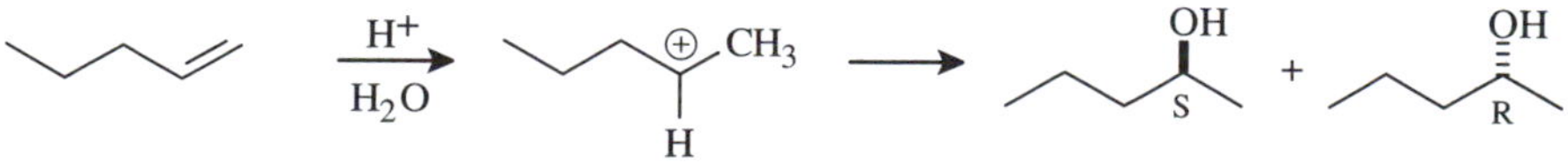

In reactions in which neither the reactants (C=O, C=C, C$^+$) nor the reagents (BH$_4^-$, EtMgBr, Br$_2$, H$_2$O, etc.) are chiral, there is no possibility for controlling which face undergoes addition (in fact, addition to either face is equivalent); thus a racemic mixture will be produced. Such processes are described as having no *enantioselectivity*.

Formation of Diastereomers

Diastereomers are defined as compounds which have the same molecular formula and sequence of bonded elements but which are non-superimposable, non-mirror images. Although Z, E isomers are one subclass of diastereomers that are achiral, the majority of diastereomeric compounds are chiral compounds which have more than one chiral center. Furthermore, it is important to recall that for a compound with *n* chiral centers, there will be 2^n stereoisomers. These will be divided into 2^n/2 pairs of enantiomers, and each pair of enantiomers will be diastereomeric with the other pairs of enantiomers. (This was discussed earlier in this chapter.)

One of the most direct ways to produce diastereomers is by addition reactions across carbon–carbon double bonds. If the structure of the olefin substrate is such that two new chiral centers are produced by the addition of a particular reagent across the double bond, diastereomers will result. For example, the addition of HBr to Z-3-chloro-2-phenyl-2-pentene produces 2-bromo-3-chloro-2-phenylpentane as a mixture of four diastereomers. Assuming only Markovnikov addition, the diastereomers are produced by the addition of a proton to C-3, followed by addition of bromide to the carbocation intermediate at C-2. Since the olefin precursor is planar, the proton can add from either face, and since the carbocation intermediate is also planar and freely rotating, the bromide can add to either face to give diastereomeric products. The possibilities are delineated schematically (but not mechanistically) in Figure 5.11.

Even though there are four possible diastereomeric products, they will not necessarily be formed in equal amounts. Diastereomers are not equal in terms of their energies; reactions which produce diastereomers reflect these energy differences and consequently proceed at different rates. Thus diastereomers are normally formed in unequal amounts. This is a very important concept, since it provides the kinetic basis for the stereoselectivity found in many different organic reactions. Restating this idea, if reactions produce diastereomers and thus proceed via diastereomeric transition states, the energy barriers for the formation of individual diastereomers will be different, the rates of formation of individual diastereomers will be dif-

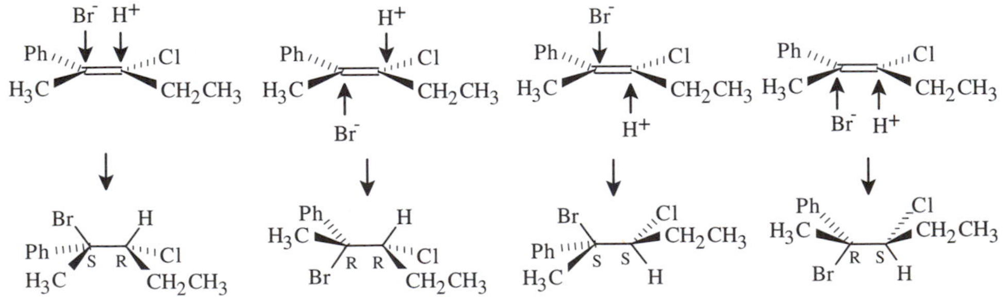

FIGURE 5.11

ferent, and the diastereomers will be formed in unequal amounts. The greater the differences in the energy barriers, the greater will be the differences in rates, and the more stereoselective will be the reaction. In contrast to HBr addition, which gives a mixture of diastereomers, a variety of other olefin addition reactions yield a single diastereomer from a starting olefin of defined stereochemistry. Furthermore, a starting olefin with a different stereochemistry will give a different single diastereomer. Such reactions are described as being *stereospecific*.

The diastereoselectivity for any process is often reported as a diastereomeric excess (de %) which is analogous to the optical purity reported for mixtures of enantiomers. The de% is defined as % de = (percentage of major diastereomer) − (percentage of minor diastereomer). For diastereospecific reactions in which a single diastereomer is produced, de % = 100, and for reactions in which there is no selectivity and diastereomers are produced in equal amounts, de % = 0.

A typical example of a stereospecific olefin addition reaction is the addition of bromine to olefins. If *cis*-2-pentene is used as the substrate, only the 2R,3R and 2S,3S pairs will be produced (they are enantiomers).

Because the addition of bromine is stereospecifically *trans* or *anti*, one bromine atom adds to each face of the olefin and can go to either carbon. If *trans*-2-pentene is used as the substrate, only the 2R,3S and 2S,3R pairs are produced (they are also enantiomers.). However, the pair from *cis*-2-pentene is diastereomeric with the pair from *trans*-2-pentene.

Stereospecificity is possible only if the inherent facial relationship of the olefinic bond is maintained throughout the addition process *and* only one bromine atom adds to each face. In the case of bromination, the electrophilic addition leads to a bridged bromonium ion which not only maintains the olefin geometry but also forces the second bromine to add from the opposite direction (*anti*).

(Contrast this to the addition of HCl or water to a double bond, where the intermediate is free to rotate so that the olefin geometry is lost, and both the proton and the nucleophile can add to either face.)

Other olefin additions which proceed via bridged intermediates should show similar stereospecificity, and addition should occur *anti*. Chlorination of olefins is an obvious analogy to bromination, but the addition of sulfenyl chlorides, oxymercuration, and epoxidation/hydrolysis

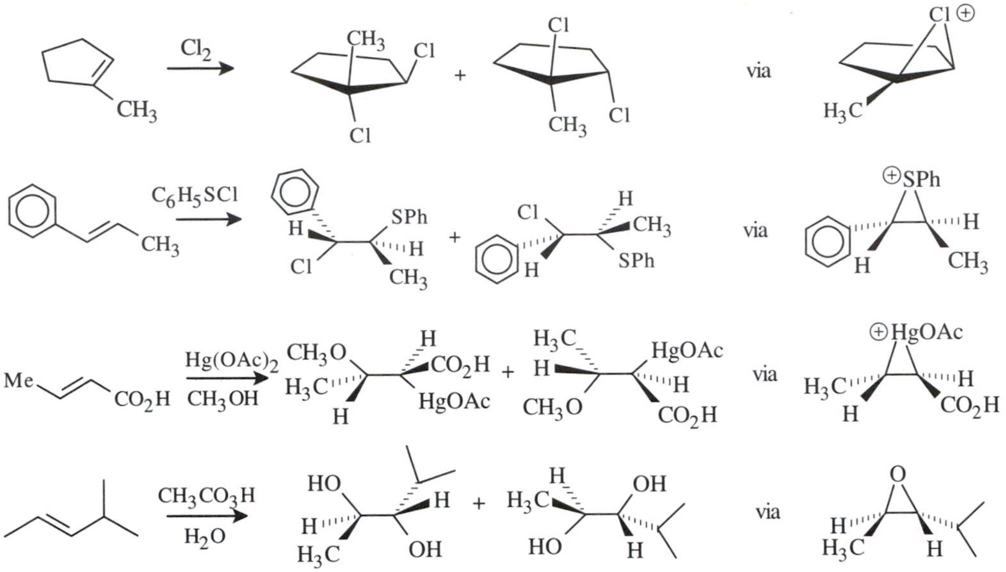

Stereochemistry and Conformation

FIGURE 5.12

all give stereospecific *anti* addition across the double bond because bridged intermediates are involved (see Figure 5.12).

There are other stereospecific olefin addition processes which occur with *cis* or *syn* stereochemistry. Common examples include catalytic hydrogenation, hydroboration/oxidation, and dihydroxylation using osmium tetroxide. The stereospecificity of these *syn* additions also requires that the facial properties of the olefinic bond be maintained throughout the addition process *and* that both new bonds be formed to the same face of the olefin. This is normally accomplished by a concerted *syn* addition to the π system (see Figure 5.13).

Stereospecificity in hydrogenation is gained by a surface-mediated delivery of the hydrogen atoms to one face of the olefin. Stereospecificity in both hydroboration/oxidation and osmium tetroxide/reduction results from a concerted addition to the π system. This mode of addition guarantees that both new bonds are formed on the same face of the olefin. Although the reagents can add to either face of the olefin, this leads only to enantiomers of a single diastereomer. The concerted addition is the key feature which assures *syn* selectivity.

Another type of stereoselectivity is possible when a new chiral center is produced in a molecule which already contains one or more chiral centers. A typical example of such a process would be addition to an aldehyde or ketone which already contains a chiral center:

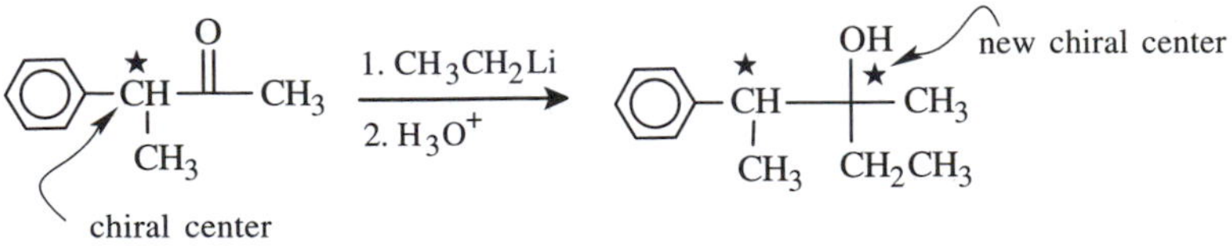

To understand the stereoselectivity that might be observed, it is first necessary to delineate the stereochemical possibilities. The existing chiral center can be R, S, or both. Addition to the

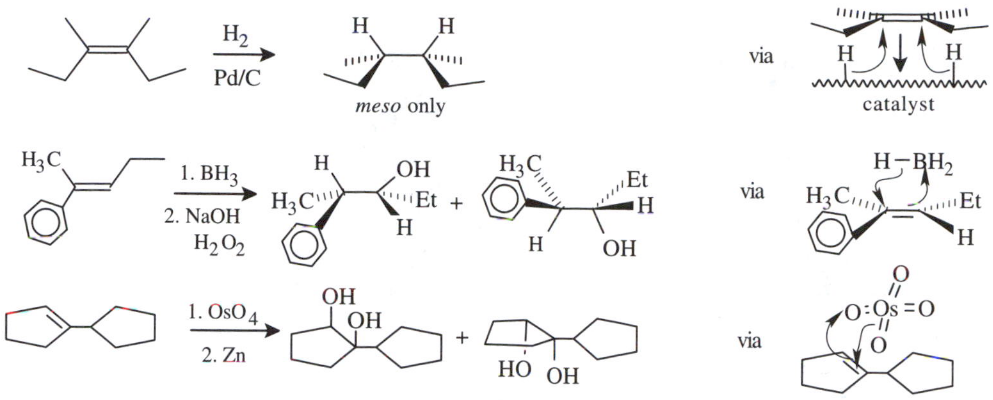

FIGURE 5.13

carbonyl group can potentially occur from either face since it is planar. Thus if the existing chiral center is of the R configuration, the products of addition can have either the R,R or R,S configuration and are diastereomers. Each diastereomer is optically active because only one enantiomer is produced. If the existing chiral center is of the S configuration, the products of addition can be either the S,S or S,R diastereomers. They will be optically active and they are enantiomers of the diastereomeric pair formed from the R configuration of the precursor. If the starting material is a racemic mixture, all four stereoisomers will be produced — two sets of enantiomeric diastereomers. (That is, the R will give R,R and R,S and the S will give S,S and S,R.) The product mixture will be optically inactive.

A given chirality of the starting material gives two diastereomers, and it is normal to find that these two diastereomers are not produced in equal amounts. Because the two diastereomeric products are of different energies, the diastereomeric transition states leading to them will be of different energies, the rates of their formation will be different and they will be produced in unequal amounts. If one diastereomer is produced in excess of the other, the reaction is diastereoselective. If only one diastereomer is produced, the reaction is diastereospecific. The same analysis would apply if the starting material is racemic. The reaction would still produce two diastereomers, each would be formed as a pair of enantiomers, and the same diastereoselectivity would be observed.

As stated previously, the addition of nucleophiles to chiral carbonyl compounds is a very common type of reaction which produces diastereomeric mixtures. The diastereoselectivity varies with the reagents and conditions. Some examples include the following:

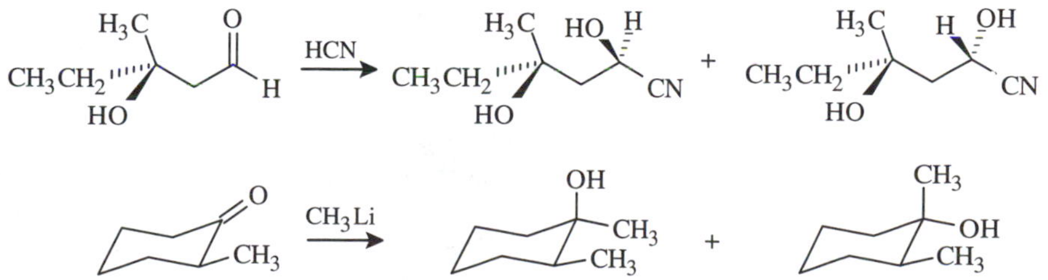

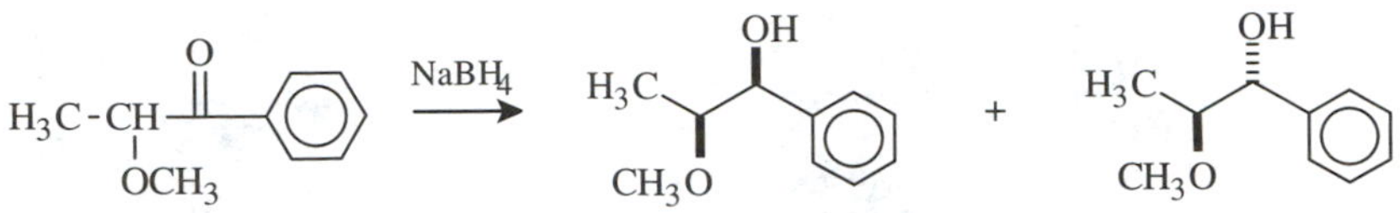

By analogy, the formation of diastereomers is observed for additions to other trigonal systems, such as olefins, which have a chiral center elsewhere in the molecule. In these cases, if optically active starting materials are used, the diastereomers will be optically active. If racemic starting materials are employed, the diastereomeric mixture will be optically inactive. In either case, it is common to find different amounts of the two diastereomers.

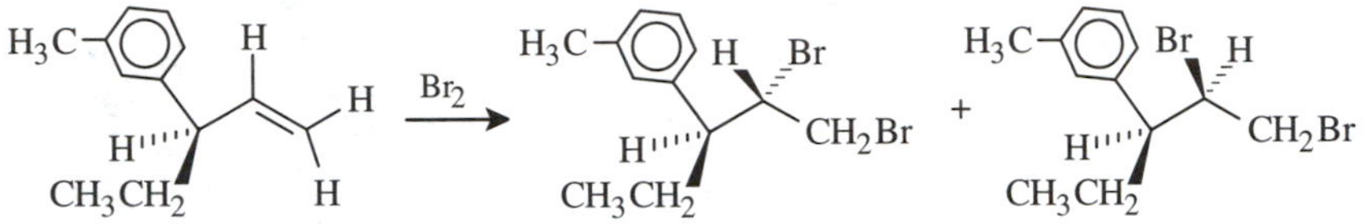

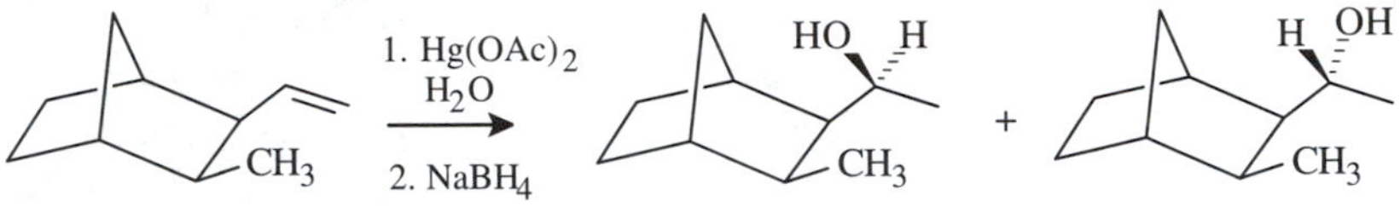

The formation of diastereomers is also possible when two new chiral centers are produced from achiral starting materials. A pertinent example is found in aldol-type reactions between enolates and carbonyl compounds. The achiral enolate and the achiral aldehyde or ketone give a product with two new chiral centers. Thus, there can be two diastereomers produced (*syn* and *anti*) and because there is no initial chirality, each diastereomer will be produced as a racemic mixture of enantiomers. The *syn* and *anti* diastereomers will not usually be produced in equal amounts.

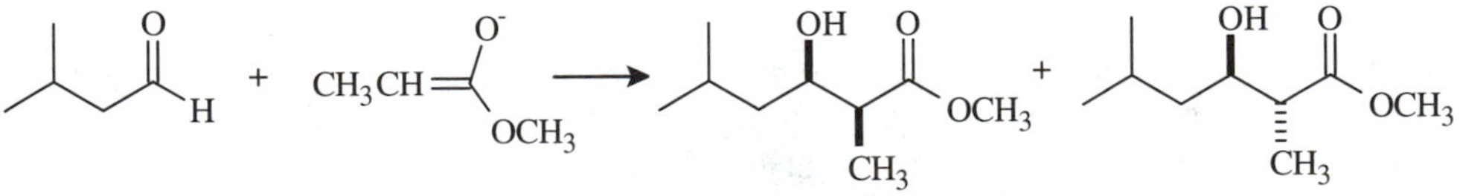

Factors which influence the stereoselectivity of organic reactions have been under intense investigation recently because of the increasing requirement and profitability of producing stereoisomerically pure compounds. A great deal of progress has been made, but even more remains to be accomplished. The specific contributors to stereoselectivity in individual reactions will be discussed as they are encountered. At this point, it is important to be aware of the stereochemical variations that are possible.

Conformational Analysis

In the most basic sense, chemical reactions are really only changes in the distribution of electrons. Such changes result in the breaking and making of chemical bonds and cause reactants to be converted to products. However, before such electronic changes can take place, molecules taking part in the reaction must approach each other within bonding distance, or they must undergo a change in geometry which permits overlap between the necessary orbitals to take place and thus results in electron redistribution. Restated another way, in order for chemistry to occur, molecules must first interact in a spatial sense. Consequently, the shapes of molecules and the surface features they display are an important influence on their interactions with other molecules.

The large number of σ bonds present in organic molecules has a direct bearing on their shapes. Since a σ bond is axially symmetric along the bond, rotations of the groups connected by a σ bond do not cause it to break. (Such cannot be said of π bonds.) Thus molecules with many σ bonds are capable of large numbers of internal rotational motions which largely determine the shape, size, and surface characteristics of the molecule. Conformational analysis is the study of rotational motions in molecules and how they affect molecular properties.

The simplest molecule capable of internal rotational motion is ethane. Ethane has two tetrahedral methyl groups connected by a carbon–carbon σ bond. As such, the methyl groups are free to rotate relative to each other. However, it is found that the various rotational positions are not equivalent spatially or energetically. In fact, there are two limiting rotational positions for ethane. The lowest energy conformation is the one in which the C—H bonds of each methyl group are staggered between the C—H bonds of the other methyl group across the σ bond. This is the lowest energy conformation because the electron clouds of the bonds are the farthest distance apart, and their repulsions are minimized.

lowest energy   staggered   highest energy   eclipsed

The highest energy conformation of ethane is the one in which the C—H bonds of each methyl group are eclipsed with the C—H bonds of the other methyl group across the σ bond. This is the highest energy because the electron clouds of the C—H bonds are as close as they can be, and their repulsions raise the energy of the molecule.

The staggered and eclipsed forms of ethane are *conformational stereoisomers* (conformational isomers, conformers) because they have the same molecular formulas and sequences of bonded elements but different spatial arrangements due to rotations around single bonds. (Actually, there are an infinite number of conformational isomers because there are an infinite number of degrees of rotation around the bond, but normally one needs only be concerned with the higher and lower energy situations.)

The difference in energy between the higher and lower energy forms of ethane is only 2.9 kcal/mol (12 kJ/mol), thus rotations around the bond are very rapid at room temperature (about 10^{11}/second). However, if one plots the change in energy as ethane rotates between the staggered and eclipsed forms, a periodic behavior is seen (Figure 5.14). Moreover, if a large number of snapshots of ethane were taken, they would show that most of the time ethane is found in the staggered conformation. The equilibrium between the staggered and eclipsed conformations favors the staggered by 99.2 percent to 0.8 percent.

A similar analysis of propane reveals analogous behavior with two major conformations—staggered and eclipsed—and periodic energy changes as rotation about a σ bond occurs (see

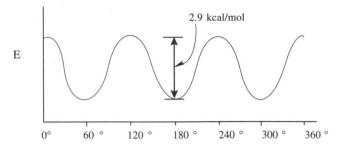

FIGURE 5.14

Figure 5.15). There is a difference from ethane, however, in that the energy difference between the staggered and eclipsed conformations is now 3.3 kcal/mol (14 kJ/mol). This increase means that a hydrogen and methyl group eclipsed across a carbon–carbon bond repel each other more than do two hydrogen atoms. This suggests that the electron cloud of the methyl group comes closer to the electron cloud of the C—H bond, so the repulsion is greater. Since the electron clouds associated with the methyl group define the space that the methyl group occupies, it follows that a methyl group occupies more space than a hydrogen; that is, it is "larger." It follows that groups in molecules have definite sizes. The sizes of these groups are factors which contribute to the overall shape of the molecule because of their influence on the preferred conformation of the molecule.

Conformational isomerism around the central bond in butane is more complex, because the various staggered and eclipsed conformations are not equivalent as they are in ethane and propane. Starting with the eclipsed conformation, with the dihedral angle between the two methyl groups at 0°, rotation around the central bond leads to two different staggered conformations and one additional eclipsed conformation (see Figure 5.16).

The most stable staggered conformation, in which the methyl groups are antiperiplanar (dihedral angle of 180°), is called the *anti* conformation. The other staggered conformation, in which the dihedral angle between the methyl groups is 60°, is called the *gauche* conformation (there are two of them, for rotations of 60° or 300°). The other eclipsed conformation is that in which the two methyl groups each eclipse a hydrogen (there are two of them, for rotations of +120° and 240°).

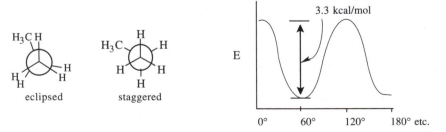

FIGURE 5.15

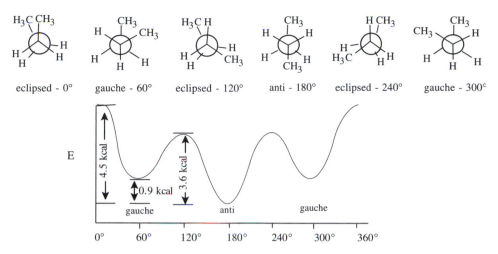

FIGURE 5.16

From the energy diagram, it is seen that the *gauche* conformation is 0.9 kcal/mol (3.7 kJ/mol) higher than the *anti*. This must be due to some residual repulsion between the methyl groups when the dihedral angle is only 60° between them. Also, the energy difference between the *anti* conformer and the highest-energy eclipsed conformer is 4.5 kcal/mol (18.8 kJ/mol). Thus the greater effective size of the methyl groups results in increased repulsion when they are eclipsed.

In addition to conformational isomerism about the 2,3 bond in butane, rotations about the 1,2 bond and the 3,4 bond are possible. The energy changes here are much smaller and are comparable to those found in propane.

The importance of conformational isomerism lies in the fact that the predominant shape that molecules adopt is dependent on the energies of the various staggered and eclipsed conformations. In combination they can be used to predict the probable shapes the molecule normally assumes, and these shapes are those which are presented to reagents in solution.

In contrast to open chain systems in which groups can rotate through 360° around σ bonds, cyclic systems can undergo conformational change through only limited ranges. Like open chain systems, however, conformational changes in rings minimize eclipsing interactions across σ bonds. Cyclopropane is a flat ring without conformational motion. Cyclobutane is not planar; if it were, all the C—H bonds around the ring would be eclipsed. The molecule undergoes a conformational change that bends the molecule out of planarity by about 35°. This reduces eclipsing and leads to a lower overall energy. A similar situation is found in cyclopentane, which adopts an envelope conformation (one ring apex out of plane) that is in equilibrium with four other envelope conformations (each apex up) to avoid the 10 C—H eclipsing interactions that would be present if the molecule were planar.

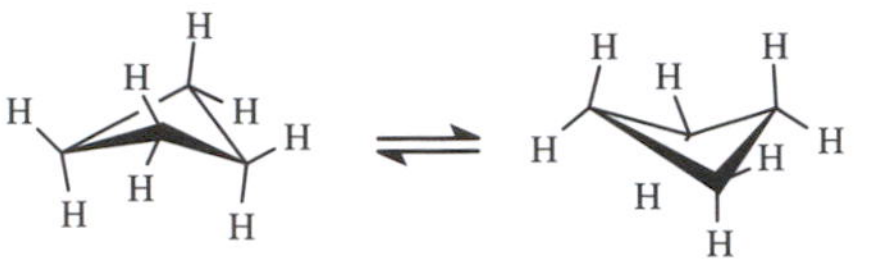

cyclobutane 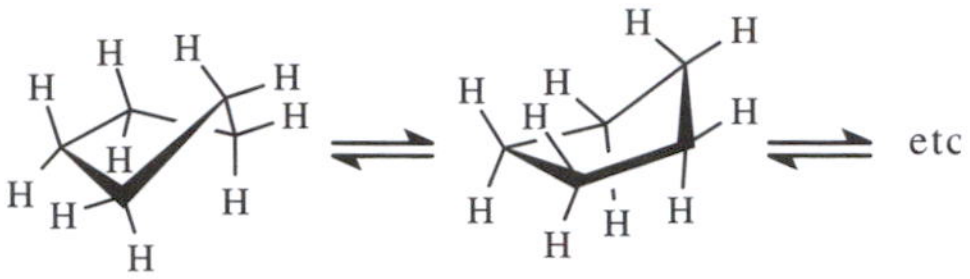   cyclopentane

Saturated six-membered rings are the most common ring systems in nature because they present an optimal conformational situation. As seen in cyclohexane, the molecule adopts a puckered shape called a *chair conformation* in order to avoid angle strain in the ring bonds. In the chair form, all bond angles are 109° *and* all the bonds are staggered.

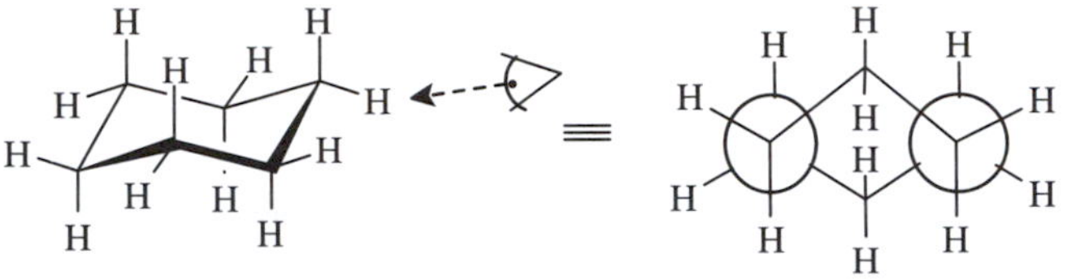

This results in a strain-free molecule whose energy is comparable to a completely staggered, open-chain alkane. This is easy to see by viewing the molecule in a Newman projection.

The chair conformer can undergo conformational isomerism to a second chair conformer which is degenerate in energy with the first. Cyclohexane is thus a dynamic molecule which exists largely in one of two chair isomers. These are the lowest-energy conformations. Other higher-energy conformations of cyclochexane include the boat form (which is 10.1 kcal/mol (42.3 kJ/mol)) above the chair form and the twist boat form, which lies 3.8 kcal/mol (15.9 kJ/mol) above the chair form.

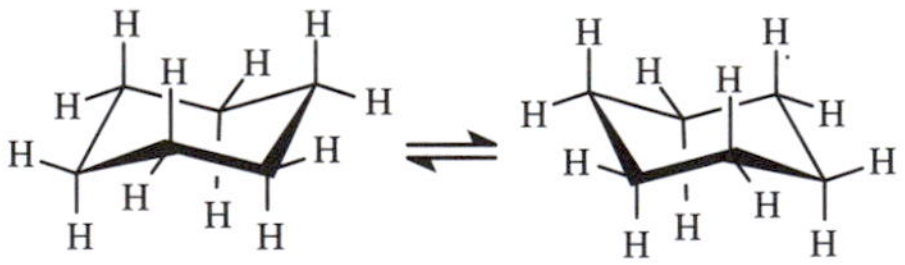

chair   chair 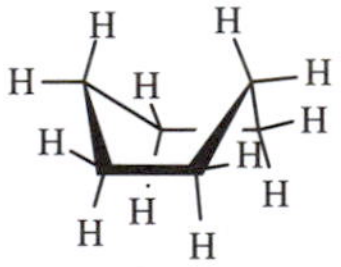   boat 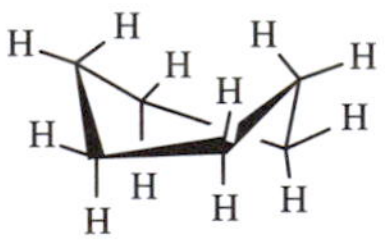   twist boat

Although these are well-defined conformational isomers, their energies are such that they are virtually unpopulated at room temperature. (The twist boat is an intermediate in the conversion of one chair form to the other.) At the same time, the conversion of one chair form to the other occurs rapidly at room temperature, and they are in rapid equilibrium.

Because cyclohexane exists in the chair form, the C—H bonds of the methylene groups are nonequivalent. There are two types of valences on each CH_2 group. One type, perpendicular to a plane loosely defined by the ring carbons, is called *axial*. The second type falls generally in the plane loosely defined by the ring carbons and is termed *equatorial*. These are shown both in combination and individually.

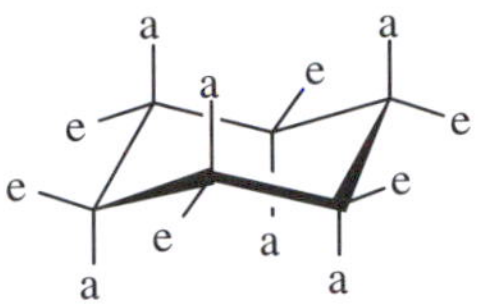

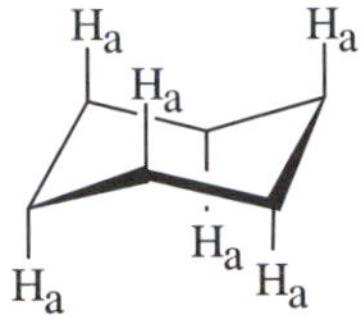

axial valences 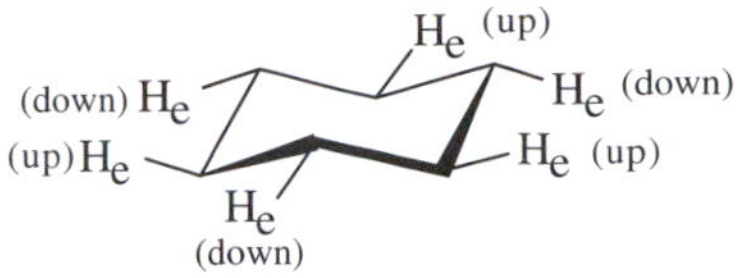   equatorial valences

Three axial valences on alternate $(1, 3)$ carbons point to one side of the ring (up) and the other three axial valences on alternate $(1, 3)$ carbons point to the other side of the ring (down). The same is true for equatorial valences. While the directionality is not so obvious for equatorial valences, they still point toward one side of the ring (up) or the other (down).

There are two chair forms and two types of valences (axial and equatorial), and the conversion of one chair form to the other interconverts the axial and equatorial valences (i.e., a valence which is axial in one chair form is equatorial in the other chair form and vice versa). In the following figure, one of the carbons is indexed with a star ($\star$) to help track it:

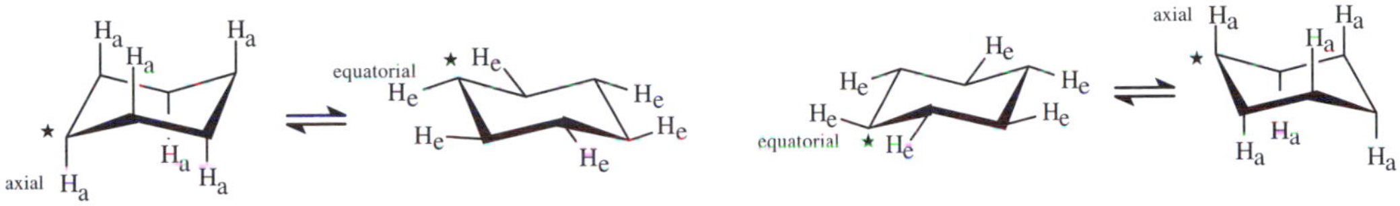

In cyclohexane itself, the chair forms have equal energy. However, if groups other than hydrogen are attached to the cyclohexane ring, the two chair forms are no longer equivalent, because in one chair isomer the group is equatorial and in the other chair isomer it must be axial. This is shown for methylcyclohexane.

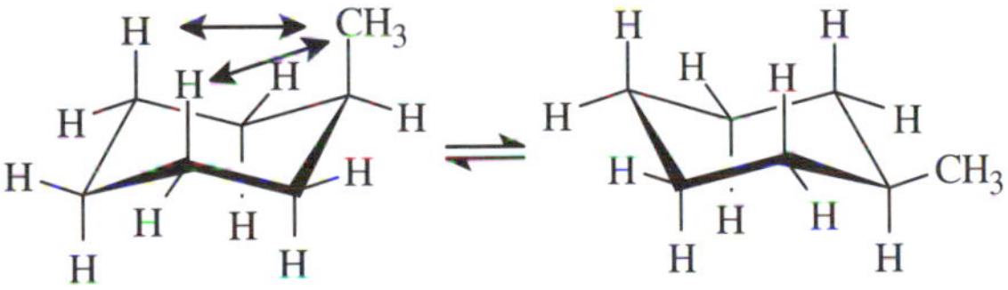

Although both conformations have the methyl group staggered between the vicinal protons, when the methyl group is axial it is sufficiently close to the *syn*–axial protons to undergo 1,3-diaxial interactions and be repelled by them. This raises the energy of the axial conformer relative to the equatorial conformer. For a methyl group, the energy difference is about 1.8 kcal/mol. (Actually, the relationship of an axial methyl group to the ring bonds is a *gauche* conformational relationship. Thus the value of 1.8 kcal/mol for an axial methyl group is the value of two *gauche* butane interactions with the ring bonds.)

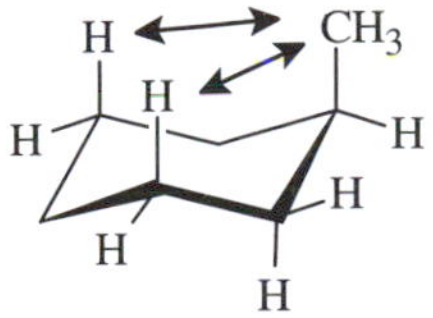

1,3-diaxial interactions 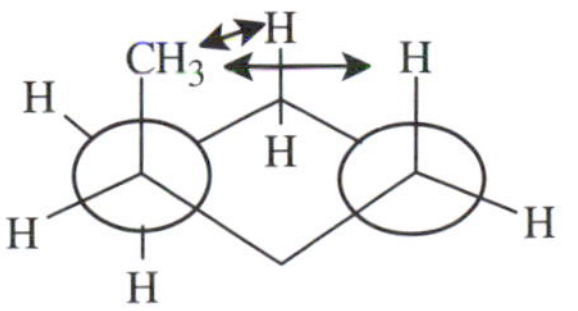   2 gauche butane interactions

Other groups would behave similarly, with the axial isomer being higher in energy (less stable) than the equatorial isomer because of 1,3 diaxial interactions. These two isomers are conformational isomers because they are interconvertible by rotations about C—C single bonds. They are also called conformational diastereomers since they have different physical properties and are non-superimposable, non-mirror images.

When more than one group is attached to cyclohexane, the stereoisomeric possibilities increase. First, structural isomers of the 1,2, 1,3, or 1,4- type are possible. Next, relative configurations (R, S) are possible for 1,2 or 1,3 disubstituted isomers. (The 1,4- isomer has a plane of symmetry.) The relative stereochemistry can be denoted as *cis* or *trans*, depending on whether the substituents point toward the same side or opposite sides of the ring. Finally, the cyclohexane ring can undergo chair–chair interconversion leading to different conformational isomers. These possibilities are shown for methylcyclohexanol (in Figure 5.17).

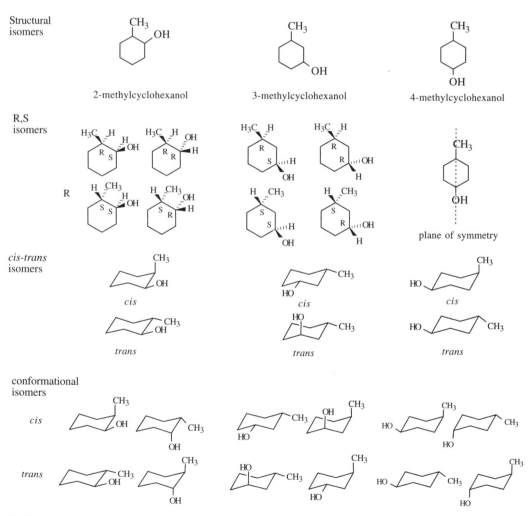

FIGURE 5.17

The first three types of isomerism are familiar and have been discussed previously. The conformational isomerism is very understandable if it is remembered that axial and equatorial valences exchange upon chair–chair interconversion. For example, in order to draw the *trans* isomer of 3-methylcyclohexanol, one of the groups must be equatorial and the other axial. The other chair form *must* have the groups in opposite valences. Similarly, *trans*-2-methyl cyclohexanol has both groups equatorial in one chair form. The other chair form must therefore have both groups axial.

Conformational Energies

The energetic consequences of chair–chair interconversions in substituted cyclohexanes are related to the interconversion of axial and equatorial valences. Because axial groups undergo 1,3 diaxial interactions which increase the energy of the molecule, the obvious energy preference is for groups larger than hydrogen to occupy equatorial positions. The relative energies of various chair conformers of multisubstituted cyclohexanes can thus be evaluated by noting the numbers of 1,3-diaxial interactions in each of the conformers.

For example, *trans*-1,2-dichlorocyclohexane has diaxial and diequatorial chair forms. The diaxial conformer should be less stable because it has two sets of 1,3-diaxial interactions between the chlorines and the axial protons. Knowing the equatorial–axial preference for a single chlorine substituent on a cyclohexane is 0.52 kcal/mol (2.2 kJ/mol), one can predict that the diequatorial isomer is favored by 1.04 kcal/mol (4.4 kJ/mol). Further, *cis*-1,3-dichlorocyclohexane has two chair forms with two equatorial chlorines and two axial chlorines, respectively. The diaxial isomer should be more than 1 kcal/mol higher in energy than the diequatorial isomer because the 1,3-diaxial interactions between two axial chlorines should be more severe than the 1,3-diaxial interactions between an axial chlorine and axial hydrogens. In contrast, *trans*-1,3-dichlorocyclohexane has a single axial chlorine in either chair conformer, so the two chair forms are of the same energy.

Similarly, *trans*-4-methylcyclohexanol has diequatorial and diaxial substituted chair forms. It is thus predicted that the former should be about 2.7 kcal/mol (11 kJ/mol) more stable than the latter, because an axial methyl is less stable by 1.8 kcal/mol (7.3 kJ/mol) and an axial OH group is less stable by 0.9 kcal/mol (3.8 kJ/mol) than the equatorial counterparts.

Similar qualitative assessments can be made for more highly substituted cyclohexanes. It is found that *cis,cis*-1,3,5-trimethyl cyclohexane exists in only one chair form. This must be the triequatorial isomer because the other chair form has three axial methyl groups interacting on the same side of the ring. This should cause severe steric interactions and should be much less stable than the all-equatorial isomer. In fact, the energy difference between the all-equatorial and the all-axial isomer should be greater than 5.4 kcal/mol (3 × 1.8kcal/mol). This can be estimated by noting that an axial methyl group is less stable by 1.8 kcal/mol when the other axial valences with which it interacts are protons. The 1,3-diaxial interactions should be even greater if the other axial valences hold groups larger than protons, thus the energy difference should be greater than for three axial methyl groups, or greater than 3 × 1.8kcal/mol.

Since 1,3-diaxial interactions are the major factor which increases the energy of conformations with axial substituents, it is reasonable to expect that larger groups would have more severe steric interactions, causing the energy difference between the axial and equatorial positions to increase. That is, the larger the group, the larger are the 1,3-diaxial interactions with axial protons, and the greater the energy difference between the axial and equatorial forms.

This effect is clearly seen by comparing methylcyclohexane and *t*-butylcyclohexane. The axial–equatorial energy difference is 1.8 kcal for the methyl group, while it is 4.9 kcal/mol for the *t*-butyl group. This is because the *tert*-butyl group is much larger than a methyl group, and 1,3-diaxial interactions are much stronger. In fact, these interactions are so large that the *t*-butyl group has been employed to anchor the particular chair conformation that has the *tert*-butyl group equatorial. The other chair form would be much higher in energy and virtually unpopulated; thus the chemistry that is observed arises from reactions of a single conformation.

A Values

Viewed another way, if the axial–equatorial energy difference is mainly a function of steric bulk, it might be used to assess the relative size of various groups. That is, if the energy difference between the two chair conformational isomers of a monosubstituted cyclohexane were measured, it might serve as a quantitative measure of the effective steric bulk of a particular group. Table 5.1 is a collection of such data. The free energy differences between equatorial and axial substituents on a cyclohexane ring are called A values and are a quantitative

TABLE 5.1 A Values: Free Energy Differences between *Axial* and *Equatorial* Conformations of Monosubstituted Cyclohexanes (kcal/mol)

Group	A	Group	A	Group	A
HgCl	−0.25	Cl	0.52	$CH=CH_2$	1.7
HgBr	0	OAc	0.71	CH_3	1.8
D	0.008	OMe	0.75	C_2H_5	1.8
CN	0.2	OH	0.94	*iso*-Pr	2.1
F	0.25	NO_2	1.1	C_6H_{11}	2.1
C≡CH	0.41	COOEt	1.15	$SiMe_3$	2.5
I	0.46	COOMe	1.3	C_6H_5	2.7
Br	0.55	COOH	1.4	*t*-Bu	4.9
OTs	0.52	NH_2	1.4		

measure in kcal/mol of the effective steric bulk of a substituent. An important point is that these A values are not a measure of the physical size of a group, but rather a measure of its steric interactions. Thus the *t*-butyl group (A = 4.9) is seen to be significantly more bulky than the trimethylsilyl group (A = 2.5), yet physically the trimethylsilyl group occupies more volume. The difference is that the carbon–silicon bond is longer than the carbon–carbon bond, so the trimethylsilyl group is farther away from the ring. Thus its effective bulk, which is the strength of 1,3-diaxial interactions, is actually less than the *t*-butyl group. The same trend is seen in the halides, where the A values for chloride (A = 0.52), bromide (A = 0.55), and iodide (A = 0.46) decrease even though the sizes of these atoms increase (Cl < Br < I). Thus A values are related to the effective steric bulk and not the actual physical size of substituents. In this respect they are very useful since it is the *effective* size of substituents which gives rise to steric effects in chemical reactions. Thus A values can be used to predict steric changes resulting from the introduction of a group into a molecule. While the effective steric bulk of a group in a different molecule may not be quite the same as when it is attached to the cyclohexane ring, the trends should be parallel. In this way, A values provide a useful way to evaluate steric effects semiquantitatively.

Stereoelectronic Effects

In addition to determining its shape and surface characteristics, the conformational preferences of a molecule can also contribute to its chemical reactivity. Many reactions require that reacting groups achieve a particular spatial relationship so that overlap of appropriate orbitals leading to the needed electron redistribution can take place. These geometry-dependent orbital interactions which influence chemical reactivity are described generally as *stereoelectronic factors*. A very well-known example of a reaction with distinct stereoelectronic requirements is the Sn2 reaction. Here, the incoming nucleophile must approach from the side opposite the leaving group. This permits electron donation into the σ^* antibonding orbital and results in the inversion stereochemistry found for these processes. Structural features which prevent the stereoelectronic requirements of the reaction from being met, such as the cage structure of the norbornyl skeleton which prevents the incoming nucleophile from approaching from the back of the leaving group, will effectively slow or stop the reaction.

unreactive

In other reactions, a particular disposition of groups in a molecule is required for the reaction to proceed efficiently. Moreover, it is often found that the proper stereoelectronic requirements of the reaction can be met only if a particular conformation of the molecule is populated. If the needed conformation is energetically accessible and thus populated, the reaction can proceed normally; if not, it is very slow and alternate processes might intervene. A classic example is the base-promoted elimination, an E2 reaction which requires an antiperiplanar relationship between the proton being removed and the leaving group which departs. In open-chain systems this rarely presents a problem, since the barriers to rotations about single bonds are low and reactive conformers are easily populated. Nevertheless, this stereoelectronic requirement can have stereochemical consequences. Hence, *d,l*-stilbene dibromide undergoes dehydrohalogenation in hot pyridine, whereas the *meso* isomer reacts much more slowly. As shown , the reactive conformation of the *d,l* isomer has the bulky phenyl groups *anti* to each other and is energetically favored, whereas the reactive conformation of the *meso* isomer has the phenyl groups *gauche* to each other and is energetically unfavorable. Thus, the reactive conformer of the *d,l* compound is populated and reacts effectively; the reactive conformer of the *meso* compound is not populated significantly and the rate of elimination is much slower. (Eclipsing of the phenyl groups in the transition state also slows the reaction of the *meso* isomer.)

In cyclic systems where conformational motions are more restricted, stereoelectronic effects can play a much larger role in determining the outcomes of reactions. For example, base-promoted elimination in *cis*-4-(*t*-butyl)-cyclohexyl tosylate occurs 70 times faster than in the *trans* isomer. The reason is that the *t*-butyl group controls the conformation of the cyclohexane ring (it occupies an equatorial valence nearly exclusively); consequently, the *cis* isomer has an axial tosylate group which is antiperiplanar to axial hydrogens in the β positions. The *trans* isomer has an equatorial tosylate group which has no antiperiplanar hydrogens.

cis- isomer - faster   *trans*- isomer- slower

The *cis* isomer meets the stereoelectronic requirements for base-promoted elimination and thus reacts significantly faster than the *trans* isomer, which does not.

The pyrolysis of acetate esters yields olefins by a concerted *syn* elimination of acetic acid. In open chain systems where rotations are facile, it is possible for the acetate group to achieve a *syn* relationship with any of the vicinal protons, and the major product primarily reflects the energies of the products (*trans* is favored).

In contrast, it is found that *trans*-1-acetoxy-2-phenylcyclohexane gives 1-phenylcyclohexene as the major product (86.5%) upon heating, whereas *cis*-1-acetoxy-2-phenylcyclohexane

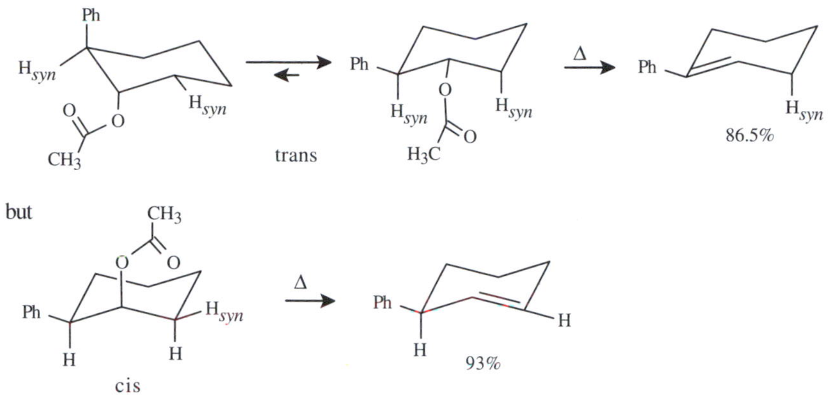

FIGURE 5.18

gives 3-phenylcyclohexene as the major product (93%). Comparing these two isomers it is calculated from A values that the *trans* diequatorial isomer is favored by 3.4 kcal/mol over the *trans* diaxial isomer (Figure 5.18). The diequatorial conformer has a proton at C-2 *syn* to the acetate group in an *axial*–equatorial disposition. This undergoes elimination to produce the more stable conjugated product. (In this case, even the less stable diaxial conformer has a *syn* proton available in an *axial*–equatorial relationship with the acetate group and gives the same product.) In contrast, the *cis* isomer has only one conformation with a *syn* proton in an *axial*–equatorial relationship with the acetate group, and that elimination gives the less stable, nonconjugated 3-phenyl cyclohexene as the major product.

A more striking example of the influence of conformation on the reaction outcome is seen in the nitrous acid deamination of 2-aminocyclohexanols (Figure 5.19). This takes place by rearrangement of a group on the carbinol carbon that is *anti* to the developing carbocation. The deamination reaction is very fast and the products reflect the population of the chair conformers. The *trans* isomer exists mainly in the diequatorial conformer, thus the only group *anti* to the amino group is a ring bond. Indeed, ring contraction is the only process observed. In the *cis* isomer, however, both chair forms are populated (the A values of OH and NH₂ are similar), so products of *both* ring contraction and hydride migration are obtained.

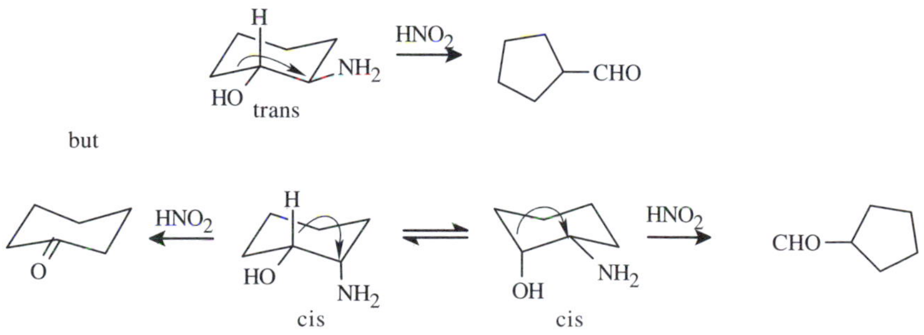

FIGURE 5.19

A particularly revealing example is seen in the reaction of 2-bromo-4-phenylcyclohexanols with silver[I]. In the secondary carbinol, reaction takes place from the higher-energy conformer because, even though its population is low, hydride migration by the hydrogen *anti* to the bromide assists the loss of axial bromide, so it is the fastest reaction. Placement of a methyl group at the carbinol carbon raises the energy of the axial bromide conformation even higher; its population is further reduced. Because of its greater relative concentration, the conformer with bromide in the equatorial position reacts faster, and leads to ring contraction.

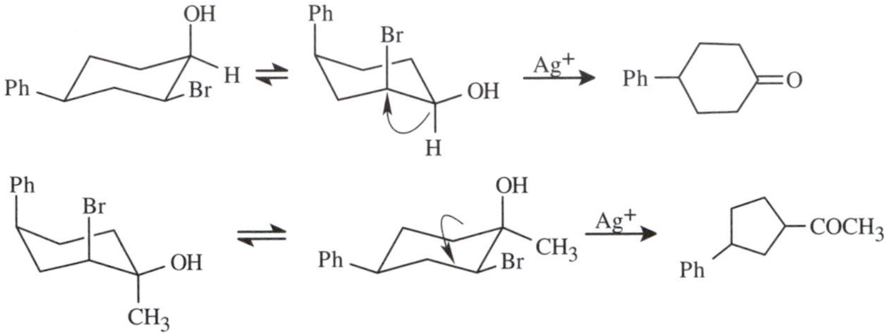

The examples presented point out some important features about organic reactions. First, many have distinct stereoelectronic requirements that must be met if the reaction is to proceed efficiently. Second, the correct stereoelectronic relationships are primarily dependent on the conformations of the substrate. Finally, the populations of various conformers determine whether stereoelectronic requirements can be satisfied and thus play a significant role in product partitioning.

Bibliography

For a good elementary discussion of priority rules see J. McMurry, *Organic Chemistry*, 3rd ed., Brooks/Cole, Pacific Grove, CA, 1992, pp. 181–86.

A succinct but complete discussion of R,S nomenclature is found in J. A. March, *Advanced Organic Chemistry, Reactions, Mechanism, and Structure*, 4th ed., Wiley, New York, 1992, pp. 109–11.

The strategies for asymmetric synthesis are found in J. A. March, *Advanced Organic Chemistry, Reactions, Mechanism, and Structure*, 4th ed., Wiley, New York, 1992, pp, 116-20.

See F. A. Carey and R. J. Sundberg, *Advanced Organic Chemistry, Part A*, 3rd ed. Plenum, New York, 1990, Chapter 2.

For a seminal discussion on conformational isomerism, see E. L. Eliel, N. L. Allinger, S. J. Angyal, and G. A. Morrison, *Conformational Analysis*, American Chemical Society, Washington, DC, 1981.

Problems

1. Indicate the chiral centers in the following molecules and give the relative configuration (R,S) of each.

(a)

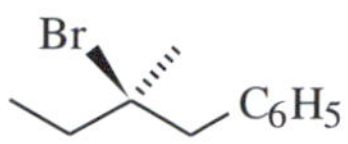

(b)

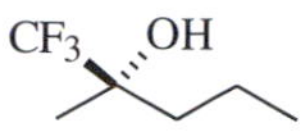

(c)

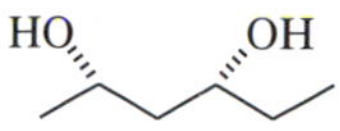

(d)

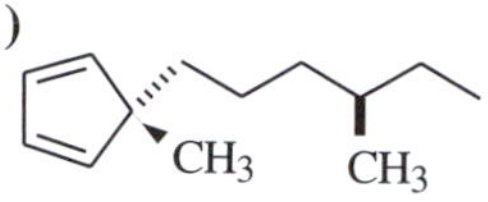

(e)

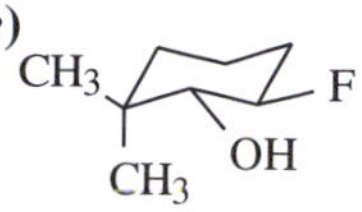

(f)

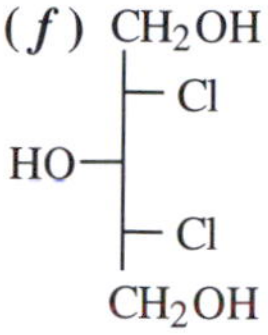

(g)

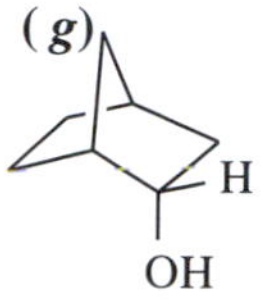

(h)

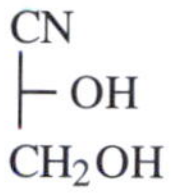

(i)

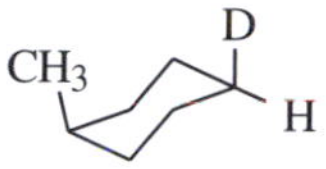

(j)

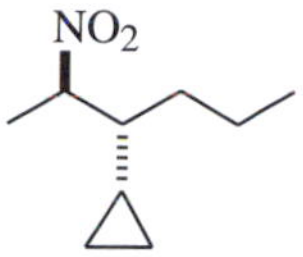

(k)

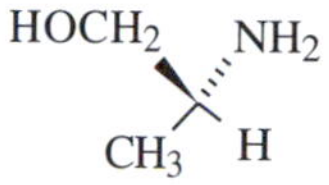

2. The following structures are representations of 3-fluoro-2-phenyl-2-pentanol. Give the stereochemical relationship of each structure to (2R, 3R)-3-fluoro-2-phenyl-2-pentanol.

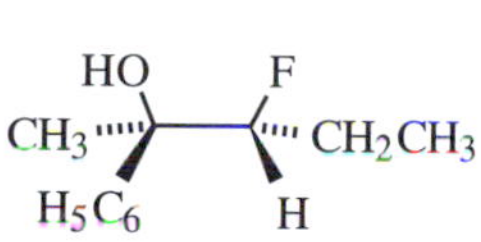

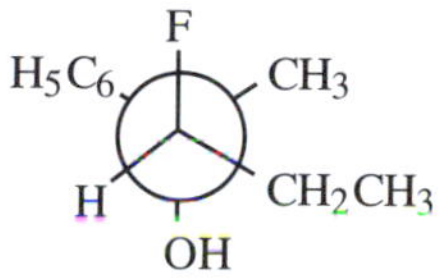

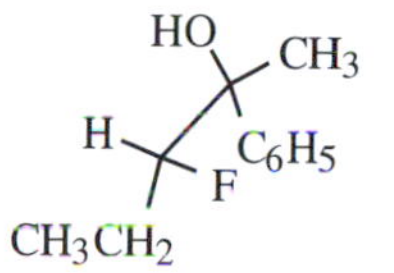

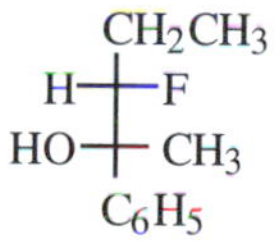

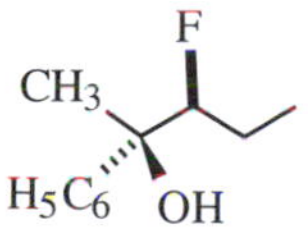

3. (a) Draw the four stereoisomers of 4-methyl-2-hexanol and give the relationship of each to the others.

(b) Draw all the stereoisomers of 3-bromo-4-methylhexane, give the R,S designation of each chiral center and give the relationship of each to the others.

(c) Using Fischer projections, draw all of the stereoisomers of 2-fluoro-3-methyl-1,4-pentanediol and give the relationship of each to the others.

(d) Draw all of the stereoisomers of 1,4-diphenyl-1,4-dibromobutane, give the R,S designation of each chiral center and give the relationship of each to the others.

4. Give the stereochemical relationship between the following pairs of compounds.

(a)

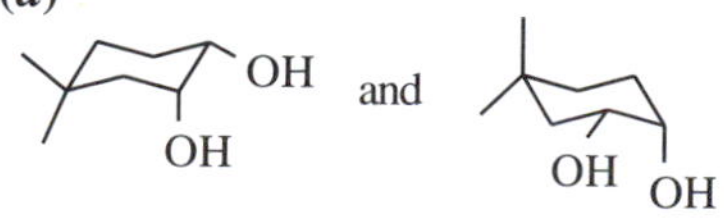

(b)

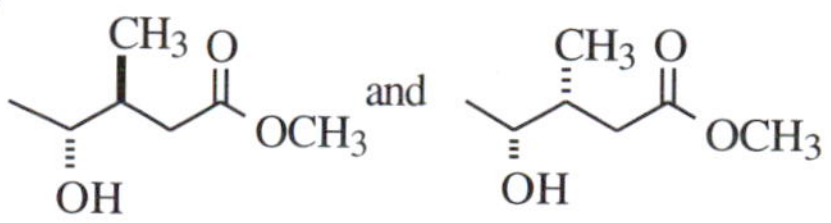

(c)

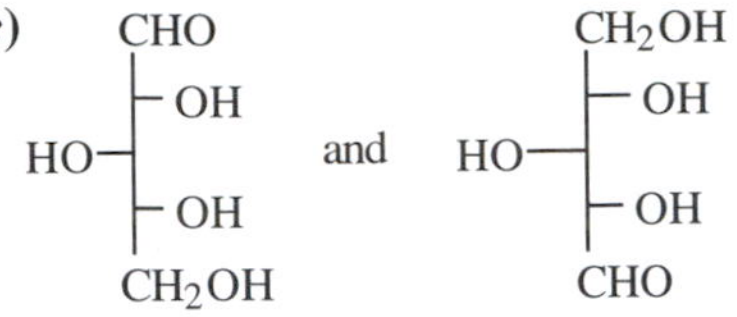

(d)

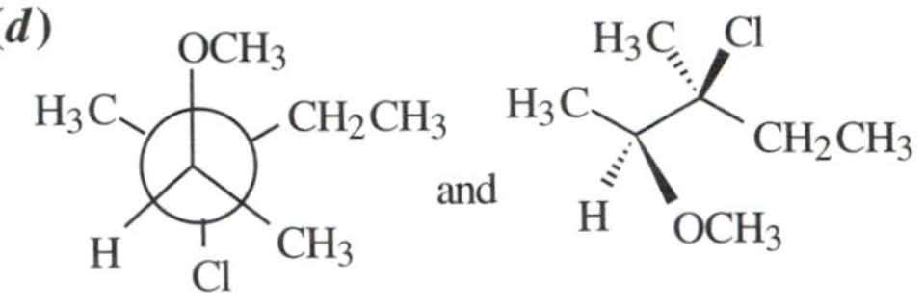

(e)

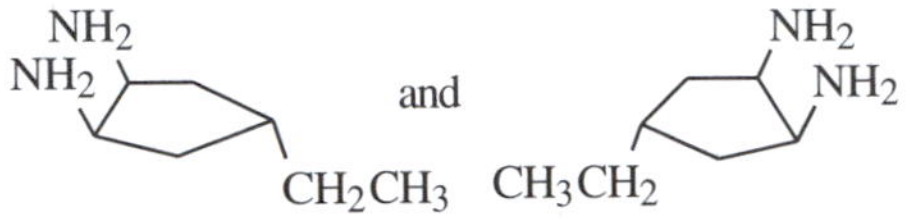

(f)

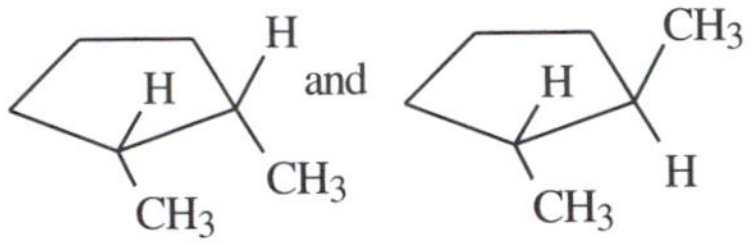

(g)

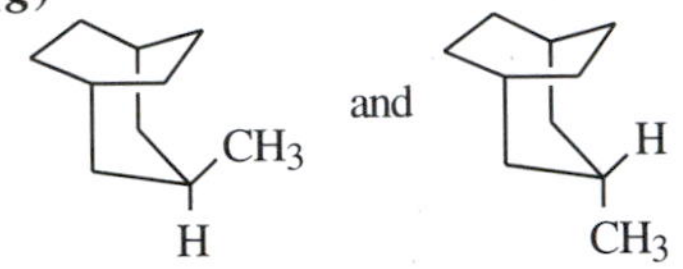

(h)

(i)

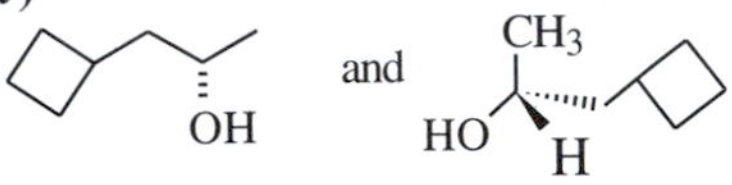

(j)

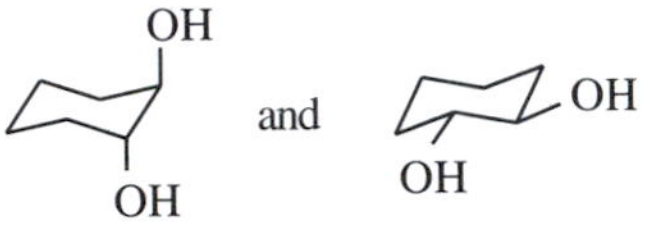

5. For the following compounds, show two chair conformations, indicate which is more stable, and give an estimate of the energy difference between the two.

(*a*) *trans*-1-ethyl-3-phenylcyclohexane

(*b*) *cis*-1-(tert-butyl)-4-isopropylcyclohexane

(*c*) *trans*-2-amino-1-cyanocyclohexane

(*d*) 2R,6S-1-bromo-2,6-dimethylcyclohexane

(*e*) *cis*-4-(tert-butyl)-2-methylcyclohexanone

(*f*) *cis*-1,1,3,4-tetramethylcyclohexane

6. Estimate the difference in energy between the chair conformations of *trans*-2-methoxycyclohexanol. The actual value is about 3.1kcal/mol. Explain this.

7. Show all of the staggered conformers of 2,3-dimethylbutane and estimate the energy differences between them.

8. The Beckman rearrangement could occur by either a stepwise or a concerted mechanism.

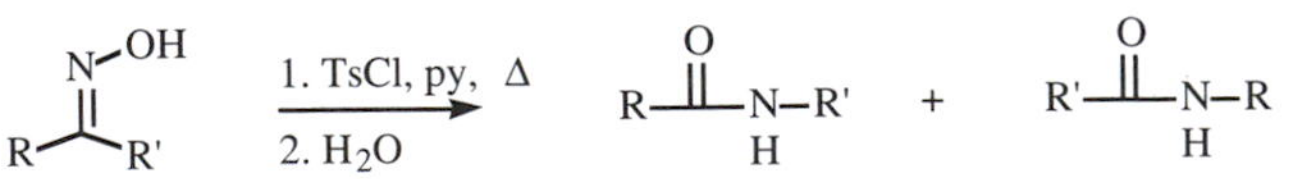

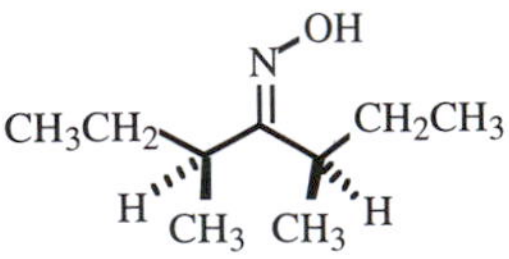

4

(*a*) Show both mechanisms using curved arrow notation.

(*b*) Suppose one made oxime, 4.
    (1) Would it rotate plane polarized light?
    (2) Label the configurations of the chiral centers in 4.
    (3) Show how 4 could be used to help distinguish the mechanisms given.

9. Explain why (1S,3R)-3-*tert*-butylcyclohexyl tosylate undergoes E-2 elimination with potassium *tert*-butoxide very slowly, while the (1R, 3R) reacts much more rapidly.

10. The reaction of *cis*-2-pentene with iodine azide (IN$_3$) in dichloromethane gives (2S,3S)-3-azido-2-iodopentane and (2R,3R)-3-azido-2-iodopentane but not any other diastereomers. Show the stereochemistry of the addition and give a curved arrow mechanism to account for it.

11. The reaction of *trans*-2-hexene with aqueous peracetic acid gives (2S,3R)-2,3-hexane diol and (2R,3S)-2,3-hexanediol but not any other diastereomers. What is the stereochemistry of the addition?

12. Heating (2S)-3-methyl-3-phenyl-2-butyl tosylate in ethanol leads to skeletal rearrangement and the formation of (3S)-2-ethoxy-2-methyl-3-phenylbutane. What does this information indicate about the stereoelectronic course of the skeletal rearrangement?

13. Treatment of *trans*-2-phenylcyclohexyl tosylate with potassium *tert*-butoxide gives mainly 3-phenylcyclohexene in a fairly slow process, whereas under the same conditions, *cis*-2-phenylcyclohexyl tosylate gives 1-phenylcyclohexene in a much shorter reaction time. Explain this regiospecificity.

FUNCTIONAL GROUP SYNTHESIS

Functional Group Manipulation

As seen in undergraduate coursework, organic chemistry is often organized around the chemistry of functional groups. Functional groups are recurring sequences of bonded elements which give typical and characteristic chemical reactions. Recall that there are characteristic reactions of ketones, esters, secondary alcohols, and so on. Consequently, it is very important to be able to introduce a particular functional group into a molecule, or to interconvert functional groups in a molecule in order to utilize particular reactions available to them.

To manipulate functional groups, it is first necessary to be able to identify them, to understand their bonding and oxidation levels, and to recognize the bonding changes that are needed to convert one functional group into another. A variety of functional group reactions are available, and much of an introductory organic chemistry course is devoted to learning reagents and reactions for carrying out functional group interconversions.

These reactions are very often traditional and illustrative, but they are not necessarily the best way to manipulate a particular functional group. Many traditional methods have been replaced, in practice, by newer reactions or reagents which offer certain advantages over older methods. In general, these advantages have to do with mild conditions, selectivity, generality, or experimental simplicity. Nevertheless, all types of functional group interconversions, new and old, are still based fundamentally on the ideas that have been developed earlier in this book.

The discussion which follows is organized in the following fashion: a common under-graduate text was used to provide a list of "standard" preparations for each of the functional groups. Most of these reactions should be familiar because they are the ones presented in a typical undergraduate course. Next, the most widely used and practical ways (or "real" ways) to introduce functional groups will be discussed. These latter methods have practical syn-thetic value and are usually the first choices in real laboratory situations, but they often differ from the standard list of preparations. The functional group order goes from highest oxidation level (carboxylic acid) to lowest oxidation level (alkanes). As might be anticipated, carboxylic acids are most often prepared oxidatively and alkanes are most often prepared reductively. Functional groups of intermediate oxidation level can be accessed either reductively or oxida-tively.

Carboxylic Acids

Carboxylic acids have a relatively high oxidation level and thus a majority of the synthetic methods to access carboxylic acids are oxidative. Traditional preparations include the follow-ing:

1. Oxidation of olefins

$$R-CH=CH-R' \xrightarrow[\text{or}]{\text{a. } KMnO_4, \text{ heat}} RCO_2H + R'CO_2H$$

$$\text{b. } 1. O_3$$
$$2. H_2O_2$$

2. Oxidation of primary alcohols

$$R-CH_2OH \xrightarrow[\text{or}]{\text{a. } KMnO_4, OH^-} RCO_2H$$

$$\text{b. } Na_2Cr_2O_7, H_2SO_4$$

3. Oxidation of alkylbenzenes

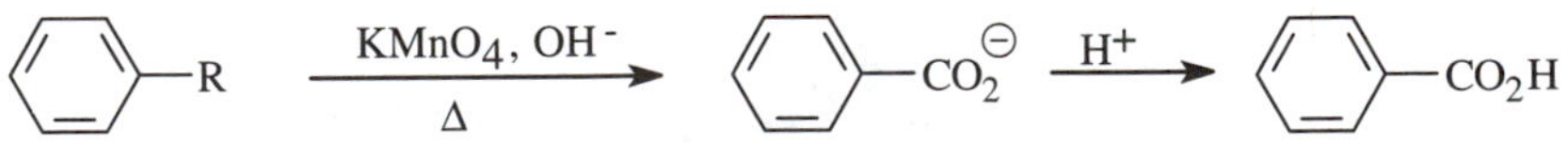

4. Hydrolysis of nitriles

$$R-C\equiv N \xrightarrow[\text{or}]{\text{a. conc } HCl, \text{ heat}} RCO_2H + NH_3$$

$$\text{b. } OH^-, H_2O, \text{ heat}$$

Although these reactions will provide carboxylic acid products, each has associated prob-lems. The cleavage of olefins to carboxylic acids (reaction 1) can be carried out either by using potassium permanganate or by ozonolysis at low temperature followed by oxidative workup with hydrogen peroxide. Neither of these methods is very useful since only symmetric olefins provide a single carboxylic acid product; unsymmetrical olefins give a mixture of two acids which must be separated. Furthermore, the most useful synthetic processes are those which build up structures, whereas these reactions are degradative in nature.

Primary alcohols can be oxidized to carboxylic acids by a variety of reagents (reaction 2). Often, potassium permanganate or sodium dichromate is given as a reagent to use in this transformation. These are powerful oxidants and many other functional groups that might be present cannot survive the reaction conditions. Milder oxidants are preferred and the best of these is chromic acid in acetone (Jones reagent). Jones reagent is a mixture of chromic acid and a stoichiometric amount of sulfuric acid which is needed in the redox process to keep the solution near a pH of 7. This technique is fast, easy, and efficient and the reagent solution is easily prepared from chromium trioxide and sulfuric acid in acetone. The oxidation can be carried out by adding the Jones reagent by burette to the alcohol. Oxidation is instantaneous and the addition can be stopped precisely when all the alcohol has been consumed. Using a stoichiometric amount of chromic acid leaves other functional groups untouched. This is the method of choice for the synthesis of carboxylic acids from primary alcohols.

The oxidation of alkyl benzenes to benzoic acids (reaction 3) is still carried out occasionally and this oxidation is most likely the only one where potassium permanganate is the reagent of choice. Any carbon group attached to the aromatic ring is degraded to the carboxylic acid group under the very vigorous conditions of this oxidation.

An interesting twist on the oxidation of aromatic compounds forms the basis of a new and very useful synthesis of carboxylic acids. Normally, the aromatic ring is resistant to oxidation and the side chains are oxidatively degraded to carboxylic acids, as in reaction 3. It has been found that ruthenium tetroxide is a mild and selective oxidant of aromatic rings, and it completely degrades the ring to the carboxylic acid but leaves aliphatic groups unoxidized. This is essentially the reverse of the chemoselectivity seen in potassium permanganate oxidations of arenes. The selectivity and mildness is seen in the following example, in which no amide or ether oxidation is observed, and there is no epimerization of either chiral center:

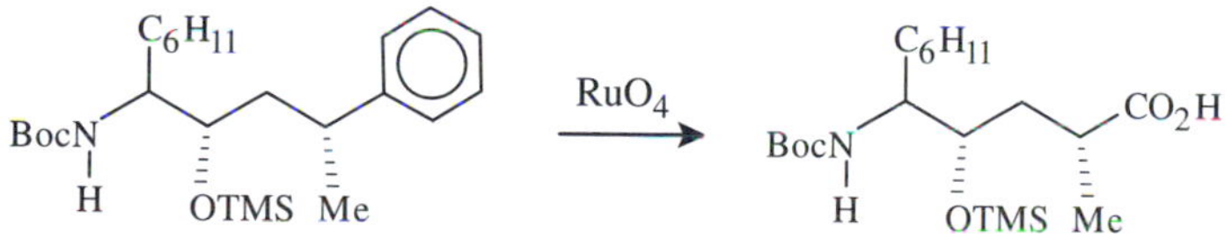

Another common way to install a carboxylic acid group is to hydrolyze a carboxylic acid derivative. Such hydrolyses do not require a change in oxidation level, but do normally require acid or base catalysis. Nitriles (reaction 4) and amides often require vigorous conditions for hydrolysis. Either concentrated hydrochloric acid or sodium hydroxide can be used to hydrolyze nitriles. The first stage of the hydrolysis produces an amide, and the amide is subsequently hydrolyzed to the acid. Each step of the hydrolysis requires strenuous conditions and is useful mainly for amides and nitriles that lack other functional groups that would be destroyed by the stringent conditions. It has been found that a mixture of sodium hydroxide and hydrogen peroxide can be used to hydrolyze nitriles and amides more efficiently than sodium hydroxide alone. This is the hydrolysis method of choice, although many other reagent combinations have been reported.

Esters, on the other hand, are very common hydrolytic precursors to carboxylic acids. The traditional reaction for the hydrolysis of esters is basic saponification using either sodium hydroxide or potassium hydroxide. Although acid catalysis can also be employed, preparative methods usually use base catalysis because formation of the carboxylate salt drives the equilibrium to the right and gives high yields of products.

$$R\overset{O}{\underset{\|}{\text{—}}}OR' \quad \xrightarrow[\text{or}]{\begin{array}{c}\text{a. NaOH, H}_2\text{O}\\ \\ \text{b. KOH, H}_2\text{O}\end{array}} \quad R\text{-}CO_2^{\ominus} \; M^{\oplus} + \; R'OH$$

Although esters are much more easily hydrolyzed than amides, traditional saponification suffers from the fact that most esters are not soluble in the aqueous base, so the rate of the hydrolysis is limited by the solubility, not by the reactivity. This limitation is overcome by the use of lithium hydroxide in aqueous THF, the reagent of choice for basic hydrolysis of esters. Methyl and ethyl esters are cleaved readily by this combination, as most esters are soluble in this solvent mixture.

$$\begin{array}{c}R\text{-}CO_2R' \\ \\ (\text{ }R' = \text{Me, Et})\end{array} \quad \xrightarrow[\begin{array}{c}\text{THF, H}_2\text{O}\\ \text{1-2 hr, 25}°\text{C}\end{array}]{\text{LiOH}} \quad \begin{array}{c}R\text{-}CO_2^{\ominus} \; Li^{\oplus} + \; R'OH\\ \\ 90\text{-}100\%\end{array}$$

If structural constraints prevent the use of basic hydrolysis of the ester group, acid hydrolysis must be used. Nowadays it is much more common, in such instances, to use *tert*-butyl esters because they are cleaved rapidly and efficiently by trifluoroacetic acid to the carboxylic acid and isobutylene. This cleavage is different from the normal acid-catalyzed hydrolysis of esters in that the alkyl–oxygen bond, rather than the acyl–oxygen bond, is broken. This change in mechanism is brought about by the stability of the *tert*-butyl cation which is produced upon alkyl–oxygen cleavage. As an added benefit, the isobutylene by-product is a gas which escapes from the reaction mixture.

If neither acidic nor basic conditions are compatible with other groups present in the ester to be hydrolyzed, β-trimethylsilylethyl esters are often prepared. Trimethylsilyethyl esters are cleaved easily by fluoride under mild, neutral conditions. Typical sources of fluoride are cesium fluoride (CsF) or the more soluble tetrabutylammonium fluoride (TBAF).

Current methods for the hydrolysis of esters are fast, efficient, and sufficiently mild to be compatible with the presence of a variety of other functional groups and stereocenters in the molecule. For example, protected amino acid esters are hydrolyzed quantitatively without either racemization or deprotection, by LiOH in aqueous THF.

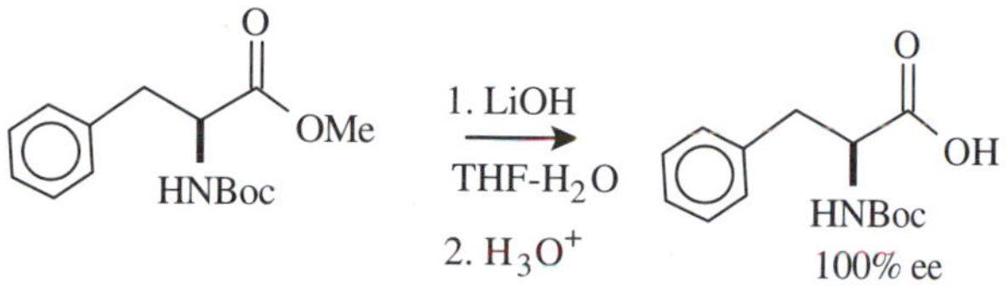

Carboxylic acid groups can also be installed in molecules using the reaction of an organometallic compound with carbon dioxide. This is a reductive method since the carbon dioxide is reduced to a carboxylic acid by formation of a new carbon–carbon bond. Both Grignard reagents and organolithium compounds work well in this reaction:

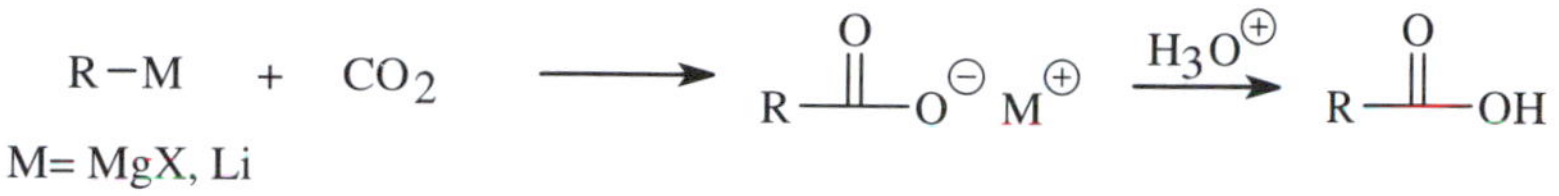

Esters

Derivatives of carboxylic acids are generally made from the carboxylic acid. The traditional routes to esters are as follows:

1. An acid-catalyzed reaction between an acid and an alcohol (Fischer esterification)

2. Reaction of an acid chloride (or acid anhydride) with an alcohol

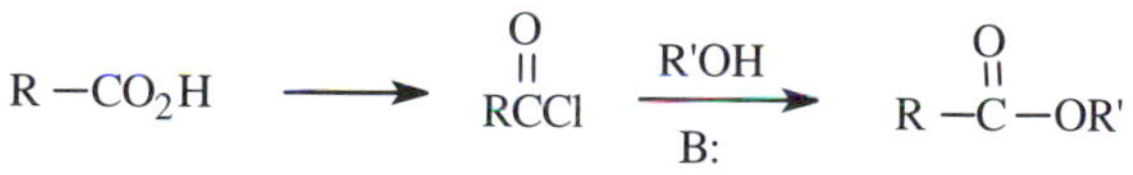

While still useful for large scale esterification of fairly robust carboxylic acids, Fischer esterification is generally not useful in small-scale reactions because the esterification depends on an acid-catalyzed equilibrium to produce the ester. The equilibrium is usually shifted to the side of the products by adding an excess of one of the reactants—usually the alcohol—and refluxing until equilibrium is established, typically for several hours. The reaction is then quenched with base to freeze the equilibrium, and the ester product is then separated from the excess alcohol and any unreacted acid. This separation is easily accomplished on a large scale where distillation is often used to separate the product from the by-products. For small-scale reactions where distillation is not a viable option, the separation is often difficult or tedious. Consequently, Fischer esterification is not widely used for ester formation in small-scale laboratory situations. In contrast, intramolecular Fischer esterification is very effective on a small scale for the closure of hydroxy acids to lactones. Here the equilibrium is driven by the removal of water and no other reagents are needed. Moreover, the closure is favored entropically and proceeds easily.

A second very common way to convert carboxylic acids to esters is by the reaction of the corresponding acid chloride with an alcohol. A tertiary amine such as pyridine or triethylamine is used to scavenge the HCl by-product. It has also been found effective to add small amounts of N,N-dimethyl-4-aminopyridine (DMAP) to the reaction mixture to promote efficient product formation. If the acid chloride is readily available, this is a very satisfactory preparation. If the acid chloride is not available, a disadvantage to this method is that a carboxylic acid must first be converted to the acid chloride which must be isolated and purified prior to the formation of the ester. If the chlorinating agent is not separated from the acid chloride, the alcohol will also react, leading to a mixture of products that may be difficult to separate.

$$R-CO_2H \longrightarrow \underset{RCCl}{\overset{O}{\parallel}} \xrightarrow[\substack{R_3N \\ DMAP\ (cat) \\ CH_2Cl_2}]{R'OH} R-\overset{O}{\overset{\parallel}{C}}-OR' \ + \ R_3\overset{\oplus}{N}H\ \overset{\ominus}{Cl}$$

For small-scale esterification reactions (< 500 mg), the best methods should occur rapidly under mild conditions and produce only by-products which are easily separated from the reaction products. Under these criteria, an extremely efficient way to convert acids to *methyl* esters is to titrate the carboxylic acid with an ethereal solution of diazomethane. The methyl ester is produced rapidly and quantitatively, and the by-product of the esterification is nitrogen. Although diazomethane is a reactive and explosive compound, solutions of diazomethane can be prepared and used safely in the laboratory from readily available reagents.

$$\underset{R}{\overset{O}{\parallel}}\text{—OH} \xrightarrow[\text{ether}]{CH_2N_2} \underset{\text{quantitative}}{R\overset{O}{\overset{\parallel}{\text{—}}}\text{OCH}_3} \ + \ N_2$$

A second method to efficiently produce *methyl* esters of carboxylic acids is to treat the acid with potassium carbonate and methyl iodide. The methyl ester is produced under mild conditions and is easily separated from the reaction by-products. This method is somewhat different, in that the ester is formed by a nucleophilic displacement of iodide by the carboxylate ion. Normally, carboxylates are not thought of as good nucleophiles, and they are not, but methyl iodide is a quite reactive electrophile which matches the poor nucleophilicity of the carboxylate satisfactorily.

$$R\overset{O}{\overset{\parallel}{\text{—}}}\text{OH} \xrightarrow[\substack{CH_3I \\ THF}]{K_2CO_3} R\overset{O}{\overset{\parallel}{\text{—}}}\text{OCH}_3 \ + \ KI$$

Besides these methods, there are many other satisfactory ways to convert acids to esters.

Amides

Amides are usually obtained from carboxylic acids or their derivatives. The traditional method of preparation of amides is to react the corresponding acid chloride with an amine. This substitution process replaces the chloride with an amine without a change in the oxidation level. This remains an excellent and efficient method. However, excess amine or another base is required to neutralize the equivalent of HCl produced by the substitution.

$$R-CO_2H \longrightarrow \underset{RCCl}{\overset{\overset{\displaystyle O}{\|}}{}} \xrightarrow[CH_2Cl_2]{R'NH_2} \underset{R-\overset{\overset{\displaystyle O}{\|}}{C}-NHR'}{} + R'NH_3^{\oplus} Cl^{\ominus}$$

In this approach, formation of an acid chloride is required to activate the carbonyl group and make it more electrophilic. This activation is required in most transformations of carboxylic acids because carboxylic acids themselves are not sufficiently electrophilic to react with most nucleophiles. Furthermore, since many nucleophiles are also basic, they can react with carboxylic acids to give the carboxylate ion which is an even poorer electrophile than the carboxylic acid itself.

Carbodiimides are increasingly being used to promote the conversion of carboxylic acids to amides. This method was originally developed for creating the amide bonds in peptides. In this method, the carboxylic acid is treated with a carbodiimide (a very common one is dicyclohexylcarbodiimide (DCC) although many others have been developed). An activated acylating agent is produced which reacts with the amine present in the mixture to produce an amide. The advantage of this approach is that the acid is activated to a reactive electrophile in situ so the activated species need not be isolated.

$$\underset{R-\overset{\overset{\displaystyle O}{\|}}{C}-OH}{} \xrightarrow[\substack{R'NH_2}]{\substack{C_6H_{11}N=C=NC_6H_{11} \\ (DCC)}} \underset{R-\overset{\overset{\displaystyle O}{\|}}{C}-NHR'}{} + \underset{\text{urea}}{C_6H_{11}NH\overset{\overset{\displaystyle O}{\|}}{C}NHC_6C_{11}}$$

Yields are normally high, and the urea by-product from the carbodiimide can be separated and removed from the amide by one of several methods.

Amides are also available from nitriles, which have the same oxidation level. Direct acid or base hydrolysis usually requires fairly severe conditions and often does not stop at the amide stage, but goes on to carboxylic acid. Treatment of nitriles with a solution of HCl in ethanol furnishes an imidate ester which is hydrolyzed in aqueous acid to the amide. Because a nitrile is the starting material, only primary amides can be produced by this process.

$$R-C\equiv N \xrightarrow[EtOH]{HCl} \underset{\substack{\text{imidate ester}}}{R-\overset{\overset{\displaystyle OEt}{|}}{C}=NH} \xrightarrow{H_3O^{\oplus}} R-\overset{\overset{\displaystyle O}{\|}}{C}-NH_2$$

Acid Chlorides

Textbook preparations of acid chlorides from carboxylic acids include the following:

$$R-\overset{\overset{\displaystyle O}{\|}}{C}-OH \xrightarrow{SOCl_2} R-\overset{\overset{\displaystyle O}{\|}}{C}-Cl + HCl + SO_2$$

$$3 \ R-\overset{\overset{\displaystyle O}{\|}}{C}-OH \quad \xrightarrow{\text{PCl}_3} \quad 3 \ R-\overset{\overset{\displaystyle O}{\|}}{C}-Cl \quad + \quad H_3PO_3$$

$$R-\overset{\overset{\displaystyle O}{\|}}{C}-OH \quad \xrightarrow{\text{PCl}_5} \quad R-\overset{\overset{\displaystyle O}{\|}}{C}-Cl \quad + \quad POCl_3 \quad + \quad HCl$$

The traditional methods utilize sulfur or phosphorus halides to convert the acid to the acid chloride. Of these methods, thionyl chloride (often with a catalytic amount of DMF) is the most useful since the by-products of the reaction are gases (SO_2, HCl) which can be easily purged from the reaction mixture with a stream of nitrogen. The acid chloride product can then be purified on a small scale by either bulb-to-bulb distillation or crystallization.

Another superior reagent for the preparation of acid chlorides is oxallyl chloride in methylene chloride. Addition of a carboxylic acid leads to the smooth evolution of gas (CO_2, CO, HCl) which can be used as a crude monitor of the reaction progress. The acid chloride is very easily purified since oxallyl chloride boils at 62°C and is easily evaporated from the product. In many instances, the crude product is sufficiently pure to be used directly.

$$R-\overset{\overset{\displaystyle O}{\|}}{C}-OH \quad + \quad Cl-\overset{\overset{\displaystyle O}{\|}}{C}-\overset{\overset{\displaystyle O}{\|}}{C}-Cl \quad \xrightarrow[\substack{25° \\ 4\text{-}5 \ hr}]{\text{CH}_2\text{Cl}_2} \quad R-\overset{\overset{\displaystyle O}{\|}}{C}-Cl \quad + \quad CO_2 + \ CO \ + \ HCl$$

Aldehydes

The aldehyde functional group is a very reactive functional group, so methods to prepare it must be mild and must allow the aldehyde group to survive the reaction conditions. Traditional methods for introduction of the aldehyde functional group include the following:

$$R-\overset{\overset{\displaystyle O}{\|}}{C}-Cl \quad \xrightarrow[\text{Pt - BaSO}_4]{\text{H}_2} \quad R-\overset{\overset{\displaystyle O}{\|}}{C}-H \quad + \quad HCl \quad \text{(Rosenmund reduction)}$$

$$\text{LiAlH(OtBu)}_3 \qquad \text{ether, -78°C}$$

or

$$R-\overset{\overset{\displaystyle O}{\|}}{C}-OR$$

$$R-C{\equiv}N$$

$$\xrightarrow[\substack{\text{hexane} \\ \text{-78°C}}]{\text{DIBAH}} \quad R-\overset{\overset{\displaystyle O}{\|}}{C}-H$$

$$R\text{-CH}_2\text{OH} \quad \xrightarrow[\text{CH}_2\text{Cl}_2]{\text{PCC}} \quad R-\overset{\overset{\displaystyle O}{\|}}{C}-H$$

$$R-C{\equiv}CH \quad \xrightarrow[\text{2. H}_2\text{O}_2, \ \text{NaOH}]{\text{1. (Sia)}_2\text{BH}} \quad R\text{-CH}_2-\overset{\overset{\displaystyle O}{\|}}{C}-H$$

Aldehydes are intermediate in oxidation level, and thus the aldehyde functional group can be installed by either reduction of carboxylic acid derivatives or by oxidation of alcohols. Aldehydes are rarely installed without a change of oxidation level. One difficulty is that they undergo both oxidation *and* reduction readily. Special methods are required to stop at the aldehyde stage rather than proceeding by further reduction or oxidation.

Reductive methods utilize carboxylic acid derivatives as starting materials and the trick is to stop the reduction at the aldehyde stage, which is normally more easily reduced than the starting material. There are a variety of reducing systems known and many employ acid chlorides as precursors. The most effective reduction method for the preparation of aldehydes is the DIBAH reduction of either esters or nitriles using a single equivalent of the reducing agent.

By using low temperatures, the intermediate anions produced by hydride addition are at the aldehyde oxidation level, but they are resistant to further reduction. Hydrolysis delivers the aldehyde. Care must be taken to maintain low temperature during both the reaction and the hydrolysis.

The oxidation of primary alcohols to aldehydes also suffers from the problem of overoxidation of the aldehyde to a carboxylic acid. Mild methods capable of stopping the oxidation at the aldehyde oxidation level are required if aldehydes are to be obtained. The most common and effective reagent for this purpose is pyridinium chlorochromate (PCC), produced by the reaction of pyridinium hydrochloride with chromium trioxide. This reagent is soluble in dichloromethane and smoothly oxidizes primary alcohols to aldehydes in high yields. Because of the mild, neutral reaction conditions and the use of stoichiometric amounts of oxidant, the aldehyde product is not further oxidized.

The activation of DMSO by electrophilic reagents, such as oxallyl chloride or trifluoroacetic anhydride (among many others), produces an oxidant capable of oxidizing primary alcohols to aldehydes in high yields. This oxidation, called the *Swern oxidation,* yields the aldehyde (oxidized product) by reductive elimination of dimethylsulfide (reduced product). This reaction proceeds under mild, slightly basic conditions. It is a second widely used and effective oxidative method for the production of aldehydes from primary alcohols.

A different oxidative approach toward the preparation of aldehydes uses the ozonolysis of vinyl groups. If a vinyl group is present in a molecule, it can be oxidatively cleaved to an aldehyde by ozonolysis. This process cleaves the carbon–carbon double bond, but it is mild and very successful in many cases.

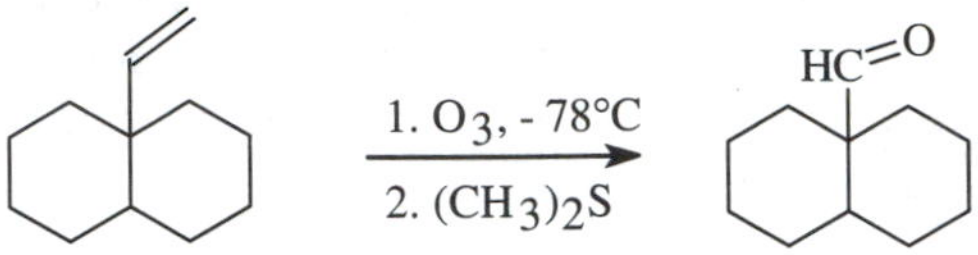

The formation of aldehydes without a change in oxidation level is not a common synthetic approach because most compounds that can be hydrolyzed to aldehydes without change in the oxidation level are formed from aldehydes in the first place. Thus acetals can be hydrolyzed rapidly to aldehydes by acidic water, but they are normally prepared from aldehydes. As such, this is a very common protection strategy for aldehydes, in which they first are converted to an acetal and later hydrolyzed back to the aldehyde when the time is right.

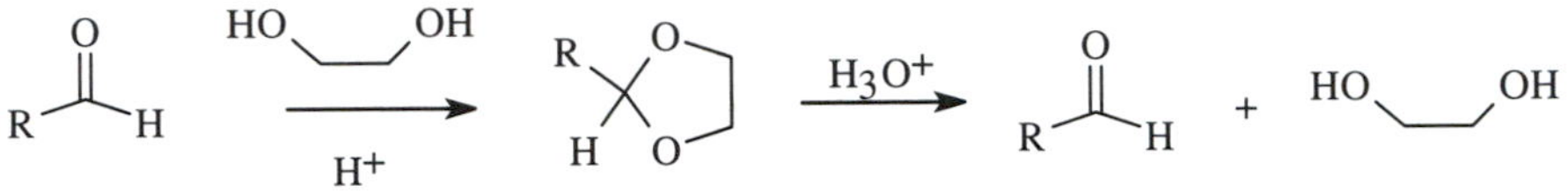

Ketones

Ketones have the same oxidation levels as aldehydes, but their preparation poses far fewer problems. Most important, they are very resistant to oxidation, so they can be prepared by any number of oxidative routes without difficulty. Textbook preparations of ketones are listed below. Many of these traditional methods remain the methods of choice for the preparation of ketones.

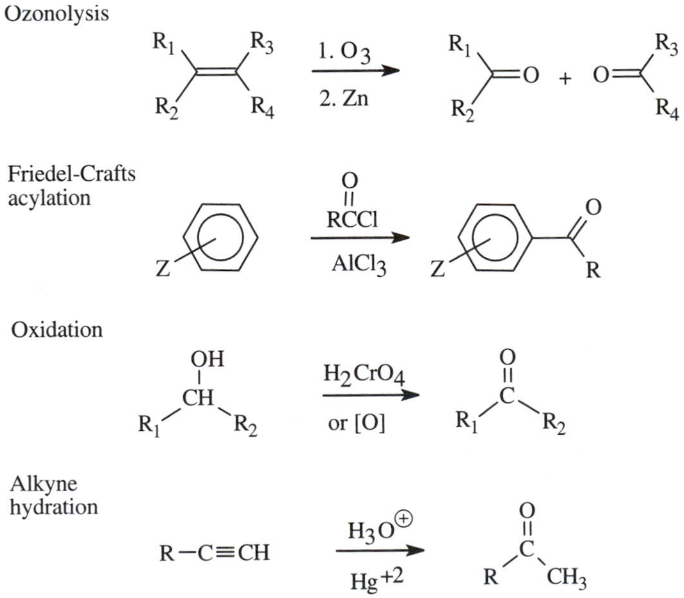

Ozonolysis

Friedel-Crafts acylation

Oxidation

Alkyne hydration

Addition to
acid chlorides

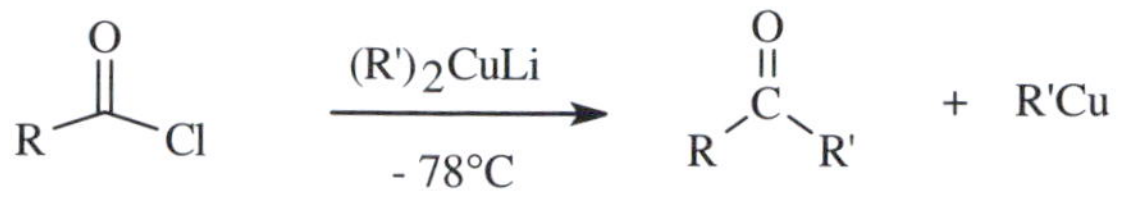

Addition to
acids

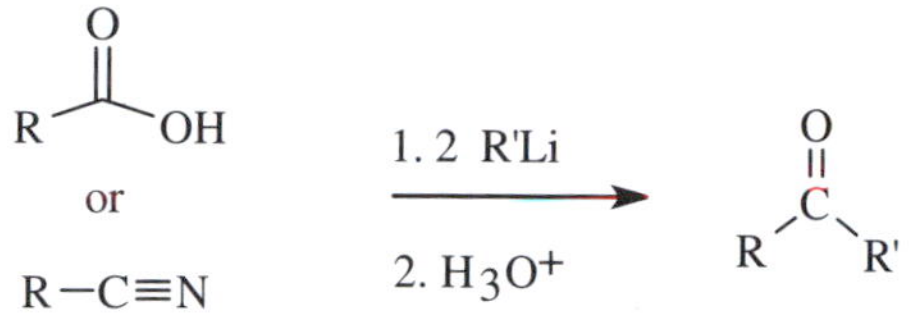

One of the most common methods for the preparation of ketones is by the oxidation of secondary alcohols. The use of chromic acid (Jones reagent) is easy, safe, and effective for the oxidation of secondary alcohols to ketones. Furthermore, Jones reagent gives a nearly neutral solution; thus it can be used with a variety of acid sensitive functional groups.

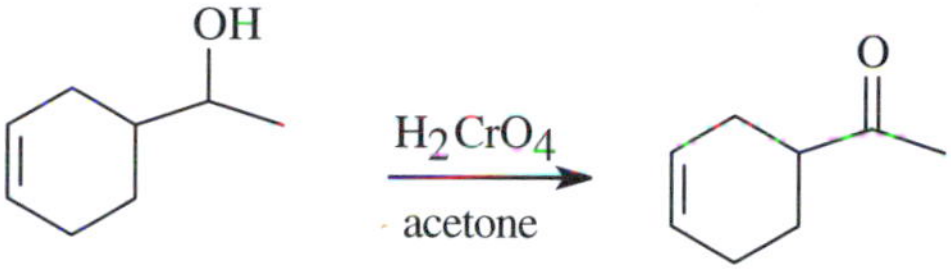

Sodium hypochlorite (household bleach) with acetic acid offers a very cheap and effective alternative to Jones reagent for the oxidation of secondary alcohols to ketones, and has been widely used for the synthesis of ketones.

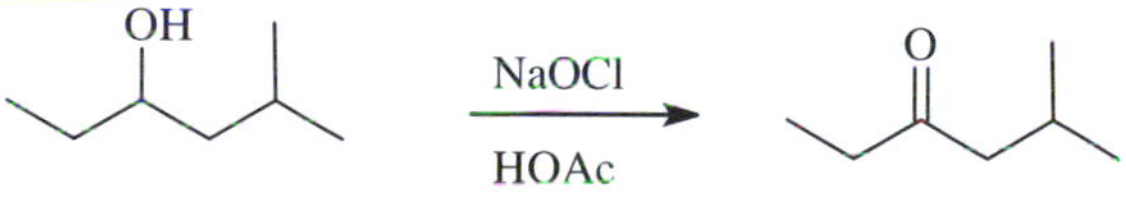

If very mild or basic conditions are required, PCC is the reagent of choice and works very well.

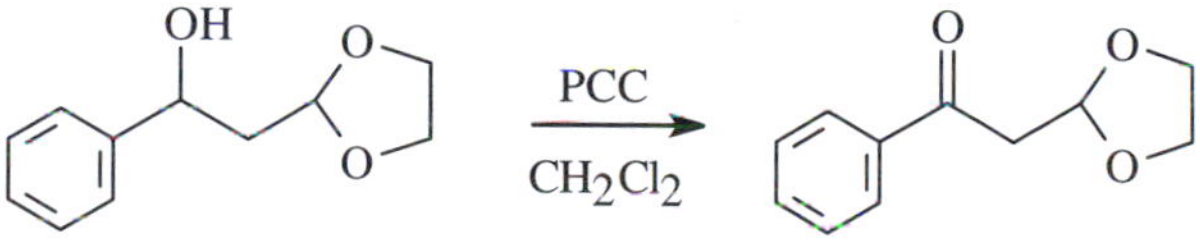

There are *many* other reported methods for the oxidation of secondary alcohols to ketones. In fact, over 140 different methods are listed in *Comprehensive Organic Transformations* by Larock. However, few are as versatile and useful as those listed here.

Conversion of carboxylic acid derivatives to ketones requires a net reduction of oxidation level. Furthermore, since the two groups attached to the carbonyl group are carbon-containing groups, it follows that a carbon nucleophile must be the reductant, usually an organometallic reagent. Carboxylic acid derivatives such as esters, acid chlorides, and acid anhydrides do not stop at the ketone oxidation level upon reaction with most organometallic reagents, but are further reduced to tertiary alcohols. (This same problem of excess reduction was seen for aldehyde preparation.) However, carboxylic acids themselves react smoothly with organo-lithium reagents to furnish ketones upon hydrolytic workup. This method is an effective way to

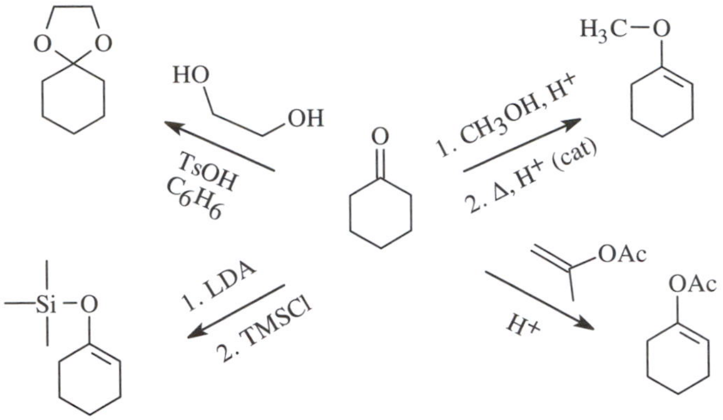

FIGURE 6.1

produce ketones. A key to the success of this reaction is the fact that the tetrahedral intermediate is a dianion which is stable to further addition. Only organolithium reagents are useful in this process, for only they are powerful enough nucleophiles to add to the very weakly electrophilic carbonyl group of the carboxylate anion.

As with aldehydes, production of ketones by nonredox processes is not a common synthetic approach. Ketone derivatives having the same oxidation level are usually produced from ketones themselves. Several examples of enol and acetal ketone derivatives are shown in Figure 6.1. All are prepared from ketones, all can be readily hydrolyzed back to the ketone in the presence of acidic water, and, with the exception of vinyl acetates, all are very stable to strong bases and nucleophiles. Acetals are often used as ketone (and aldehyde) protecting groups, and enol derivatives are versatile synthetic intermediates.

One hydrolytic method that is useful for the preparation of ketones is the hydrolysis of dithianes. These dithioacetal compounds can be prepared by alkylation methods and then converted to ketones by hydrolysis. Dithioacetals are much more resistant to hydrolysis than acetals, and thus Hg^{2+} is often used to promote efficient hydrolysis.

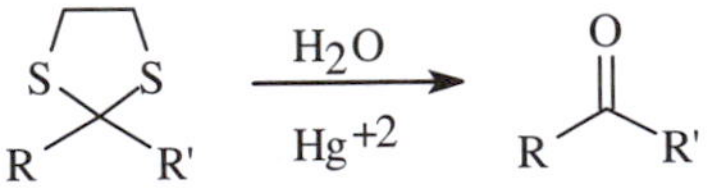

Imines and Imine Derivatives

Nitrogen analogs of aldehydes and ketones having the same oxidation level are imines and imine derivatives. In almost every instance, these compounds are prepared by an exchange reaction between an amine derivative and a carbonyl compound.

$$R-\overset{\cdot\cdot}{N}H_2 \quad + \quad \underset{R_2}{\overset{O}{\underset{}{\overset{\|}{C}}}}\!\!-R_1 \quad\longrightarrow\quad R-\overset{\cdot\cdot}{N}=C\overset{R_1}{\underset{R_2}{\big\langle}} \quad + \quad H_2O$$

For simple imines, R (= alkyl, aryl) water removal, either by a dehydrating agent (KOH or molecular sieves) or by azeotropic distillation, is often employed to drive the reaction to completion. A wide variety of R groups are possible, and all give the C=N product (see Figure 6.2). However, most substitutions on the nitrogen atom reduce the nucleophilicity of the amino group so acid catalysis is used to facilitate the reaction.

Alcohols

The alcohol functional group is a very important functional group in organic chemistry. Not only do many important compounds and pharmaceuticals contain the alcohol group, but the alcohol group can be prepared *from* many other groups and converted *to* many functional groups. Alcohols thus occupy a central position in functional group manipulations. What's more, most of the traditional methods are still among the most useful methods for the preparation of alcohols.

$$R_1\!-\!\overset{O}{\overset{\|}{C}}\!\!-OR_2 \quad\xrightarrow[\text{2. }H_2O]{\text{1. LAH}}\quad R_1CH_2OH \;+\; R_2OH$$

R_2= H, alkyl, aryl

$$R_1\!-\!\overset{O}{\overset{\|}{C}}\!\!-OR_2 \quad\xrightarrow[\text{2. }H_2O]{\text{1. }BH_3,\ THF}\quad R_1CH_2OH$$

R_2= H, but <u>not</u> alkyl, aryl

Alcohols are at a fairly low oxidation level compared to other oxygen-containing functional groups, and they are consequently readily prepared by reduction. Large numbers of reductive methods have been reported for the preparation of alcohols. Carboxylic acids and esters react vigorously with lithium aluminum hydride (LAH) to produce primary alcohols. Carboxylic acids, but not esters, are also reduced easily by borane which is the only reducing agent that reacts faster with carboxylic acids than with esters or other acid derivatives.

$$\underset{R\quad OH}{\overset{:O:}{\|}} + BH_3 \;\xrightarrow{-H_2}\; \underset{R\quad O-B\cdots H}{\overset{:O:}{\|}}_H \;\longrightarrow\; \underset{H}{\overset{R}{\diagdown}}\!\!\underset{O}{\overset{O}{\diagdown}}\!B\!-\!H \;\longrightarrow\; \underset{RCH_2O}{B}\overset{O}{\overset{\|}{\diagdown}} \xrightarrow{H_2O} RCH_2OH$$

Its unique reactivity comes from the fact that borane first forms a Lewis acid–base complex with the acid and then a boron–carboxylate intermediate which increases the reactivity of the boron hydride and delivers the hydride by an intramolecular reaction. As such, it provides a selective way to reduce acids and produce alcohols in the presence of most other functional groups.

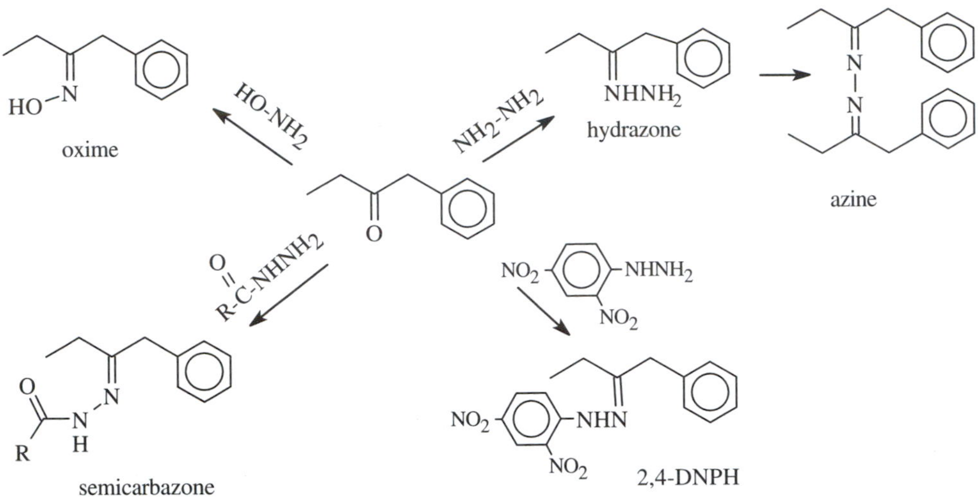

FIGURE 6.2

Aldehydes and ketones are conveniently reduced by sodium borohydride, which is much milder than LAH and does not require aprotic conditions (an alcohol is often the preferred reaction solvent). Aldehydes give primary alcohols, and ketones give secondary alcohols.

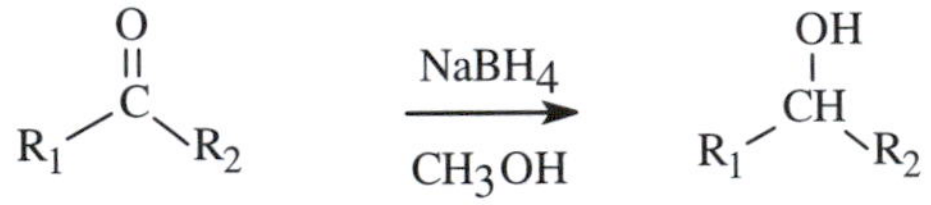

R_2 = H, alkyl, aryl

Alcohols and olefins are the same oxidation level and are interconvertible without a change in oxidation level. Addition of water across an olefinic double bond is thus a common method for the preparation of alcohols. However, simple acid-catalyzed addition of water is often the least desirable alternative. Instead, methods are used which are milder and permit much greater control. For example, the preferred way to produce the Markovnikov alcohol from an olefin is to use hydroxymercuration for the addition step, followed by reductive removal of mercury with $NaBH_4$. In this process, mercury serves as a surrogate for the proton during the addition step and is replaced by hydrogen in the reduction. The advantage is that the mercuric electrophile forms a bridged or partially bridged intermediate. This is stabilized and hence gives higher yields and cleaner product mixtures because rearrangements are suppressed.

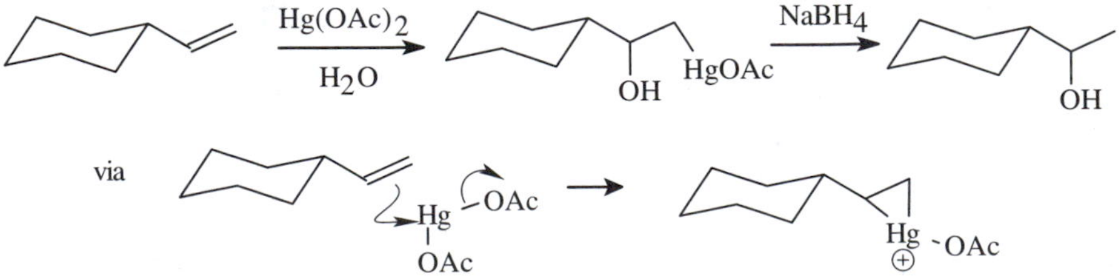

Other nucleophiles such as alcohols, acids, and hydrogen peroxide can also be employed as the cation trap.

Hydroboration is widely employed to obtain an anti-Markovnikov alcohol from an olefin. Addition of diborane to the double bond produces an organoborane intermediate. Three equivalents of the olefin are needed to consume the BH_3 and a trialkylborane is produced. Reaction with basic H_2O_2 converts the carbon–boron bond to a carbon–oxygen bond. This process is effective and widely used.

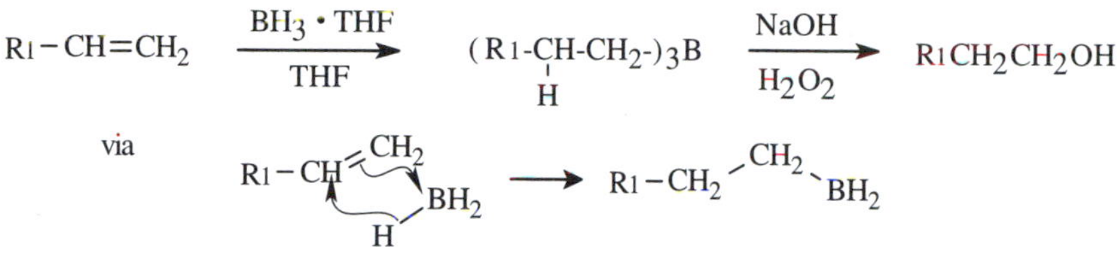
R1−CH=CH2
BH3 • THF
THF
(R1-CH-CH2-)3B
H
NaOH
H2O2
R1CH2CH2OH
via
R1−CH
CH2
BH2
H
R1−CH2
CH2
BH2

The initial addition step occurs in a concerted fashion so that the hydrogen and boron are added to the same side of the planar olefin. Furthermore, the cleavage of the carbon–boron bond occurs with retention of configuration. These features can be used advantageously to prepare stereochemically pure alcohols.

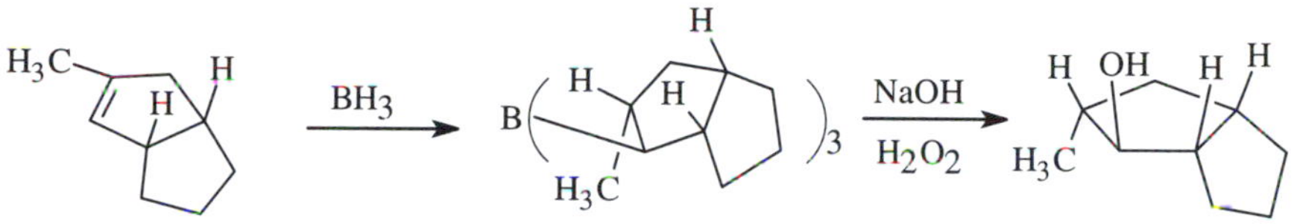
H3C
H
H
BH3
B
H
H
H
H3C
3
NaOH
H2O2
H OH
H
H
H3C

Amines

Amines are saturated nitrogen-containing functional groups that are widely encountered. Because the nitrogen atoms of amines are basic, nucleophilic, and oxidizable, some constraints on the preparation of amines result. A collection of textbook amine preparations includes the following:

Amonolysis $\quad NH_3 + R\text{-}X \longrightarrow R\text{-}NH_2 + \text{polyalkylation}$

Azide substitution $\quad N_3^{\ominus} + R\text{-}X \longrightarrow R\text{-}N_3 \xrightarrow[\text{or LAH, or } H_2/Pd]{Na/EtOH,} R\text{-}NH_2$

Gabriel synthesis

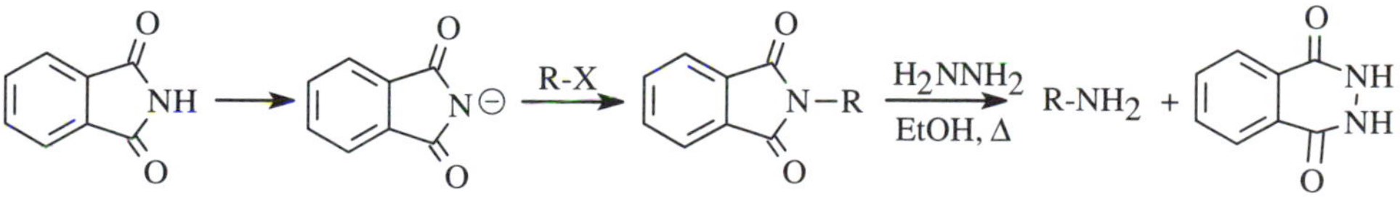
O
NH
O
O
N⊖
O
R-X
O
N−R
O
H2NNH2
EtOH, Δ
R-NH2 +
O
NH
NH
O

Nitro group reduction $\quad Ar\text{-}NO_2 \xrightarrow[\text{or Fe / HCl}]{H_2/Pd} Ar\text{-}NH_2$

Reductive amination

$$R\overset{O}{\underset{}{\|}}R' + R''\text{-}NH_2 \xrightarrow[\text{or NaCNBH}_3]{\text{H}_2 \text{ / Ni}} \underset{R'}{\overset{R}{\rangle}}NH\text{-}R''$$

R, R', R" = H, alkyl
aryl

Reduction

$$R\overset{O}{\underset{}{\|}}NR_2' \quad \text{or} \quad R\text{-}C\equiv N \xrightarrow[\text{ether}]{\text{LAH}} R\text{-}CH_2\text{-}NR_2'$$

R' = H, alkyl, aryl

Amines are at the same low oxidation level as alcohols and consequently are easily pre-pared by reduction. Amides and nitriles are reduced efficiently by LAH to amines. Nitriles give only primary amines, whereas amides give 1°, 2°, or 3° amines, depending on the number of carbon substituents on the amide nitrogen. The advantage of this method is that amides are easy to prepare from acid chlorides and amines while nitriles are available by displacement reactions.

$$R\overset{O}{\underset{}{\|}}NHCH_2CH_3 \xrightarrow[\text{2. H}_2\text{O}]{\text{1. LAH, ether}} RCH_2NHCH_2CH_3$$

$$\text{CN} \xrightarrow[\text{2. H}_2\text{O}]{\text{1. LAH, ether}} CH_2{\diagup}NH_2$$

One problem with this method is that the workup must be done carefully, as the amine products tend to complex tenaciously with the aluminum salts formed upon workup and are not easily recovered. There are standard workups which avoid these issues, but these should be followed carefully. Reduction of azides by either catalytic reduction, phosphine or phosphite reagents, or tin II chloride are all effective methods. The azides are also available from displacement reactions and give primary amines upon reduction.

$$H_3C\text{...}N_3 \xrightarrow[\substack{\text{or P(C}_6\text{H}_5)_3, \text{ H}_2\text{O} \\ \text{or SnCl}_2, \text{MeOH}}]{\text{H}_2 \text{ / Pd or Pt}} H_3C\text{...}NH_2$$

Aromatic amines are readily prepared by the reduction of aromatic nitro groups by Fe/HCl or Sn/HCl or by catalytic hydrogenation.

$$Z\text{—}NO_2 \xrightarrow[\substack{\text{or Fe / HCl} \\ \text{or Sn / HCl}}]{\text{H}_2 \text{ / Pd or Pt}} Z\text{—}NH_2$$

Displacement reactions are rarely used for the preparation of amines, as polyalkylation re-duces yields and makes product mixtures more complex. However, reaction of primary amines with primary and secondary sulfonates can provide good yields of monoalkylated product if care is taken to control both the conditions and the mode of addition. Benzylamine is particularly common as a primary amine nucleophile since the benzyl group can be removed by hydrogenolysis to give a primary amine.

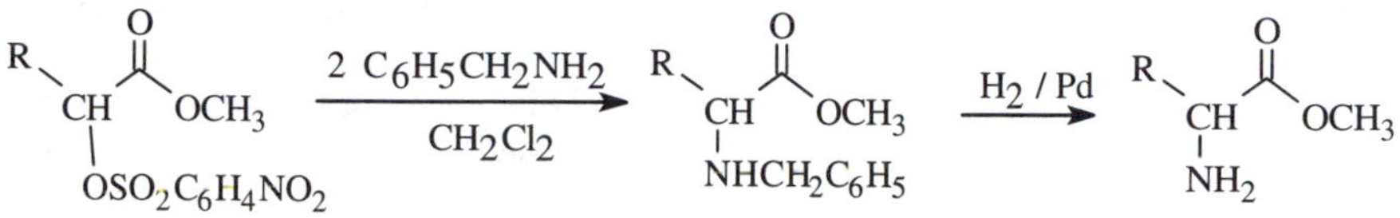

Alkenes

Traditional preparations of alkenes include the following:

Elimination

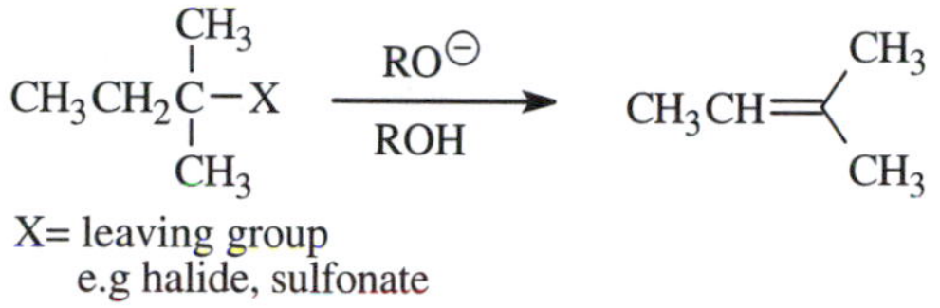

Dehydration

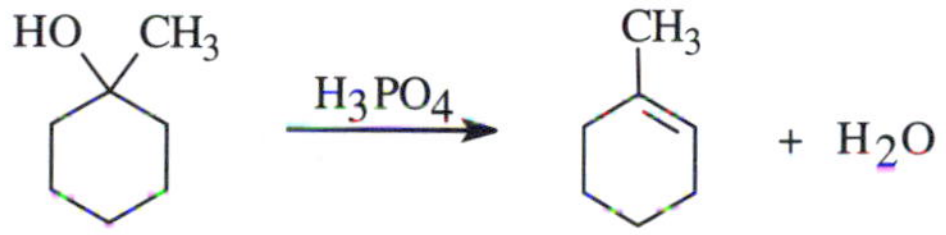

Dibromide reduction

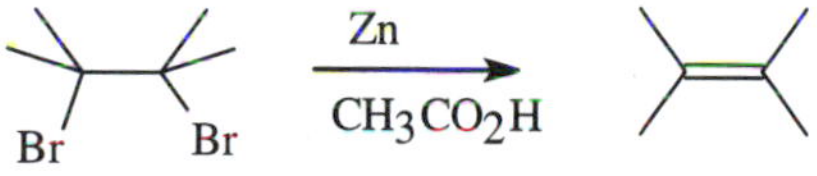

Alkyne reduction

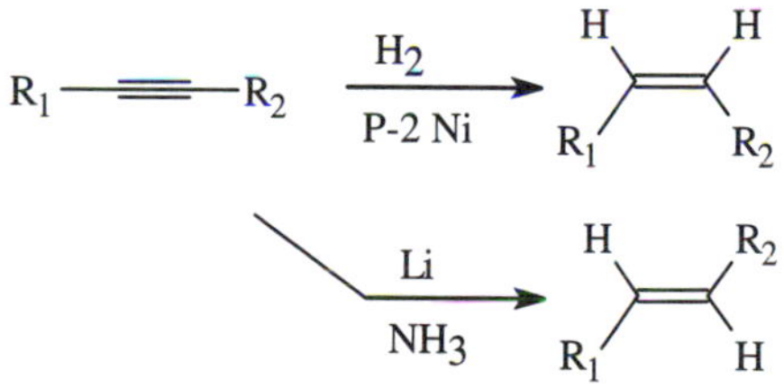

Alkenes are relatively low oxidation level hydrocarbons. The most common way to prepare alkenes is to carry out the elimination of a small molecule from between vicinal carbon atoms. However, this is a viable strategy only if the regiochemistry of elimination can be controlled. That is, traditional dehydrohalogenations or dehydrations often are regioselective but not regiospecific, so that mixtures of structurally isomeric olefins are formed.

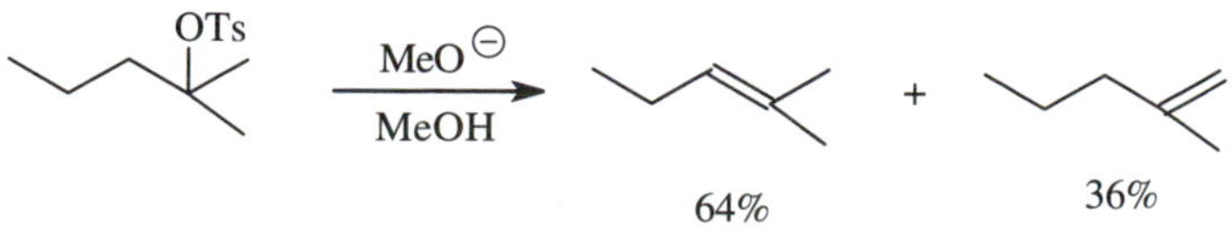

The formation of regioisomers is due to the presence of several sets of nonequivalent vicinal hydrogens of similar, but not identical, reactivity. The resulting mixture of similar products must be separated if only one of the regioisomers is desired.

Several strategies have been developed to control the elimination regiochemistry. These include placement of the leaving group, steric bulk of the base, and establishment of thermodynamic control. By placing the leaving group at the end of a chain, only terminal olefins can be produced by elimination because there is only one set of vicinal hydrogens that can be removed by the base.

sole product

but

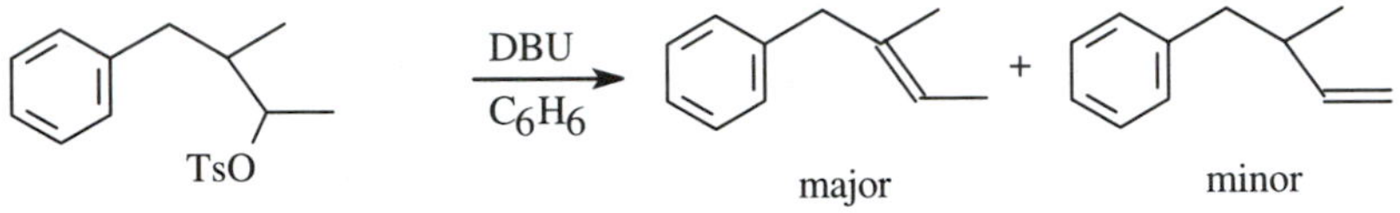

major   minor

By using very bulky alkoxide bases (*t*-butoxide or amyloxide), attack of the base occurs at the least hindered position—usually at the end of chains, if possible. For example,

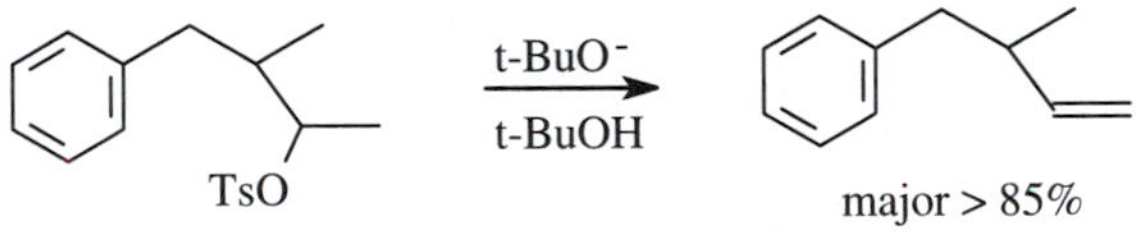

major > 85%

Finally, when eliminations which give conjugated systems are possible, they are favored significantly by the greater stability of the conjugated π system.

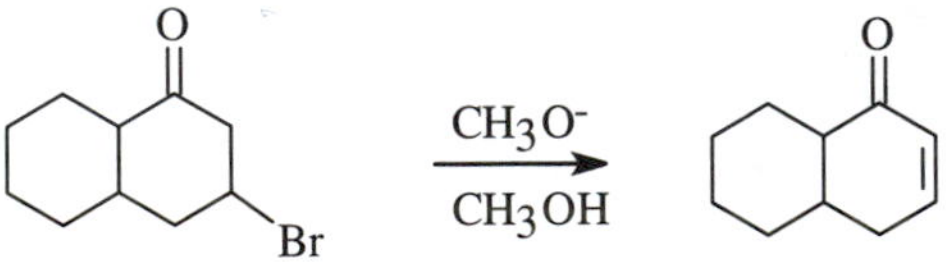

only

Dehydrations produce olefins from alcohols by the acid-catalyzed elimination of a water molecule from between two carbons. Acid-catalyzed dehydrations often give mixtures of products because the intermediate carbocation is prone to cationic rearrangements to more stable carbocations prior to formation of the olefin product. Moreover, even when the intermediate carbocation is not subject to skeletal rearrangement, as in the case of tertiary alcohols, mixtures of regioisomers are often produced during the loss of a proton from the carbocation.

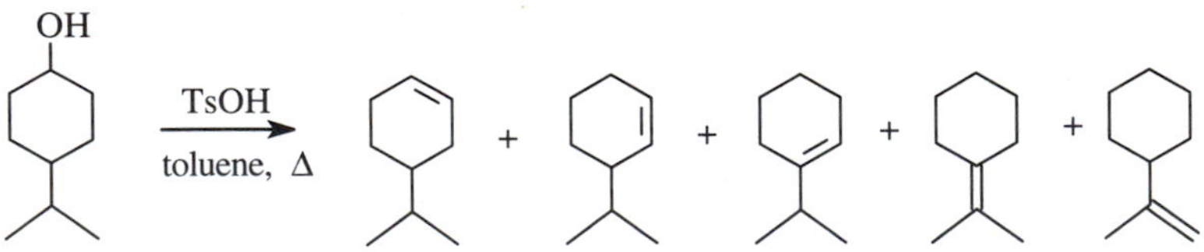

As a consequence, the acid-catalyzed dehydration of alcohols is generally not a viable synthetic method.

Many other methods are used currently for carrying out 1,2 eliminations to give olefins. Several are particularly useful and widely used. Selenoxide eliminations are frequently used to install the double bond of α,β-unsaturated carbonyl compounds. They occur by concerted, cyclic, *syn* processes.

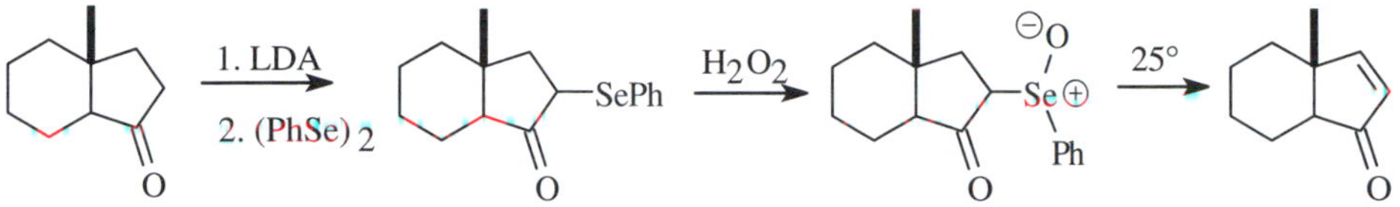

Silyloxide eliminations (Petersen olefination) also proceed readily, and regiospecifically, to give olefins:

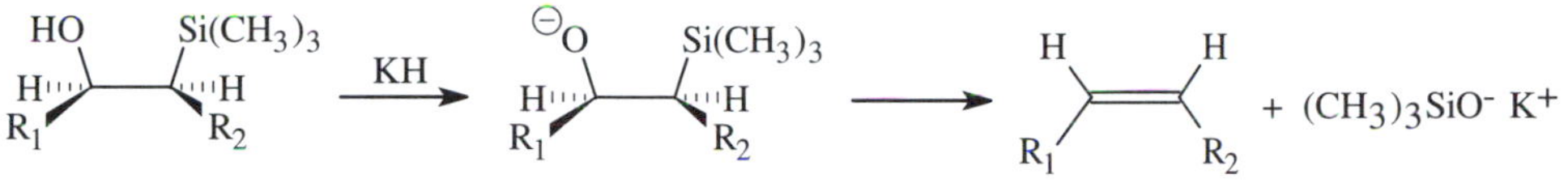

Similarly phosphine oxide eliminations (Wittig reaction) occur very readily to give olefins:

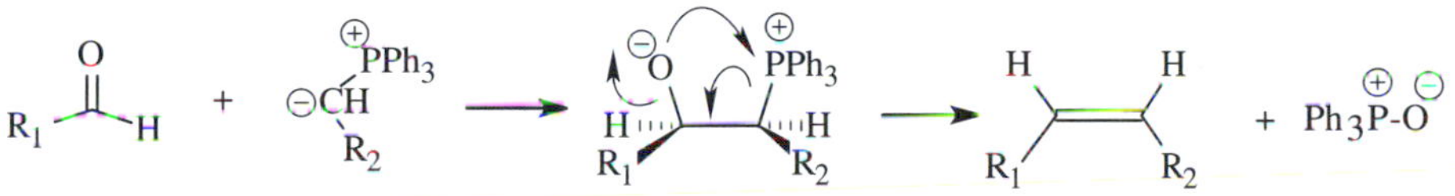

Each of these latter two methods of elimination is part of a longer sequence of reactions that produce olefins. Initial formation of a single bond to a carbonyl carbon is followed by elimination to an alkene. Thus the alkene is a condensation product of two smaller units. Schematically, where X is an element (Si or P) that can remove oxygen to the alkene:

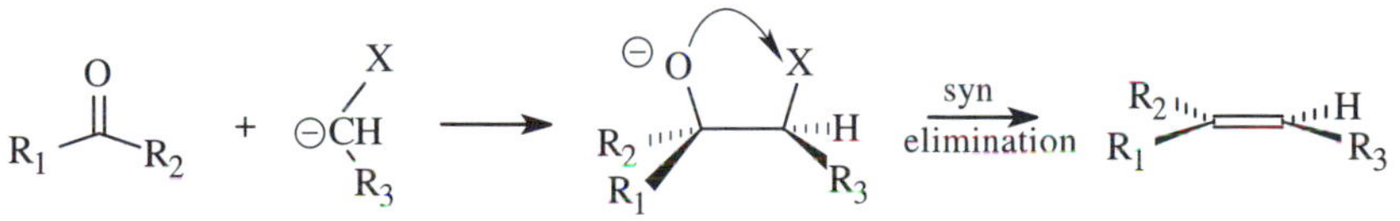

It should also be noted that both of these eliminations proceed with *syn* stereochemistry between the oxyanion and the heteroatom; thus, the stereochemistry of the intermediate dictates the geometry of the olefin product.

Alkenes can also be produced effectively by the reduction of alkynes. The reduction can be carried out stereospecifically to give either *cis* or *trans* olefins, as desired. This is a very useful method because of the stereocontrol. The P-2 nickel catalyst for the *cis* hydrogenation is produced in situ by the reduction of nickel [II] acetate with sodium borohydride, and the reaction is carried out at atmospheric pressure, making this a very simple method for the preparation of *cis* olefins. The lithium in liquid ammonia reduction of alkynes to the *trans* olefin is also very straightforward experimentally.

$$R_1 \equiv R_2 \xrightarrow[\text{P-2 Ni}]{H_2} \underset{R_1 \quad\quad R_2}{\overset{H \quad\quad H}{\diagdown\diagup}}$$

$$\xrightarrow[\text{NH}_3]{\text{Li}} \underset{R_1 \quad\quad H}{\overset{H \quad\quad R_2}{\diagdown\diagup}}$$

Alkanes

Alkanes are the most highly reduced of all organic compounds. As a consequence, virtually all preparations of alkanes are reductive. Alkenes and alkynes can both be reduced to alkanes by catalytic hydrogenation. While many catalysts can be employed, palladium on carbon is by far the most common.

$$\begin{array}{c} R_1 \equiv R_2 \\ \text{or} \\ R_1 \equiv R_2 \end{array} \xrightarrow[\text{Pd / C}]{H_2} R_1\text{-CH}_2\text{CH}_2\text{-R}_2$$

Primary and secondary alcohols can be converted to alkanes by conversion to tosylates, followed by reduction with LAH. This reduction is valuable because deuterium can be easily introduced into the alkane by the use of lithium aluminum deuteride (LAD) instead of LAH.

$$R_1\text{-CH}_2\text{-OH} \longrightarrow R_1\text{-CH}_2\text{-OTs} \xrightarrow{\text{LAH}} R_1\text{-CH}_3$$

Ketones can be reduced directly to alkanes by the Wolff–Kishner reduction. In this reduction, the ketone is converted to the hydrazone which is treated in situ with sodium hydroxide. An internal redox reaction occurs in which the carbon is reduced and the hydrazine is oxidized to nitrogen. The best experimental conditions include the use of NaOH and ethylene glycol as solvent to carry out the reduction.

$$\underset{R_2}{\overset{R_1}{\diagdown}}{=}O \xrightarrow{H_2NNH_2} \underset{R_2}{\overset{R_1}{\diagdown}}{=}NNH_2 \xrightarrow[\text{HO} \quad \text{OH} , \Delta]{\text{NaOH}} \underset{R_2}{\overset{R_1}{\diagdown}}CH_2 + N_2$$

The reduction of ketones to alkanes can also be done by the Clemmensen reduction using zinc and HCl. This reaction is specific for aromatic ketones, however.

$$\underset{\text{Ar} \quad\quad R_1}{\overset{O}{\diagdown\diagup}} \xrightarrow[\text{HCl}]{\text{Zn}} \text{Ar-CH}_2\text{-R}_1$$

Alkyl halides (Cl, Br, I) can be converted to alkanes by two types of reactions. The halogen can be reduced off using lithium or zinc metal most effectively. This procedure works best with bromides and iodides.

$$\text{RX} \xrightarrow[\text{or Li / EtOH}]{\text{Zn / HOAc}} \text{RH}$$

Alternatively, alkyl halides undergo coupling reactions with lithium organocuprates (which are themselves prepared from alkyl halides) to give alkanes by carbon–carbon bond formation. Other metals can be used to promote the same kind of coupling, but the use of cuprates is the most efficient and general.

$$R_1X \ + \ (R_2)_2CuLi \ \longrightarrow \ R_1\text{-}R_2 \ + \ R_2\text{-}Cu$$

It is clear that there are many different ways to carry out the installation of a particular functional group in a molecule. The ones discussed here are often the most general and practical, and they are often the first ones that are tried in the laboratory. However, it is also common that a particular substrate will not give good results with any of the common reagents. For this reason, new methods of functional group manipulation are constantly being sought that are even more general, more selective, milder, cheaper, easier, and use more readily available starting materials than other methods.

Bibliography

Most undergraduate textbooks cover the preparation of functional groups but don't necessarily give the best or most widely used methods. See R. T. Morrison and R. N. Boyd, *Organic Chemistry,* 6th ed., Wiley, New York, 1993; M. A. Fox and J. K. Whitesell, *Organic Chemistry,* Jones & Bartlett, Boston, 1994; and J. McMurry, *Organic Chemistry,* 3d ed., Brooks/Cole, Pacific Grove, CA, 1992.

The index of J. A. March, *Advanced Organic Chemistry, Reactions Mechanism, and Structure,* 4th ed., Wiley Interscience, New York, 1992, has a listing of functional groups and references to reactions and page numbers where they are prepared.

The best comprehensive listing of functional group preparations is in R. C. Larock, *Comprehensive Organic Transformations,* VCH Publishers, New York, 1989.

Problems

1. Show how to prepare each of the following products from the given starting material. For each *overall* transformation, indicate 1. the starting and ending functional group and 2. what change in oxidation level (if any) must be accomplished. Where more than one step is required, show each step distinctly, including reagents and conditions needed to effect the conversion.

(a)

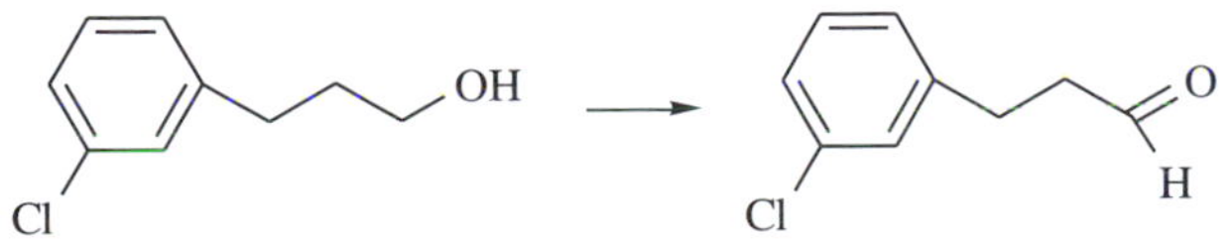

(b)

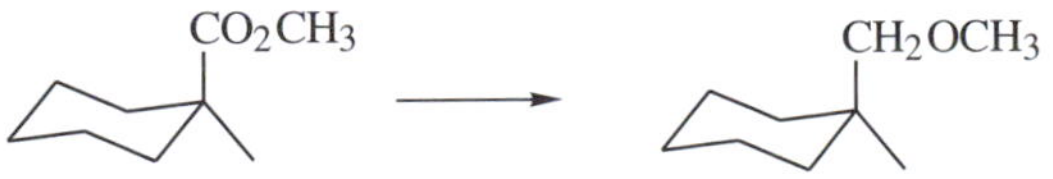

(*c*) 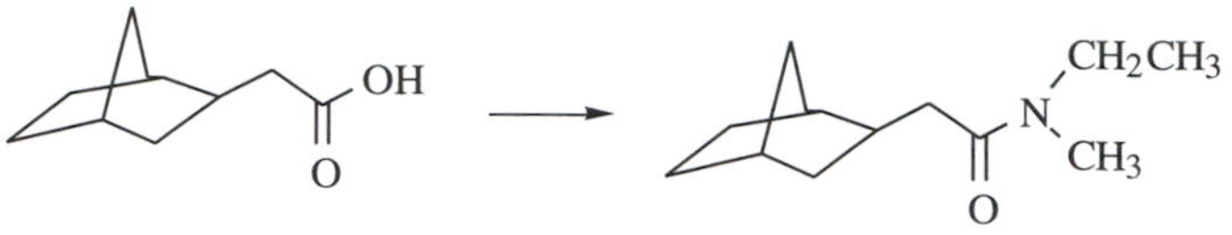
OH
O
CH_2CH_3
N
CH_3
O

(*d*) 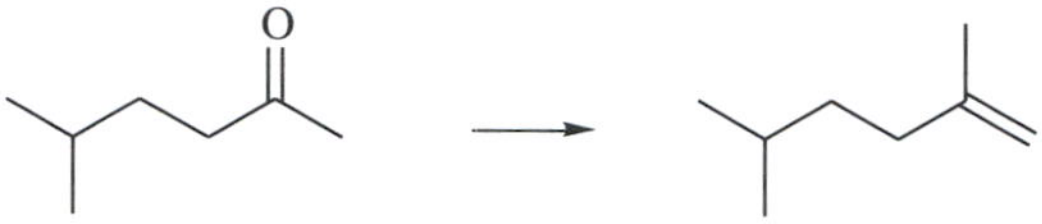
O

(*e*) 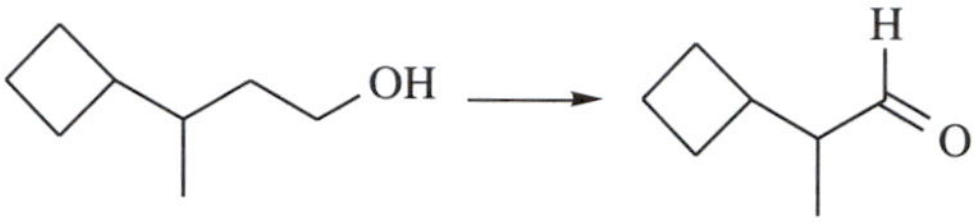
OH
H
O

(*f*) 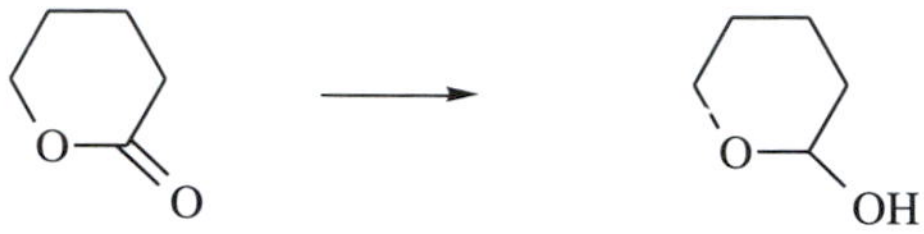
O
O
O
OH

(*g*) 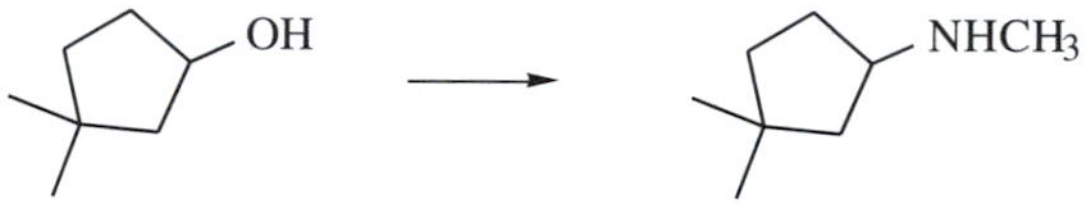
OH
$NHCH_3$

(*h*) 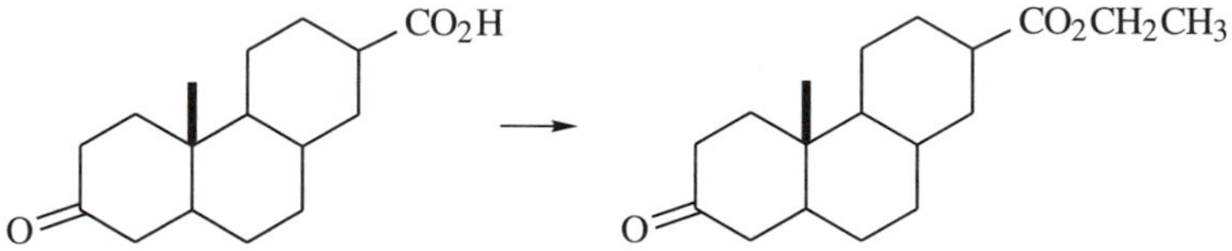
CO_2H
O
$CO_2CH_2CH_3$
O

(*i*) 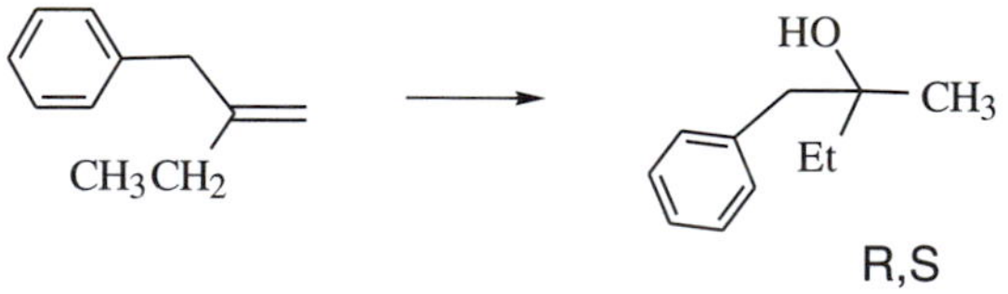
CH_3CH_2
HO
CH_3
Et
R,S

(*j*) 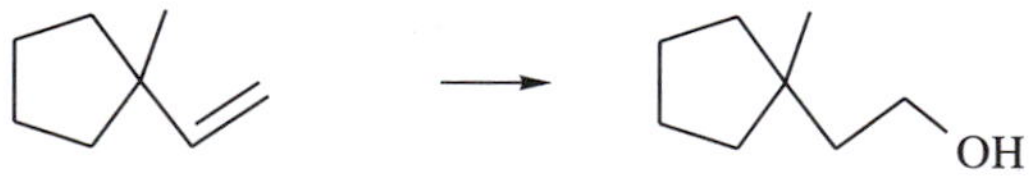
OH

(*k*) 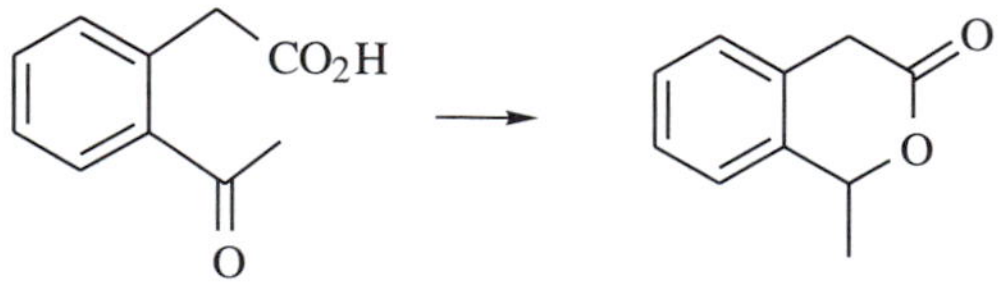
CO_2H
O
O
O

(*l*)

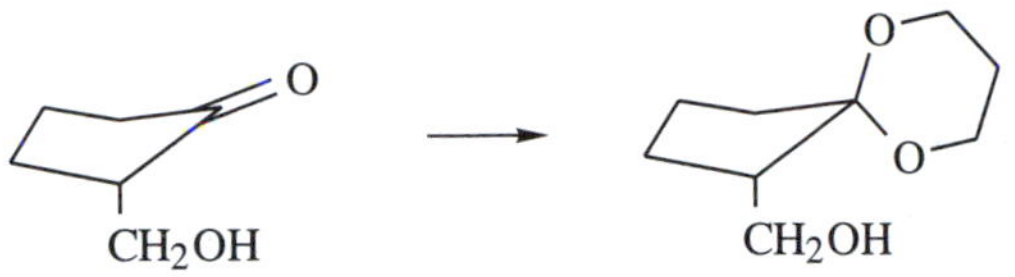
O
CH2OH
O
O
CH2OH

(*m*)

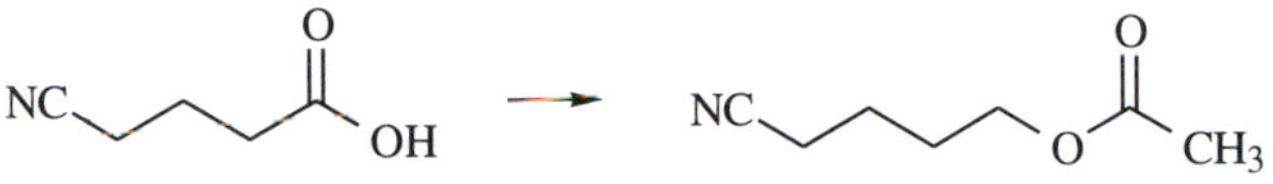
O
NC
OH
O
NC
O
CH3

(*n*)

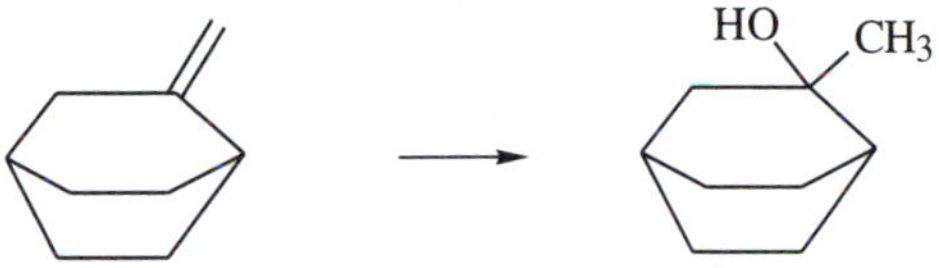
HO
CH3

(*o*)

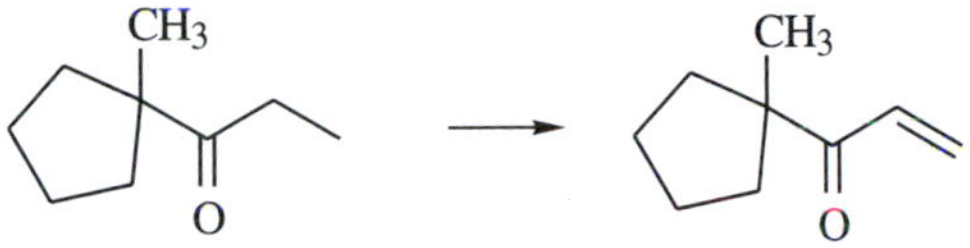
CH3
O
CH3
O

(*p*)

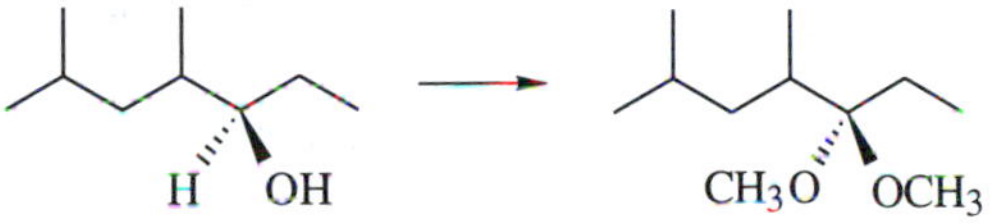
H
OH
CH3O
OCH3

(*q*)

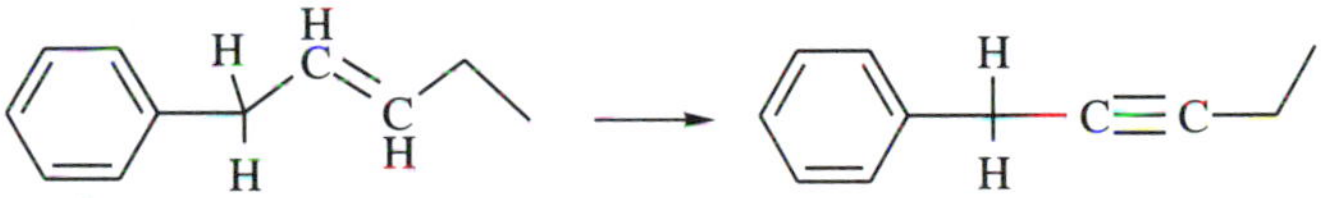
H
H
C
C
H
H
H
H
C
C
H

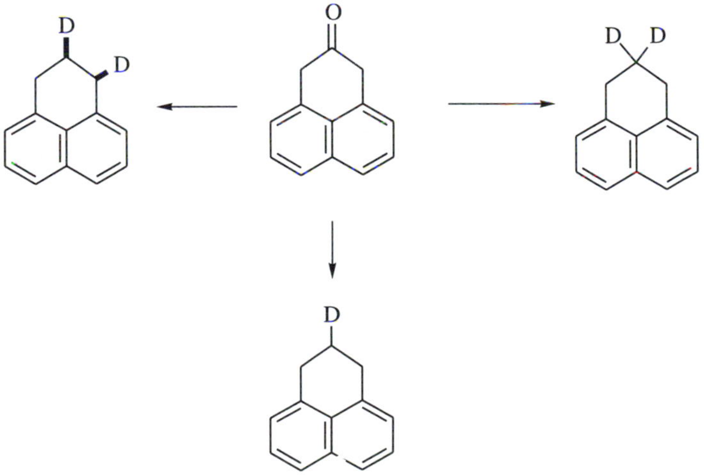
D
D
O
D
D
D

(*r*)

(s) CH₃ ⟶ CO₂H ⟶ (amide product with C₆H₅, CH₃, N-H groups)

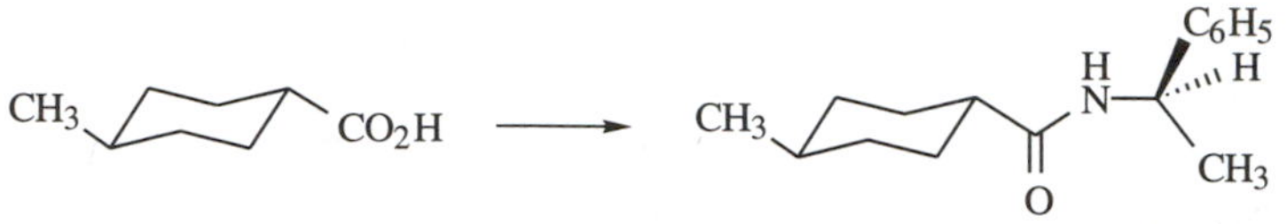

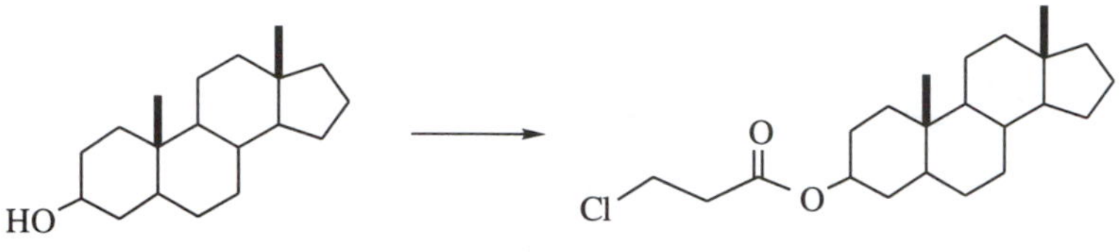

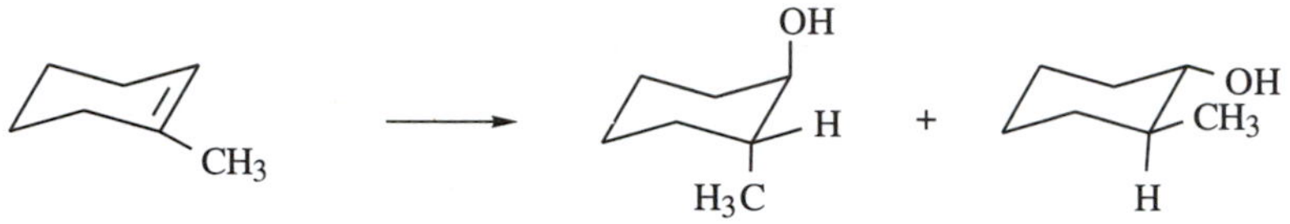

(t) HO— (steroid) ⟶ Cl— (ester-linked steroid product)

(u) (cyclohexene with CH₃) ⟶ (product with OH, H, H₃C) + (product with OH, CH₃, H)

(v) (Br compound) ⟶ (diene product)

2. Show *two* different starting materials from which one could obtain each of the following compounds. Show the reagents and conditions necessary to convert each of the starting materials into the desired product.

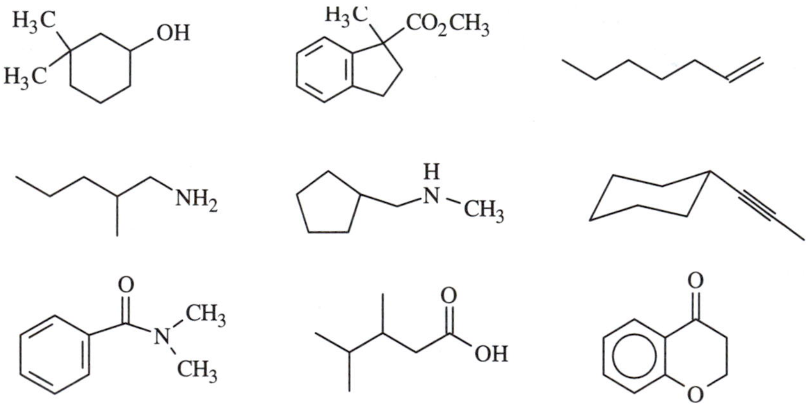

H₃C, H₃C— (cyclohexane with OH)

H₃C, CO₂CH₃ (indane ester)

(branched alkene)

(branched chain with NH₂)

(cyclopentane with N-H-CH₃)

(cyclohexane with alkyne)

(benzamide with N(CH₃)₂)

(branched carboxylic acid)

(chromanone)

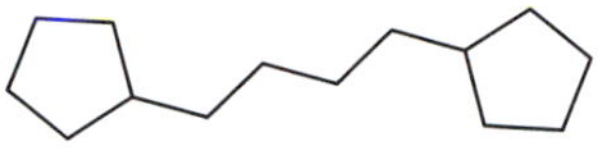

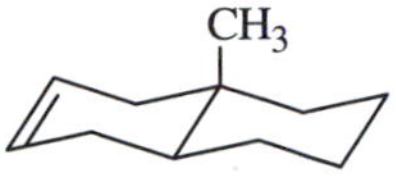

CH₃

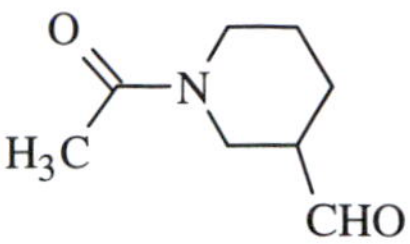

H₃C

CHO

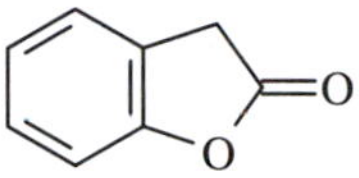

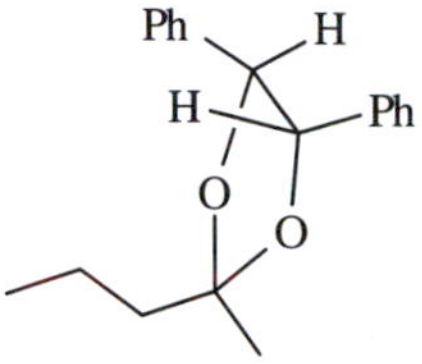

Ph   H

H   Ph

O   O

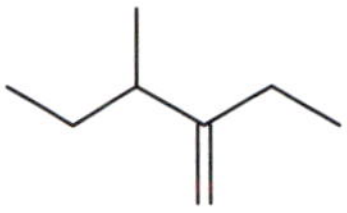

CARBON–CARBON BOND
FORMATION BETWEEN CARBON
NUCLEOPHILES AND CARBON
ELECTROPHILES

Synthetic Strategy

The object of the game in organic synthesis is to assemble a particular molecule, the target, which has the correct carbon skeleton and the proper array of functional groups distributed on the skeleton. There are two general strategies for accomplishing this synthetic challenge. The first is to start with a molecule having the desired carbon skeleton and manipulate the functionality on the skeleton to that of the desired compound. The previous chapter as well as many texts deal with functional group preparations and thus provide methods for the installation and manipulation of functional groups on the skeleton.

The second general synthetic strategy is to assemble the proper carbon skeleton and then adjust the functionality in the resulting molecule to that of the target. Obviously, assembling the carbon skeleton from smaller building blocks is a more versatile and convergent approach, because the carbon skeleton of the target may not be available (or at least not readily available). Moreover, once the synthetic sequence needed to assemble the carbon skeleton has been developed, it can potentially be adapted to produce a series of structurally related molecules. This is particularly useful when such a series of compounds is needed to test for particular physical, chemical, or biological properties.

In order to assemble the needed carbon skeletons from smaller units, it is absolutely crucial that one can form carbon–carbon bonds between them. Consequently, carbon–carbon bond-forming reactions are among the most important organic reactions. Since any particular carbon–carbon bond is merely a covalent link between two skeletal fragments, synthesis of a target, such as R_3C-CR_3', can be accomplished by forming a carbon–carbon bond between the two skeletal fragments, R_3C and CR_3'. Since the bond to be made contains two electrons that are shared, there are three modes by which these electrons can be distributed between the two fragments to be joined:

$$R_3C\oplus \quad \text{and} \quad \ominus CR'_3$$
$$R_3C\ominus \quad \text{and} \quad \oplus CR'_3 \quad \longrightarrow \quad R_3C-CR'_3$$
$$R_3C\cdot \quad \text{and} \quad \cdot CR'_3$$

There are two ionic modes of bond formation where one carbon fragment is nucleophilic (e^--rich) and the other is electrophilic (e^--poor). There is one free radical mode, in which each fragment has a single, unpaired electron which becomes shared upon bond formation. In order to understand carbon–carbon bond-forming reactions that take place by ionic or two-electron processes, one needs to know what species, or compounds, can serve as carbon-centered nucleophiles, and what species, or compounds, can serve as carbon-centered electrophiles. (Free radical reactions will be considered later.)

Nucleophilic Carbon

Carbon-centered nucleophiles are those compounds, or intermediates, which contain an electron-rich carbon atom and are thus capable of donating an electron pair from that carbon atom to an electrophile. The electron pair that is donated is found in a filled orbital in the nucleophilic carbon, and the electrons are not tightly bound. Donation to the electrophilic carbon occurs by overlap of the filled orbital of the donor with an unfilled orbital of the acceptor. The most common carbon nucleophiles fall into three main classes:

1. Organometallic molecules contain a carbon–metal bond which is polarized toward carbon. As a result, the carbon is very electron-rich and reacts vigorously with electrophiles.

$$\underset{\delta-|\quad\ \delta+}{-C-M} \qquad M = MgX, Li, Cu, Zn, \text{etc.}$$

Examples of organometallic compounds that are most commonly used as carbon nucleophiles include Grignard reagents (RMgX), organolithiums (RLi), organocuprates (R$_2$CuLi), and occasionally organocadmiums (R$_2$Cd) and organozinc reagents (RZnBr).

2. Enolates are anionic derivatives of carbonyl compounds formed by removal of a proton from a position adjacent to the carbonyl group.

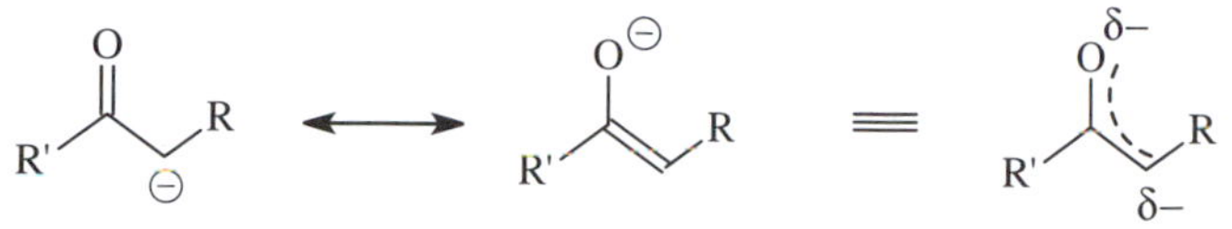

Resonance delocalization of the negative charge with the carbonyl group stabilizes enolate anions and makes them somewhat less reactive than organometallic compounds. They are, however, reactive carbon nucleophiles. Examples of enolates include anionic derivatives of aldehydes, ketones, acid derivatives, and dicarbonyl compounds.

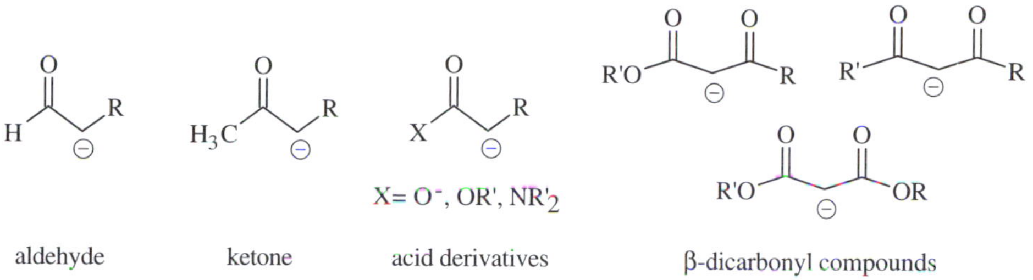

X= O$^-$, OR', NR'$_2$

| aldehyde | ketone | acid derivatives | β-dicarbonyl compounds |

Structurally related to enolates are anionic derivatives of imines and nitro compounds. The former are less stable (more reactive) than enolates because nitrogen cannot support a negative charge as well as oxygen, and thus resonance stabilization is diminished compared to enolate anions. Nitronate anions are much more stable (less reactive) than enolates because resonance with the nitro group transfers the negative charge to oxygen, where it is stabilized by the formal positive charge on nitrogen.

3. A third type of molecule, capable of functioning as a nucleophilic carbon equivalent, is one that contains an electron-rich π bond. Such species are usually uncharged and function as carbon nucleophiles because the π electrons are less tightly bound and can consequently be donated to good electron acceptors. Enol derivatives are neutral derivatives of carbonyl compounds which have an oxygen or nitrogen substituent attached to a carbon–carbon double bond. They are covalent analogs of enolate ions and because they are neutral, they are much less reactive electron donors than enolate anions. In fact, most enol derivatives are stable compounds which can be isolated. In spite of being neutral, however, resonance interaction of the lone pairs of the heteroatom with the π system increases the electron density at the β position of the double bond, and the double bond is electron-rich. Besides the parent enol, examples of enol derivatives of carbonyl compounds include the following:

| enol | enol ether | enol acetate | silyl enol ether | enamine |

Besides a heteroatom substituent, which renders a double bond electron-rich by resonance interaction of the lone pairs with the π system, other substituents can also result in π bonds being electron-rich and thus reacting as electron donors. Attachment of substituents less electronegative than carbon to the double bond increases its π electron density significantly by an inductive effect. Vinylsilanes and vinyl stannanes can both be considered to have electron-rich π bonds and have been used as π electron donors. Allylsilanes, by virtue of hyperconjugation between the allylic carbon–silicon bond and the π system, also function as good π electron donors in many reactions.

| vinyl silane | vinyl stannane | allyl silane |

Electrophilic Carbon

Carbon-centered electrophiles are compounds, or intermediates, which are electron-poor and thus capable of accepting electrons from electron donors. In order to be an electron acceptor, an electrophile must have an unfilled orbital on carbon available for overlap with a filled orbital of the donor. Unfilled atomic p orbitals or antibonding orbitals (both $\sigma*$ and $\pi*$) are the most common types of acceptor orbitals. The most common carbon electrophiles fall into four major categories.

1. Cationic carbon electrophiles are the most reactive because of the positive charge they carry. They can, however, have a variety of structures, depending on the hybridization of the carbon acceptor. Trialkyloxonium tetrafluoroborates (Meerwein salts) $R_3O^+BF_4^-$, for example, are sp^3-hybridized carbon electrophiles and are extremely reactive toward nucleophiles. The acceptor orbital is an antibonding C—O $\sigma*$ orbital which is low in energy because of the positive charge on oxygen.

Triphenylmethyl (trityl) tetrafluoroborate, on the other hand, is sp^2-hybridized, but it is also extremely reactive toward electron donors. The acceptor orbital of the trityl cation is an unfilled 2p atomic orbital on the charged carbon. These carbon electrophiles are isolable compounds, but they are extremely reactive with any sort of electron donor (H_2O vapor is a common culprit).

Many other cationic carbon electrophiles cannot be isolated, but they can be generated in situ in a reaction mixture by Bronsted or Lewis acid–base reactions. In the presence of Bronsted acids, carbonyl compounds are protonated and produce positively charged oxonium ions. Compared to the carbonyl compound itself, the $\pi*$ orbitals of oxonium ions are much stronger electron acceptors. Protonation of the carbonyl group is an equilibrium process and the extent of protonation (the position of the protonation equilibrium) is dictated by the pK_as of the Bronsted acid and the oxonium ion. While this equilibrium usually lies far to the left, the reactivity of the oxonium ion is often sufficiently great that only small amounts of the oxonium ion are needed to react effectively with the electron donor.

Addition of a strong Lewis acid, such as $TiCl_4$, BF_3, or $SnCl_4$, to a carbonyl compound is another common method to produce a very powerful cationic electrophile in solution. Complexation between the Lewis acid and the lone pairs of electrons on the carbonyl oxygen gives a species which, although formally neutral, behaves as a cationic carbon electrophile in the same fashion as a protonated carbonyl group. These are strong electrophiles that react with many types of nucleophiles. Aldehydes and ketones are common carbonyl components which are activated with Lewis acids. However, esters and amides also yield strongly electrophilic species with Lewis acids.

2. Aliphatic compounds, with good leaving groups attached to primary or secondary carbon atoms, are very commonly used as carbon electrophiles. The leaving group is an electronegative group attached by a polarized σ bond.

The bond polarity makes the carbon atom electron deficient and capable of accepting electrons into the $\sigma*$ antibonding orbital from carbon electron donors. The carbon electron donors are usually termed *nucleophiles* in this type of process. Population of the $\sigma*$ orbital by electron donation weakens the bond to the leaving group. Ultimately, the leaving group is cleaved from the molecule and retains the pair of electrons from the connecting bond. Examples of such compounds include alkyl halides, alkyl sulfonates, and alkyl sulfates.

The electrophilicity of such compounds is largely related to the leaving ability of the leaving group. The leaving ability of a group is, in turn, related to both its bond strength to carbon and its ability to accept the bonded pair of electrons and become electron-rich (most often, negatively charged). Leaving abilities range from excellent (triflate) to moderate (chloride). The electrophilicity of the acceptor molecule can be adjusted by changing the leaving group.

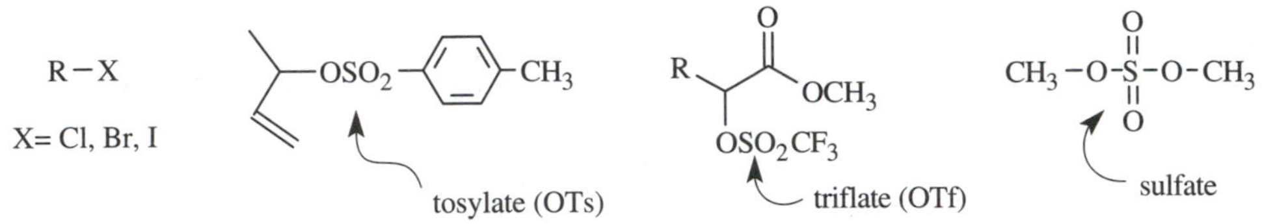

R—X

X= Cl, Br, I

tosylate (OTs)

triflate (OTf)

sulfate

3. Carbonyl compounds are very common carbon electrophiles, by virtue of the polarized carbon–oxygen π bond. Electron donation into the $\pi*$ orbital of the carbonyl carbon breaks the C—O π bond and produces a tetrahedral adduct which can then proceed to products.

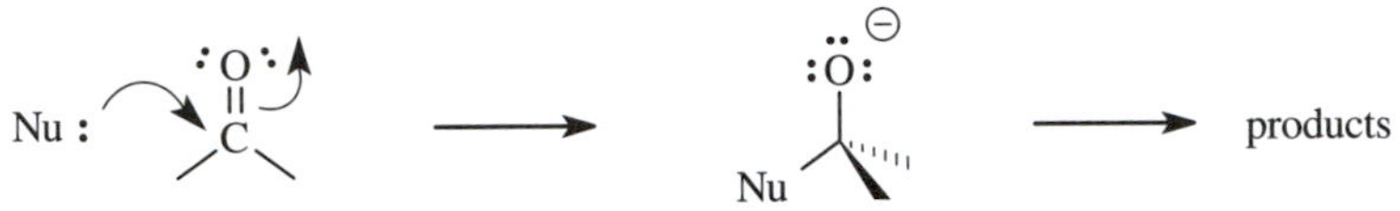

products

This is a very general process for carbonyl compounds. However, the electrophilic reactivity of the carbonyl group is very dependent on the groups attached to it. The reactivity is ranked in the following order:

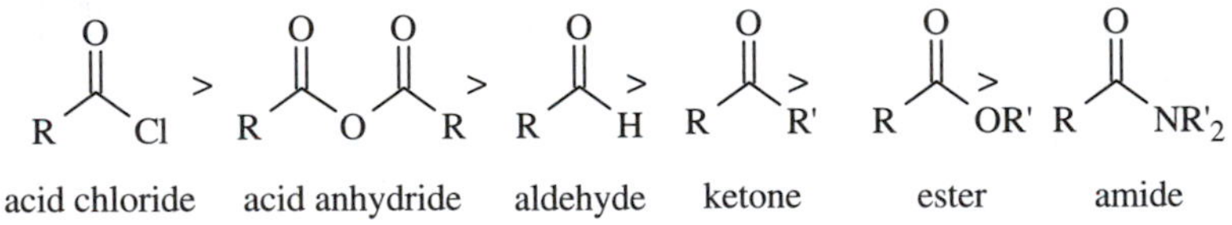

acid chloride   acid anhydride   aldehyde   ketone   ester   amide

Electron-withdrawing groups (Cl, RCO_2) increase the electrophilicity, while resonance-donating groups ($-OR$, $-NR_2$) decrease the reactivity toward electron donors. Steric effects are also a significant influence on carbonyl reactivity. The trigonal carbonyl reactant goes to a more crowded tetrahedral intermediate upon addition of the nucleophile. Thus bulky groups attached to the carbonyl carbon lead to more crowded transition states and result in much slower addition reactions. This steric rationale explains the greater reactivity of aldehydes over ketones. Extremely sterically hindered ketones, such as di-tert-butyl ketone, undergo carbonyl addition by nucleophiles at negligible rates for most nucleophiles.

4. α, β-unsaturated carbonyl compounds can act as electrophiles under certain conditions, and are bidentate in that both the carbonyl carbon and the β carbon are electron deficient. Thus nucleophiles can attack at either position.

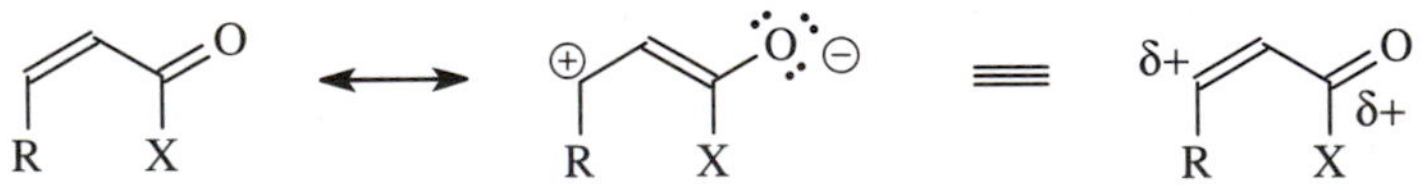

X= H, alkyl, OR

The regioselectivity of nucleophilic addition is a function of the type of nucleophile employed. In many instances, this can be controlled to give Michael addition to the β carbon.

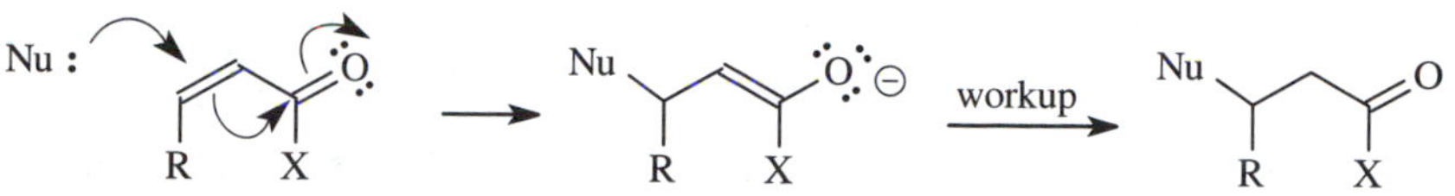

X= H, alkyl, OR

Reactivity Matching

Having defined the types of commonly used carbon nucleophiles and carbon electrophiles, it would seem that reacting any of the carbon nucleophiles (electron donors) with any of the carbon electrophiles (electron acceptors) should form a carbon–carbon bond. Although this is theoretically true, it is unworkable from a practical point of view. If, for example, a carbanion nucleophile were reacted with a cationic electrophile, it is unlikely that the desired carbon-carbon bond formation would be detected, even after the smoke cleared. Or if a silyl enol ether nucleophile were reacted with an α, β-unsaturated ester, no reaction could be observed to take place in any reasonable time frame:

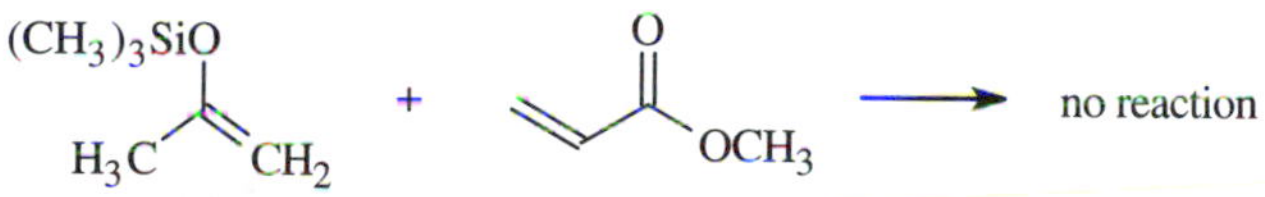

In order for carbon–carbon formation to be successful, the reactivities of the nucleophile and electrophile must be matched so that the reaction occurs at a reasonable and controllable rate. Thus one must be able to easily generate both carbon nucleophiles and carbon electrophiles in order to be able to choose the appropriate partners for successful C—C bond formation.

Generation of Nucleophilic Carbon Reagents

The three major classes of nucleophilic carbon species are (1) organometallic compounds, (2) enolate derivatives (and related carbanionic compounds), and (3) neutral enol derivatives.

Organometallic compounds which contain a carbon–metal bond are the most reactive carbon nucleophiles. In most cases, they are also powerful bases and must be prepared and used under strictly anhydrous and aprotic conditions. A very common way to produce organometallic compounds is to reduce alkyl halides with active metals. Grignard reagents and organolithium compounds are routinely produced in this manner. The transformation is a two-electron reduction of the alkyl halide to a carbanion equivalent; the metal is oxidized:

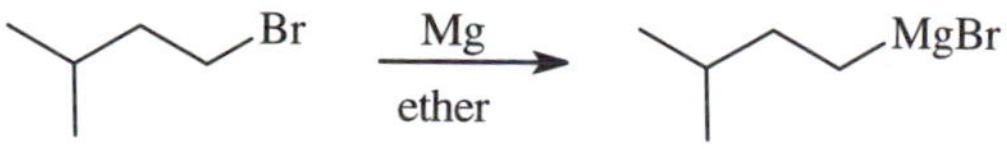

$$\text{(cyclohexylmethyl)Br} \xrightarrow[\text{hexane}]{\text{2 Li}} \text{(cyclohexylmethyl)Li} + \text{LiBr}$$

This procedure works well for alkyl, vinyl, and aryl halides, and provides a convenient source of organomagnesium halides and organolithium compounds. In addition, a variety of other metals can be exchanged for lithium in organolithium compounds to give different organometallic compounds of modified reactivity. Reaction of two equivalents of an organolithium compound with a cuprous halide gives a lithium organocuprate, in which the carbon–lithium bonds of the organolithium reactant are converted to carbon–copper bonds in the anionic organocuprate. Lithium merely serves to balance the charge of the organocuprate. By a similar exchange, dialkylmercury compounds can be prepared from organolithiums and mercury[II] halides:

$$2\ \text{RLi}\ +\ \text{CuX} \longrightarrow \underset{\substack{\text{lithium}\\\text{organocuprate}}}{\text{R}_2\text{CuLi}}\ +\ \text{LiX}$$

$$2\ \text{RLi}\ +\ \text{HgX}_2 \longrightarrow \underset{\substack{\text{organomercury}\\\text{compound}}}{\text{R}_2\text{Hg}}\ +\ 2\text{LiX}$$

A second way to make organometallic compounds (for use as carbanion nucleophiles) is to use halogen–metal exchange. In this process, an alkyl halide and an organometallic compound undergo a metathesis reaction to give a new organometallic compound and a new alkyl halide. This process is thought to take place by nucleophilic attack on the halogen atom by the organometallic reagent.

$$\text{R-Li}\ +\ \text{R'-Br} \rightleftarrows \text{R'-Li}\ +\ \text{R-Br}$$

One requirement is that the pK_a of the new organometallic compound be lower than the pK_a of the starting organometallic. This means the equilibrium is driven to products by the formation of a more stable anion. This method is commonly used to make vinyl lithiums from vinyl halides, and alkyl lithiums and aryl lithiums from aryl halides and alkyl lithiums. It is used because the electron pair in an sp^2 orbital of a vinyl or aryl lithium compound is more stable than the electron pair in an sp^3 orbital of an alkyl lithium.

$$\text{(vinyl)Br}\ +\ \text{BuLi} \longrightarrow \text{(vinyl)Li}\ +\ \text{BuBr}$$

A method often employed to drive the halogen–metal exchange equilibrium to completion is to employ tert-butyl lithium as the organolithium component. This the most basic organo-lithium compound because of the tertiary substitution; conversion of the tert-butyl lithium byproduct to isobutylene also occurs under the reaction conditions and drives the exchange equilibrium to completion. Note that two equivalents of tert-butyl lithium are required, as one equivalent is used in the halogen–metal exchange and one equivalent is consumed in converting tert-butyl bromide to isobutylene.

$$\text{Ph-Br}\ +\ 2\ \text{(t-Bu)Li} \longrightarrow \text{Ph-Li}\ +\ \text{(isobutylene)}\ +\ \text{LiBr}$$

Enolates and related carbanionic nucleophiles are routinely generated by removal of an acidic proton in a molecule with a base. Carbonyl groups somewhat acidify their α protons and make their removal by base a common process. However, structural features other than carbonyl groups can also acidify protons bound to carbon and thus facilitate their removal by bases. For example, pK_a values for structurally acidified C—H protons include

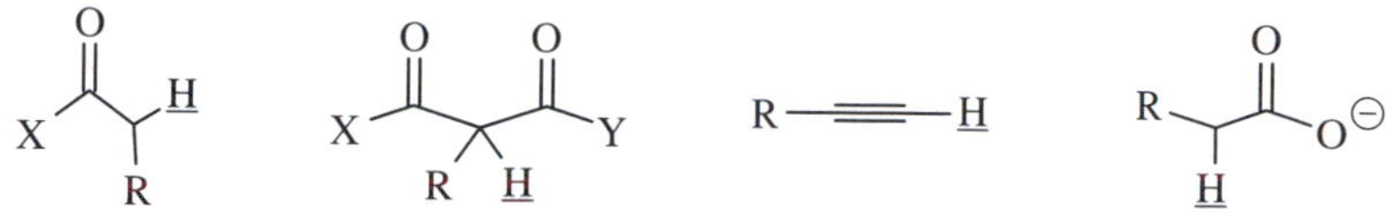

pK_a = 20-25   10-14   25   26

Some pK_a's of commonly used bases include

pK_a 12   15-19   35   > 40

By knowing (or estimating) the pK_a of a proton to be removed, it is possible to choose a base with a higher pK_a, in order to have essentially complete conversion to the anionic carbon nucleophile. When these conditions are met, proton exchange occurs readily and a carbon nucleophile is produced. It must be remembered, however, that many bases can themselves serve as nucleophiles. If the structural feature which acidified the C—H proton is itself an electrophile, a nucleophilic base cannot be used. For example, butyl lithium ($pK_a > 45$) converts phenylacetylene ($pK_a \sim 25$) smoothly to its conjugate base by proton removal. However, it reacts as a nucleophile with the carbonyl group of acetophenone, in spite of the fact that the α protons of acetophenone have $pK_a = 21$ and are thus more acidic than the terminal proton in phenylacetylene.

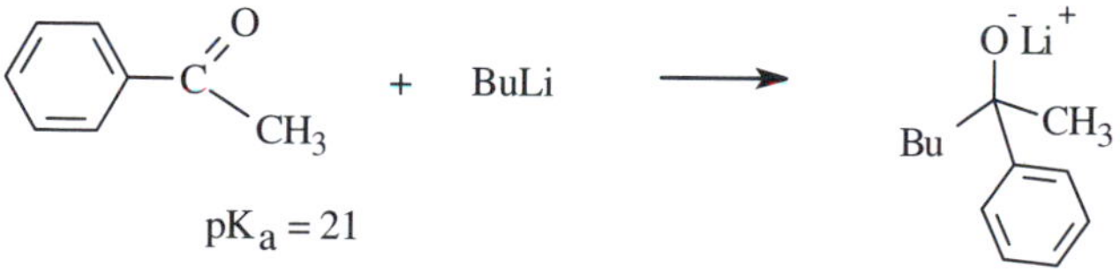

To circumvent problems of nucleophilicity, lithium diisopropylamide (LDA), potassium hexamethyldislylamide (KHMDS), and KH are often employed for proton removal, since they are very strong bases ($pK_a > 35$) but are relatively poor nucleophiles. Hence they remove protons from acidic C—H bonds but normally do not attack carbonyl groups or other electrophilic centers.

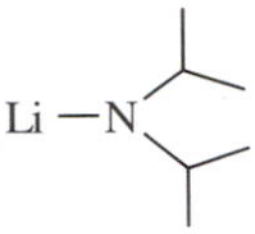

Lithium Diisopropyl amide   Potassium   Potassium Hydride
                               Hexa methyl disilylamide

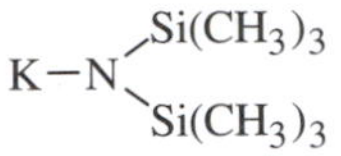

         LDA   KHMDS

If the C—H proton is highly acidified, as in a β-dicarbonyl compound ($pK_a \sim 10$–14) or nitro compounds (pK_a 9–12), weaker bases such as alkoxides ($pK_a \sim 17$) can be used to convert the material completely to its conjugate base. Thus, aprotic conditions are no longer required. However, a common protocol to convert dicarbonyl compounds to their enolates in a clean, controllable manner is to use sodium hydride in dry THF:

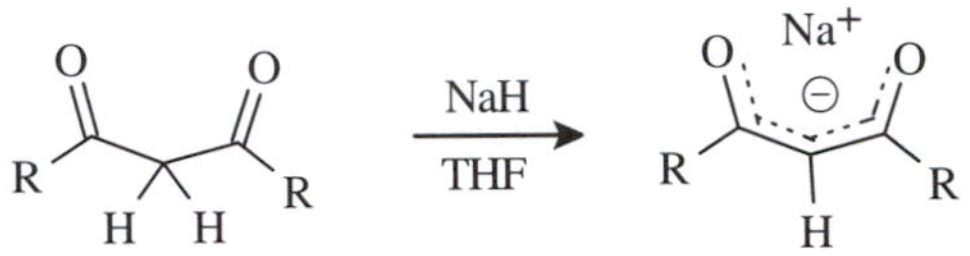

A third major class of carbon nucleophiles is enol derivatives. In general, these are stable compounds that are prepared by one of the functional group transformations outlined in the previous chapter.

Generation of Electrophilic Carbon Reagents

Electrophilic carbon species are usually stable compounds with an electrophilic functional group present. Since they are stable molecules, they need not be generated as transients in the reaction mixture. The functional types which are good electrophiles have been defined earlier in this chapter, and the preparations of these functional groups are outlined in the previous chapter.

Matching Nucleophiles with Electrophiles

Having defined nucleophilic and electrophilic carbon species, and having demonstrated how to produce a variety of them, the next step is to match the reactivities of the nucleophiles and electrophiles so that carbon–carbon bond formation can occur in a controllable and selec-

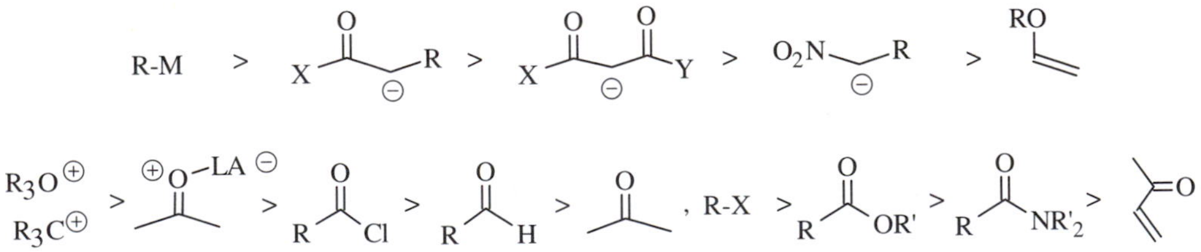

FIGURE 7.1

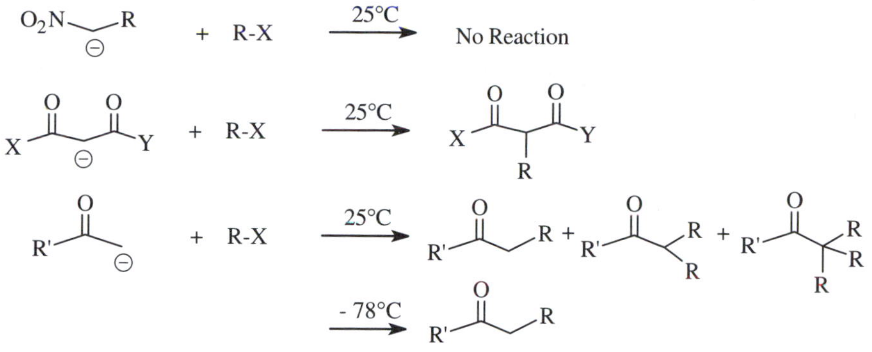

FIGURE 7.2

tive fashion. Qualitatively, the order of reactivities for nucleophiles and electrophiles used in carbon–carbon bond-forming reactions are shown in Figure 7.1. In general, many of the most useful carbon–carbon bond-forming reactions take place with nucleophiles and electrophiles in the middle ranges of reactivity. Highly reactive electrophiles and nucleophiles are often difficult to control, while nucleophiles and electrophiles of low reactivity often fail to react effectively. Nevertheless, it is reactivity matching that is most important in producing useful reactions.

When stabilized (and consequently less reactive) anions are employed as nucleophiles, more reactive electrophiles are needed for successful carbon–carbon bond formation. Nitronate anions, which are highly resonance stabilized, fail to react with simple alkyl halide electrophiles. On the other hand, β-dicarbonyl compounds react effectively with primary, and some secondary, alkyl bromides and iodides to give monoalkylated products (see Figure 7.2). Under the same conditions, simple enolates react vigorously with alkyl halides (which must be primary) to give mono- and polyalkylated products. The reactivity of the simple enolate is greater and cannot be controlled at room temperature. However, if the alkylation is carried out at low temperature, the reaction can be controlled and smooth monoalkylation of simple enolates can be achieved. The same is true for the alkylation of acetylide anions, which must be carried out at low temperature for successful alkylation.

$$R-C\equiv C\ominus \quad + \quad CH_3CH_2\text{-}I \quad \xrightarrow{-78°C} \quad R-C\equiv C-CH_2CH_3$$

Enolates

Enolates are important nucleophiles which react nicely with a variety of carbonyl compounds. In this case, the nucleophilic reactivity of the enolate and the electrophilic reactivity of the carbonyl group are well matched, and a wide variety of products can be made. The type of enolate (ketone, ester, etc.) and the type of carbonyl electrophile (aldehyde, ketone, ester, etc.) determine what the structure of the final product will be. Furthermore, these reactions are often named according to the two partners that are reacted and the type of product produced from them.

The aldol condensation is the reaction of an aldehyde or ketone enolate with an aldehyde or ketone to give a β-hydroxy aldehyde or ketone. A simple aldol reaction is one in which the enolate nucleophile is derived from the carbonyl electrophile. Very often, the β-hydroxy carbonyl product dehydrates to give an α, β-unsaturated carbonyl compound. However, the aldol nature of the dehydration product can be discerned by disconnection of the double bond of the unsaturated product.

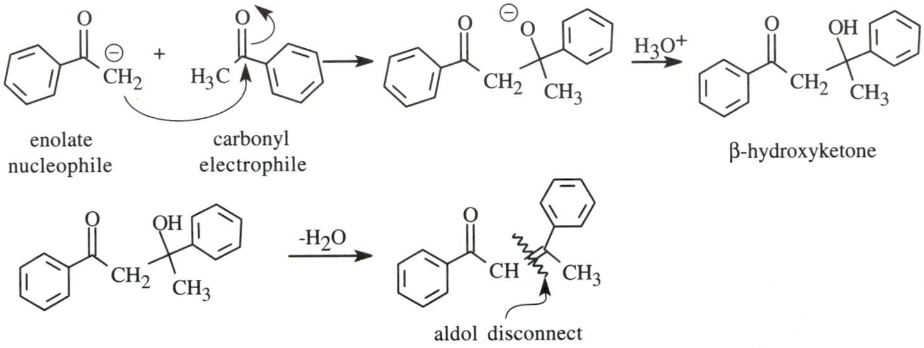

enolate nucleophile   carbonyl electrophile β-hydroxyketone

aldol disconnect

If the enolate nucleophile is derived from an aldehyde or ketone different from the carbonyl electrophile, a crossed aldol condensation results. Normally, best success is achieved if the carbonyl electrophile employed for the crossed aldol condensation is more reactive than the carbonyl electrophile from which the enolate is derived. For example, ketone enolates react with aldehydes effectively, but aldehyde enolates do not give the crossed aldol with most ketones; they self-condense instead.

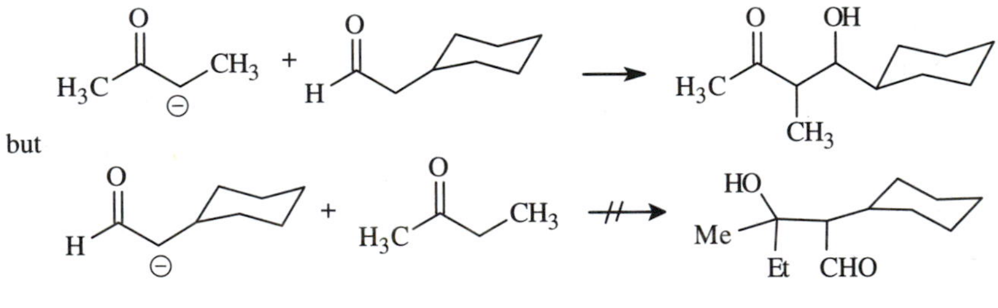

The Claisen condensation is the reaction of the enolate of an ester with an ester electrophile. The product is a β-ketoester, since the tetrahedral intermediate collapses by expulsion of an alkoxide.

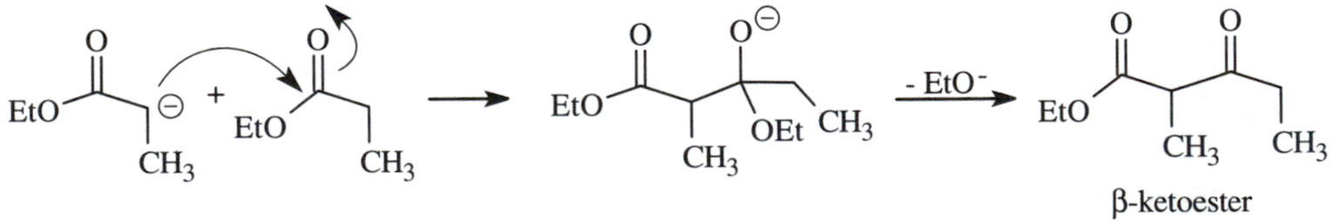

β-ketoester

A crossed Claisen is the reaction of an ester enolate with an aldehyde or ketone to produce a β-hydroxy ester. This works well because aldehydes and ketones are more reactive electrophiles than are esters, and the ester enolate reacts faster with the aldehyde or ketone than it condenses with itself. Therefore product mixtures are avoided. Moreover, the aldehyde or ketone should not have α hydrogens, so that proton transfer to the more basic ester enolate is avoided. This would lead to the formation of an aldehyde or ketone enolate in the mixture, and an aldol reaction would be a major competing reaction.

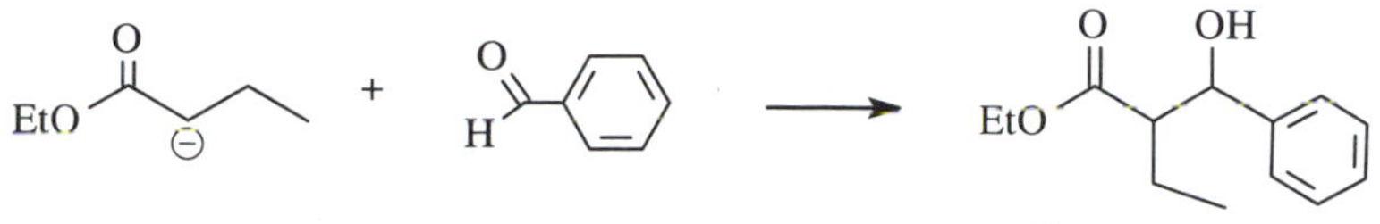

β-hydroxyester

For the same reason, it is generally not feasible to carry out a crossed Claisen reaction between the enolate of one ester and a second ester which has α protons. This is due to the fact that, if nucleophilic addition to the carbonyl group is not fast, proton exchange can occur to give a mixture of enolates, thus a mixture of products (see Figure 7.3).

There are many other named reactions that follow these same general features but differ as to the type of enolate or the carbon electrophile. These include the Reformatski reaction, the Darzens reaction, and the Dieckmann ring closure. They were in widespread use for many, many years and were named as a convenient way to characterize the reactants employed and the type of product which results. The reason there are so many variations on the same theme is that control of the reaction products depends on the ability to generate a particular enolate nucleophile and react it with a particular carbonyl electrophile. In earlier times, alkoxide bases were the strongest bases routinely available to synthetic chemists. Since alkoxides have $pK_a = 15$–19, while protons α to carbonyl groups have $pK_a = 20$–25, the reaction of an alkoxide base with a carbonyl compound produces only a small amount of the enolate at equilibrium. This is produced in the presence of the unreacted carbonyl compound, which is itself an electrophile. For simple aldol and Claisen reactions, this is the ideal situation for self-condensation.

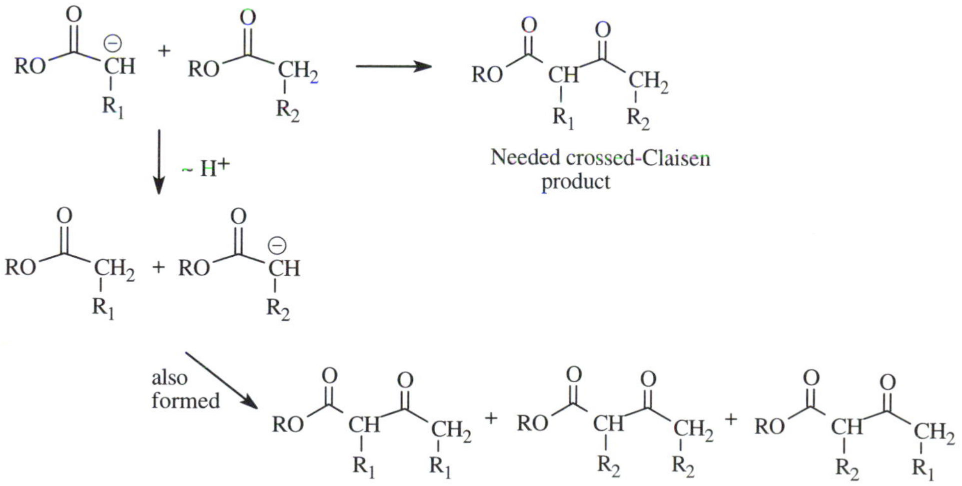

FIGURE 7.3

X= H, alkyl, OR   enolate   unreacted

If, however, it is necessary to generate a crossed product, by the reaction of an enolate (derived from one carbonyl compound) with a second carbonyl compound as the electrophile, things can go bad rapidly. Because both carbonyl groups must be present in solution at the same time, and each can form enolates to some extent, there can be four possible products from the various combinations of enolates and carbonyl compounds. This problem was illustrated for the crossed Claisen condensation (Figure 7.3). The number of products can be minimized if one carbonyl component lacks α protons, cannot form an enolate, and is a more reactive electrophile than the second carbonyl component. If these conditions are met, crossed condensations can be carried out successfully using alkoxide bases. Many of the named reactions were developed so that product mixtures could be avoided.

Today reactions of enolates are usually carried out much differently, by utilizing very strong, non-nucleophilic bases for generating the enolate nucleophile. Instead of having only small equilibrium concentrations of an enolate produced in solution, the use of strong, non-nucleophilic bases, such as LDA, KHMDS, and KH, that have $pK_a > 35$ permits carbonyl compounds (whose α protons have pK_a 20–25) to be converted completely to enolate anions. This completely converts the carbonyl compound into a nucleophile which cannot condense with itself and is stable in solution. This enolate can then be reacted with a second carbonyl compound in a subsequent step to give product:

Step 1

X= H, alkyl, OR   100%

Step 2

As long as nucleophilic addition of the preformed enolate to the second carbonyl component is rapid, and the carbonyl electrophile is added *after* the enolate is formed, the product is predictable and is not a mixture. The rule of thumb is that the carbonyl electrophile should be more reactive than the carbonyl compound from which the enolate is derived. If this condition is met, the carbonyl electrophile can have α protons and the structural possibilities are increased tremendously. Typical enolate–carbonyl pairs that have been condensed by this methodology are shown in Table 7.1.

Acetylides can also react as nucleophiles toward aldehydes and ketones to give propargylic alcohols. This provides a simple way to install the triple bond in molecules:

$$R-C\equiv C^{\ominus} \quad + \quad \underset{R_1 \quad R_2}{\overset{O}{\parallel}} \quad \longrightarrow \quad R-C\equiv C-\underset{R_2}{\overset{OH}{\underset{|}{\overset{|}{C}}}}-R_1$$

Esters, amides, and nitriles are relatively weak electrophiles. They react sluggishly, or not at all, with enolates. Esters are more electrophilic than amides and nitriles, and react readily with carbanionic-type reagents such as organolithiums or Grignard reagents. As seen previously, two equivalents of the organometallic are added, and tertiary alcohols are produced. Tertiary amides and nitriles react with organolithiums (but not Grignard reagents) to give ketones after hydrolytic workup. A single nucleophilic addition occurs to give an anionic intermediate which is stable to further nucleophilic addition. The oxidation level is that of a ketone which is unmasked upon hydrolysis.

$$R_1\text{-Li} \quad + \quad R_2-C\equiv N \quad \longrightarrow \quad \underset{R_1 \quad R_2}{\overset{N-Li}{\parallel}} \quad \xrightarrow{H_3O^+} \quad \underset{R_1 \quad R_2}{\overset{O}{\parallel}}$$

$$R_1\text{-Li} \quad + \quad \underset{R_2}{\overset{O}{\underset{\parallel}{C}}}-N\overset{R}{\underset{R}{\diagdown}} \quad \longrightarrow \quad \underset{R_1 \quad R_2}{\overset{R}{\underset{N}{\diagup}}}\overset{OLi}{\diagdown} \quad \xrightarrow{H_3O^+} \quad \underset{R_1 \quad R_2}{\overset{O}{\parallel}}$$

Carbonyl electrophiles are obviously a very important group of electrophiles that react successfully with a spectrum of carbon nucleophiles. Among carbonyl electrophiles, however, large differences in reactivity are observed. Acid chlorides are very reactive electrophiles, whereas esters and amides are much weaker and fail to react with several classes of carbon nucleophiles. Aldehydes and ketones are probably the most widely utilized groups of carbonyl electrophiles and exhibit moderate electrophilic reactivity. No matter what carbonyl electrophile is used, however, it reacts by nucleophilic addition to the carbonyl carbon to produce a tetrahedral intermediate. The ultimate reaction product reflects subsequent chemistry of the tetrahedral intermediate.

While the use of strong bases has changed the way in which many condensation reactions are carried out, it is important to remember the types of products that are produced from them. Recall that the aldol condensation yields β-hydroxy aldehydes or ketones which are easily dehydrated to α, β-unsaturated aldehydes or ketones.

TABLE 7.1 Typical Enolate–Carbonyl Pairs

Enolate	Carbonyl Compound	Product Type
Ester enolate	Aldehydes, ketones	β-hydroxy ester
Ester enolate	Acid chorides	
	or	β-ketoester
	Carbonylimidazole	
Ketone enolate	Aldehyde	β-hydroxyketone
Ketone enolate	Acid chloride	β-diketone

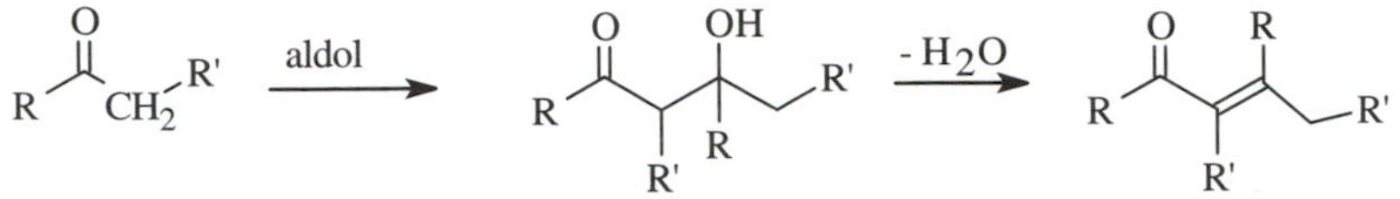

It is thus possible to look at a molecule such as **A** in the following diagram and recognize that it is a β-hydroxy ketone that could be formed in a crossed aldol reaction between enolate **B** and aldehyde **C**. Likewise, **D** could potentially be produced by dehydration of the aldol product of cyclohexanone.

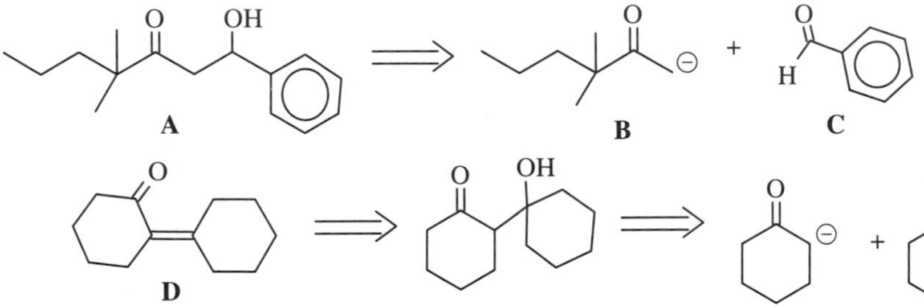

In addition to these intermolecular processes, intramolecular versions of the Claisen (Dieckmann reaction) and the mixed Claisen and the aldol reaction (Robinson annulation) are well known. In all cases, the same structural classes of products are formed.

Enolate Regioisomers

Enolates are commonly used as the nucleophilic component in carbon–carbon bond-forming reactions. Through the use of strong, non-nucleophilic bases, both esters and ketones are easily converted to their enolates. Ketones, however, are problematic with regard to the regioselectivity of enolate formation if they are unsymmetric. As seen in the following example, two regioisomeric enolates can be produced by the removal of the nonequivalent α protons of 2-pentanone by base; thus two regioisomeric aldol products are possible.

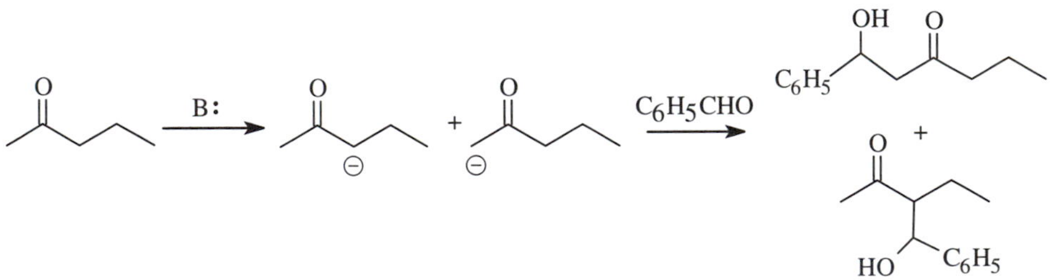

Regiochemical control of enolate formation is an important consideration when planning ways to construct a carbon–carbon bond using a ketone enolate.

There are several strategies for controlling the regiochemistry of proton removal. The first is to take advantage of the fact that the less-substituted α position has slightly more acidic protons. If a ketone is added *slowly* to a cold solution of LDA, the more acidic proton will be removed preferentially. The resulting enolate is termed the kinetic enolate because the more acidic proton is removed *faster* than the less acidic proton. Both steric and electronic factors contribute to the more rapid removal of protons from the less highly substituted α carbon.

The enolate that is the most stable usually has the most highly substituted double bond and is called the thermodynamic enolate. If a slight excess of the ketone is used, or a trace of protic impurities is present, equilibrium between the enolates is established and isomerization to the more highly substituted enolate occurs.

kinetic enolate

thermodynamic enolate

The thermodynamic enolate is lower in energy, so it is favored if equilibrium is achieved. For this reason, great care must be taken in the preparation and reaction of the kinetic enolate so that equilibration does not occur. On the other hand, preparation and reaction of the thermodynamic enolate is much easier and demands less rigorous reaction conditions.

Besides the direct formation of kinetic or thermodynamic enolates of ketones, other strategies can be employed to produce regiospecific products. An older and extremely valuable strategy for making the synthetic equivalent of a particular regiospecific enolate utilizes some group (G) to acidify a proton α to a ketone so that it is removed preferentially by base. The resulting enolate is used as a carbon nucleophile and the group (G) is then removed. In this way, it appears that one α position of a ketone has been regioselectively transformed when, in fact, the group G has guided the chemistry in the reactant but is not present in the product.

more acidic

synthon for RCH$_2$CCH$_2$

The most common group G is an ester function (although many other groups have been employed). The starting β-ketoester, which can be prepared easily by a Claisen-type reaction of an ester enolate and an acid chloride, has a very acidic α proton (p$K_a \sim$ 9–10) which is easily removed (i.e., 1 equiv. NaH or 1 equiv. EtO$^-$). The resulting enolate is used as a nucleophile to form a new carbon–carbon bond. The ester is then hydrolyzed, and CO$_2$ is thermally ejected to provide an α-substituted ketone. This strategy is simple, efficient, and convenient, and is widely used.

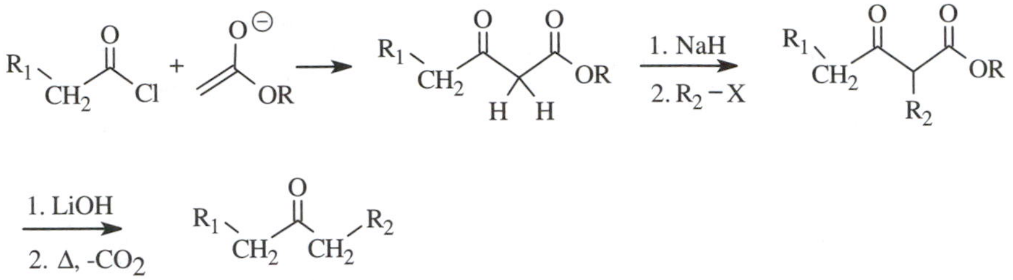

A third strategy for controlling enolate formation is to convert the carbonyl group to an *N,N*-dimethylhydrazone. The hydrazone is less reactive than the carbonyl group, and removal of an α proton by a strong base takes place at the least hindered α position. Alkylation, followed by hydrolysis, gives back carbonyl product that is the same as the result of kinetic control of enolate formation. However, this method does not have the problems of equilibration found for simple enolate formation. The regioselectivity of proton removal from the hydrazone is probably related to the geometry of the hydrazone. The dimethylamino group is pointed toward the least hindered α position for steric reasons, and directs the base to that position by coordination with the lone pairs on nitrogen.

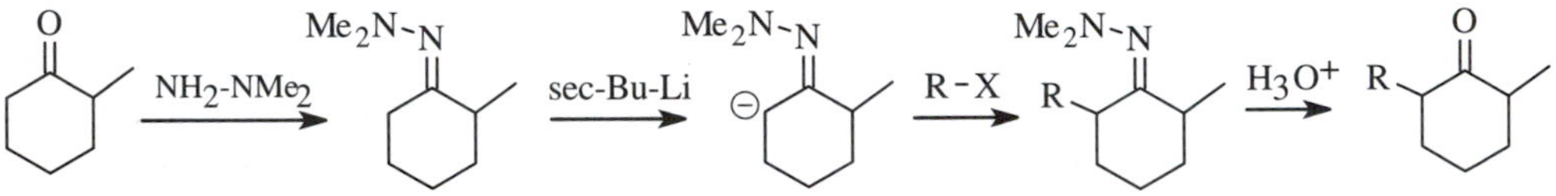

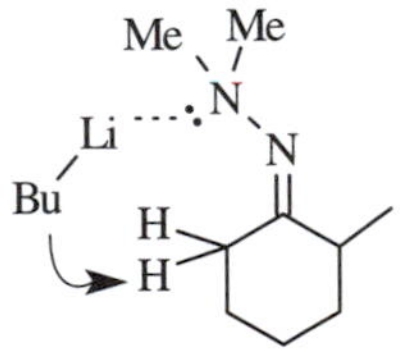

The use of hydrazones is particularly important to form the enolate equivalents of aldehydes. Aldehydes are quite reactive as electrophiles, so as soon as some enolate has been formed, it reacts with the unreacted aldehyde present in solution. Conversion of the aldehyde to its *N,N*-dimethyl hydrazone lowers the electrophilicity so that α proton removal can take place and the electrophile of choice can then be added. Hydrolysis gives back the aldehyde. In this case, the geometry of the hydrazone is unimportant, since aldehydes have only one α position from which protons can be removed by base.

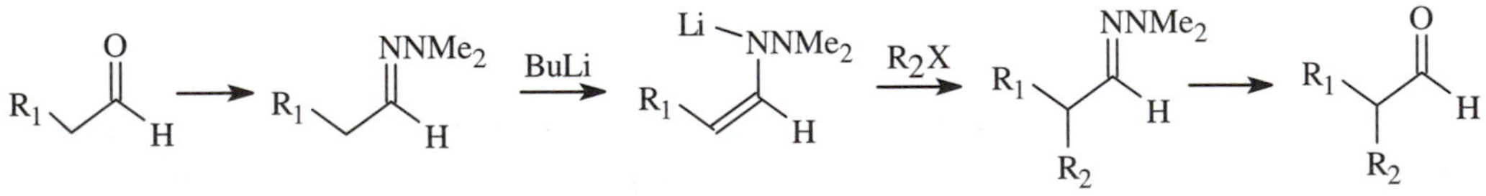

Diastereoselection in Aldol Reactions

The reaction of enolates with aldehydes or ketones to produce β-hydroxy carbonyl derivatives is a very common and very useful way to make carbon–carbon bonds. A fundamental stereochemical feature of the reaction is that two new chiral centers are produced from achiral starting materials. Hence *syn* and *anti* diastereomers will be produced, each as a pair of enantiomers. This is shown schematically for the reaction of a propionate enolate with isobutyraldehyde. Because they have different energies, the *syn* and *anti* diastereomers will be produced in unequal amounts, but each will be produced in racemic form, since both starting materials are achiral.

syn - R,S and S,R   anti - R,R and S,S

The diastereoselectivity of the reaction results from a combination of three factors. First, the carbonyl electrophile can undergo addition on either its Re or Si face. Second, the enolate nucleophile is planar, and can attack the carbonyl group from either of its faces. Third, the enolate geometry can be either Z or E. To control the diastereoselectivity, it is first necessary to use a single isomer of the enolate. In general, the E enolate is the kinetic enolate and the Z enolate is thermodynamically favored. Methods are available to produce either as the major isomer by α proton removal from carbonyl compounds with strong bases. This is particularly true of esters and amides. Pure Z and E enolates can also be prepared by first converting the carbonyl compound to a Z and E mixture of silyl enol ethers, then separating these isomers and regenerating the Z and E enolates with methyl lithium. Suffice it to say that there are known ways to produce either Z or E enolates in pure form.

LDA	9	91
LDA/HMPT	84	16

The stereoelectronic requirement for carbonyl addition is that electron donation occurs by interaction of the donor with the $\pi*$ orbital of the carbonyl group. To meet the stereoelectronic requirements and explain the diastereoselectivity, the Zimmerman–Traxler model is used. Interaction of the lithium cation with both the oxygen of the enolate and the carbonyl electrophile leads to a six-membered chairlike transition state. If the geometry of the enolate is fixed, the only variable is the orientation of the electrophile. The preferred orientation has the larger substituent in a pseudoequatorial position. This preferred orientation produces the major diastereomer. An example is shown for the Z enolate of ethyl propionate, reacting with isobutyraldehyde. This model predicts that the *anti* diastereomer should be favored (and it is). A similar analysis predicts that the *E* enolate should give the *syn* diastereomer as the major product (and it does).

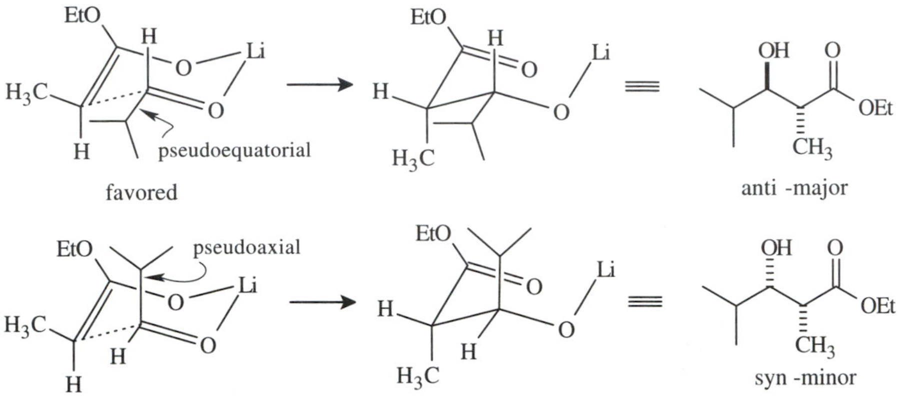

This model is extremely useful in understanding the stereochemical outcomes of aldol processes. It also provides a framework for influencing the diastereoselectivity in a rational way. For instance if the ethoxy group is changed to a much bulkier group, increased transannular interactions in the pseudoaxial transition state would make it even higher in energy. This results in increased selectivity for the *anti* isomer.

Even greater diastereoselectivity in the aldol reaction can be achieved using boron enolates as the carbon nucleophile. Boron enolates are easily prepared from aldehydes and ketones, and the *syn* and the *anti* isomers can be separated as pure compounds. They react with aldehydes and ketones to give aldol products, by a similar transition state. The difference is that boron–oxygen bonds are shorter than lithium–oxygen bonds, and steric interactions in the transition state are therefore magnified and result in greater diastereoselectivity:

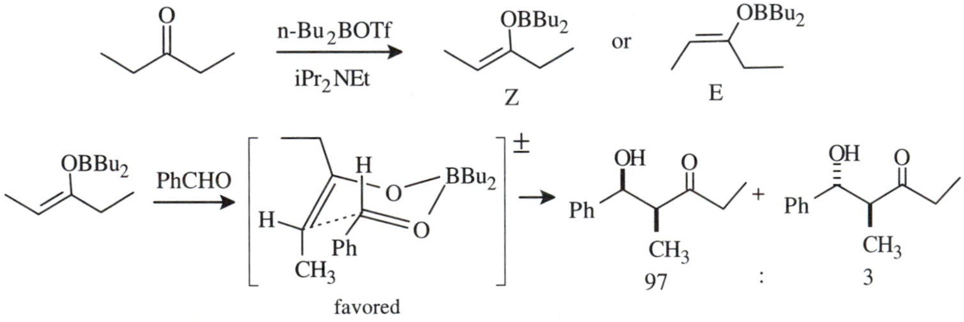

Organometallic Compounds

Organometallic compounds such as Grignard reagents and organolithium reagents are very powerful nucleophiles which react with a wide variety of carbonyl compounds. In general, organolithium reagents are more reactive than Grignard reagents. Both types of reagents react rapidly with aldehydes and ketones to yield secondary and tertiary alcohols after aqueous workup. The new carbon–carbon bond joins the organometallic fragment with the carbonyl carbon.

$$R\text{-}M \quad + \quad \underset{R_1 \quad\quad R_2}{\overset{O}{\|}} \quad \longrightarrow \quad \underset{R_2}{\overset{OH}{R\text{---}R_1}}$$

M= MgX, Li

The tetrahedral intermediate produced by addition of organolithiums or Grignard reagents to esters collapses to a ketone which is more reactive than the original ester electrophile. A second equivalent adds to give a tertiary alcohol product. This process cannot be controlled by temperature or by mode of addition, since the intermediate product is more reactive than the starting material.

$$R\text{-}M \quad + \quad \underset{R_1 \quad\quad OEt}{\overset{O}{\|}} \quad \longrightarrow \quad \underset{R_1}{\overset{O^-M}{R\text{---}OEt}} \quad \xrightarrow{-\ MOEt} \quad \underset{R_1 \quad\quad R}{\overset{O}{\|}} \quad \xrightarrow[2.\ H_3O^+]{1.\ R\text{-}M} \quad \underset{R}{\overset{OH}{R\text{---}R_1}}$$

M= MgX, Li

Collapse of the tetrahedral intermediate can be prevented, however, by reaction of a carboxylic acid with two equivalents of an organolithium (or a carboxylate salt with one equivalent of the organolithium). Addition to the carboxylate gives a dianionic intermediate which has the ketone oxidation state, but is stable to further addition under the reaction conditions. The ketone is revealed only upon hydrolysis. This is a widely used method for the preparation of ketones by the formation of a carbon–carbon bond. However, it is restricted to organolithium reagents as the carbon nucleophile. Grignard reagents do not react with carboxylates, illustrating their slightly reduced nucleophilicity relative to organolithiums.

$$R_1CO_2H \quad \xrightarrow{R_2\text{-}Li} \quad R_1CO_2^{\ominus}\ Li^{\oplus} \quad \xrightarrow{R_2\text{-}Li} \quad \underset{OLi}{\overset{OLi}{R_1\text{---}R_2}} \quad \xrightarrow{H_2O} \quad \underset{R_1 \quad\quad R_2}{\overset{O}{\|}}$$

Neutral Carbon Nucleophiles

Other types of carbon nucleophiles, such as uncharged enol derivatives, are only weakly nucleophilic. Consequently, they are unreactive toward normal carbon electrophiles such as carbonyl compounds, alkyl halides, or sulfonates. In order for them to be used effectively as nucleophiles, strong electrophiles must be used to match their reduced nucleophilicity. This can be accomplished by increasing the reactivity of a normal electrophile. Typically, this is done by treating an electrophile with a Lewis acid. Coordination of lone pairs of electrons with the Lewis acid increases the electrophilicity markedly. Examples include the following:

$$R-Cl \quad \xrightarrow[\substack{or\ AlCl_3 \\ ZnCl_2 \\ SnCl_4}]{TiCl_4} \quad \overset{\delta+}{R}\cdots Cl\cdots M \quad\quad \text{equivalent to } [R^+]$$

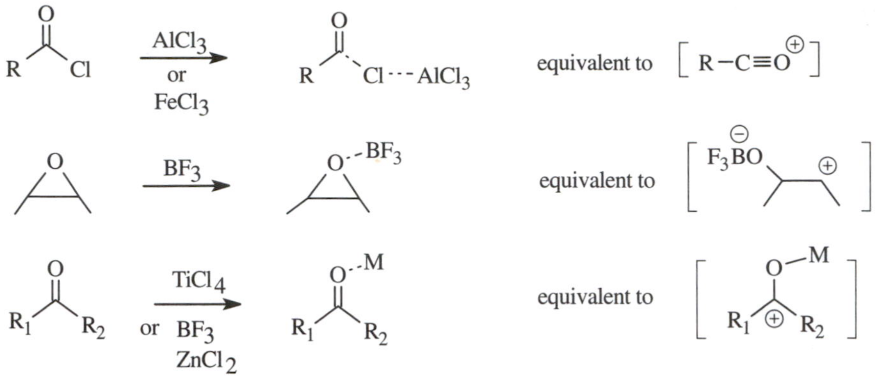

A variety of Lewis acids can be used. Commonly used ones are $AlCl_3$, $TiCl_4$, $SnCl_4$, $ZnCl_2$, BF_3, and TMSOTf. The choice of Lewis acid is often critical to the success of the reaction, and is usually made by referring to similar transformations that have been successfully reported in the literature. Often it is not possible to rationalize why one Lewis acid works and another one does not, so the initial choice of catalyst is normally made by literature precedent. With the arsenal of Lewis acids available, a suitable catalyst can usually be found.

By this strategy, reactive carbon electrophiles can be generated for successful reaction with a variety of weak carbon nucleophiles. More important examples include the Freidel–Crafts reaction, in which aromatic compounds (nucleophiles) react with alkyl and acyl halides (electrophiles in the presence of Lewis acids).

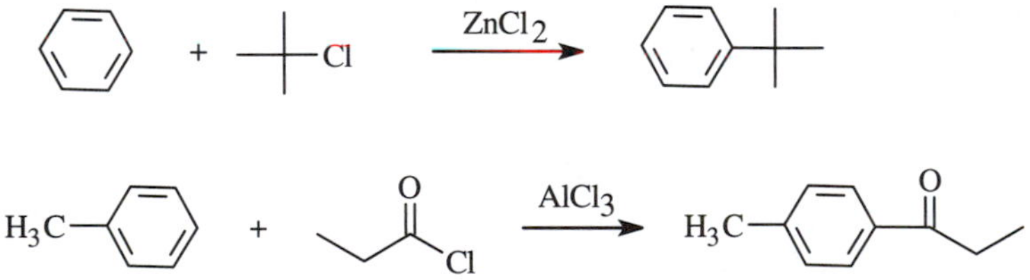

The Mukaiyama reaction is a versatile, crossed aldol type of reaction that uses a silyl enol ether of an aldehyde, ketone, or ester as the carbon nucleophile, and an aldehyde or ketone (activated by a Lewis acid) as the carbon electrophile. The product is a β-hydroxy carbonyl compound. The advantage to this approach is that it is carried out under acidic conditions, and elimination does not usually occur.

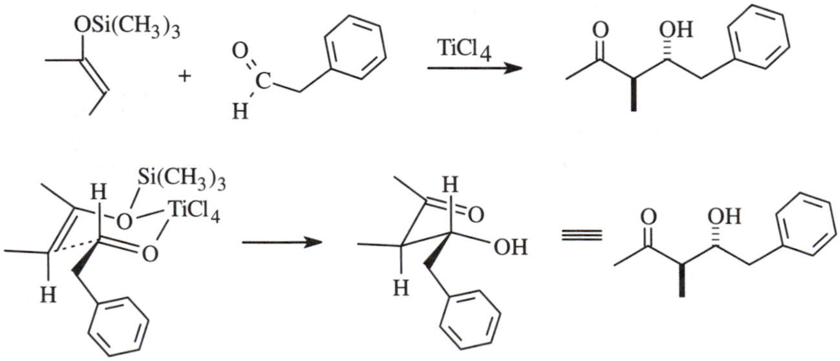

chair-like transition state

The transition state is thought to be an open structure. Assuming that a particular silyl enol ether geometry is used, the substituents will tend to occupy opposite faces of the transition state and give a particular diastereomer (*syn–anti*) preferentially. Because of the open transition state geometry, the diastereoselectivity is not high.

The reaction of allyl silanes with aldehydes and ketones activated as electrophiles by Lewis acids is a very useful method for preparing homoallylic alcohols. Since allyl silanes are only modestly nucleophilic, strong electrophiles are needed to insure a good reactivity match.

The same considerations apply to intramolecular versions. For example, although epoxide (*E*) is a stable compound, treatment with a Lewis acid activates the epoxide as an electrophile, and cyclization with the olefinic π system occurs to give the steroid ring system. Note that the silyl enol ether group of *E* functions as the terminating group. This polyene cyclization is similar to the biosynthesis of steroids. The cyclization is triggered by the generation of an electrophile sufficiently strong to react with the weakly nucleophilic π bond.

C=C Formation

In addition to connecting skeletal fragments by forming carbon–carbon single bonds, it is also possible to utilize reactions which give carbon–carbon double bonds to assemble carbon skeletons. It should be recognized that, although the final products of such reactions contain a carbon–carbon double bond, they are generally sequential processes in which a single carbon–carbon bond is formed first and the π bond is formed in a subsequent elimination step.

An elementary approach to this process is the reaction of an organometallic reactant with a ketone (or aldehyde), followed by dehydration of the resulting alcohol to the olefin. This is truly a sequential process, in that the product alcohol is dehydrated in a second, independent reaction step. It suffers as a useful synthetic method because regioisomers are often formed in the elimination step.

Alternatively, it is possible to have both steps (addition and elimination) occur spontaneously if appropriate reagents are employed. There are two common strategies in use. The Wittig olefination uses a phosphorus-stabilized carbanion (ylid) as a nucleophile, and a carbonyl compound as an electrophile. Typically the ylid is generated in situ from a triphenylphosphonium salt and a strong base such as LDA or an alkyl lithium.

$$R\text{-}CH_2\text{-}Br \;+\; :P(C_6H_5)_3 \longrightarrow \underset{\text{phosphonium salt}}{RCH_2\overset{\oplus}{-}P(C_6H_5)_3 \;\; \overset{Br\,\ominus}{}} \xrightarrow{LDA} R\overset{\ominus}{C}H\overset{\oplus}{-}P(C_6H_5)_3$$

$$RCH{=}P(C_6H_5)_3$$
ylid or phosphorane

The ylid is a neutral compound which is resonance-stabilized by phosphorus. However, it has an electron-rich carbon next to the phosphorus and is a good carbon nucleophile which adds to carbonyl groups to form a new carbon–carbon bond. The cyclic intermediate (oxaphosphetane) spontaneously loses triphenylphosphine oxide at room temperature, to give an olefin.

In sum, a new olefinic link is produced, but by an addition–elimination sequence. This is a very important method for olefin formation. The stereochemistry about the new carbon–carbon double bond is the *Z* (or less stable) isomer. This unusual stereoselectivity indicates that product formation is dominated by kinetic control during formation of the oxaphosphetane.

By adding a strong base to the cold solution of the oxaphosphetane before it eliminates, the oxaphosphetane equilibrates to the more stable *anti* isomer, and the *E* olefin is produced upon elimination. This so-called Schlosser modification, in conjunction with the normal Wittig reaction, enables either the *Z* or *E* isomer of the olefin to be prepared selectively.

The Wittig–Horner reaction is the Wittig process applied to carbonyl activated ylids; it uses trimethylphosphite as the phosphorus reagent. Reaction with a bromoester gives a phosphate intermediate. Deprotonation with a base, such as sodium hydride, and addition of an aldehyde or ketone gives (after elimination of a phosphonate), an α, β-unsaturated ester. In this case, the intermediate betaine is acidic and undergoes equilibration prior to elimination so that only the more stable *E* regioisomer is produced.

A recent alternative to the Wittig reaction uses silicon as the atom which promotes oxygen loss. This reaction, called the Peterson olefination, uses an α-silyl anion as the carbon nucleophile and a carbonyl compound (aldehyde or ketone) as the electrophile. Thus ethyl

α-trimethylsilylacetate can be converted to an enolate and can react with an aldehyde to give an α, β-unsaturated ester. The driving force for elimination is the formation of an extremely strong silicon oxygen bond, which converts the oxygen atom into a much better silyloxy leaving group. Only the more stable olefin isomer is produced, since equilibration occurs in the enolate intermediate.

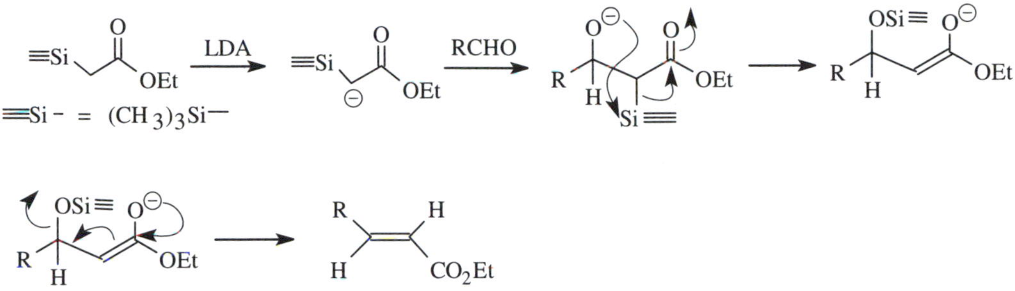

Another common α-silyl anion is produced by the halogen exchange from a methyl (but not other group) attached to silicon. Other α-silyl carbanions can be generated by other processes.

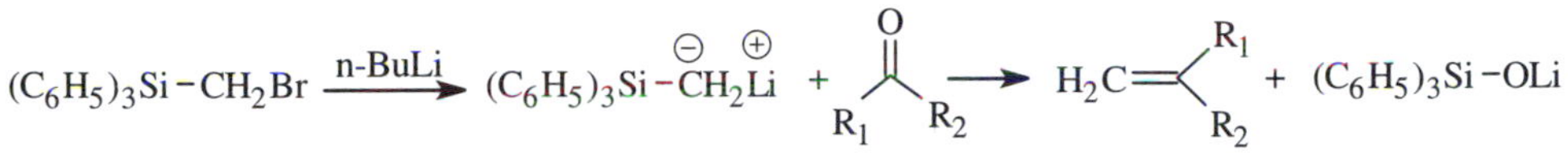

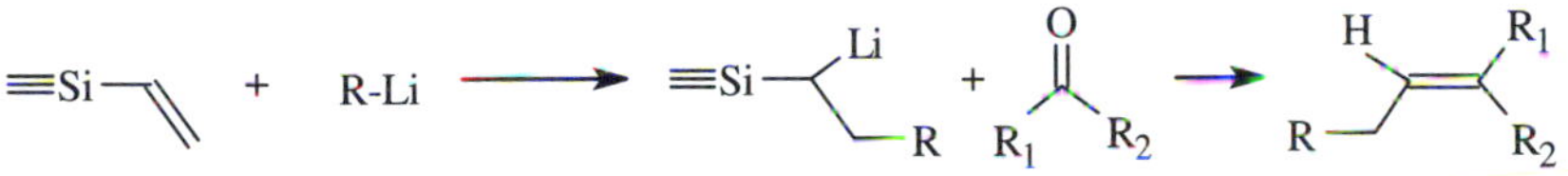

Cyclopropanation Reactions

Formation of two single bonds from a carbon atom is also a well-known method for building up carbon skeletons. In this process, a three-membered ring is formed by reaction of a difunctional carbon atom with an olefin.

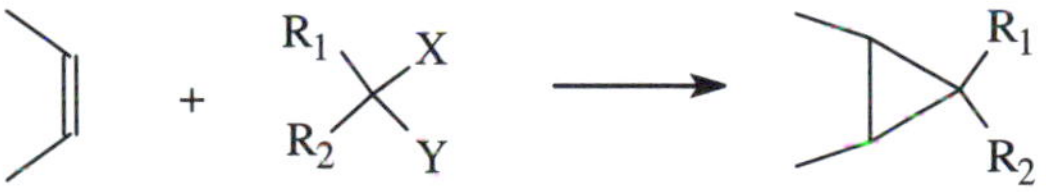

Because of the strain of three-membered rings, their synthesis is not trivial, and a small number of reactions which effectively append three-membered rings to molecules are important and widely used.

The Simmons–Smith cyclopropanation utilizes methylene diiodide and a zinc–copper couple to produce a carbenoid intermediate. This intermediate reacts with olefins to give cyclopropanes. The geometry of the double bond is preserved in the cyclopropane.

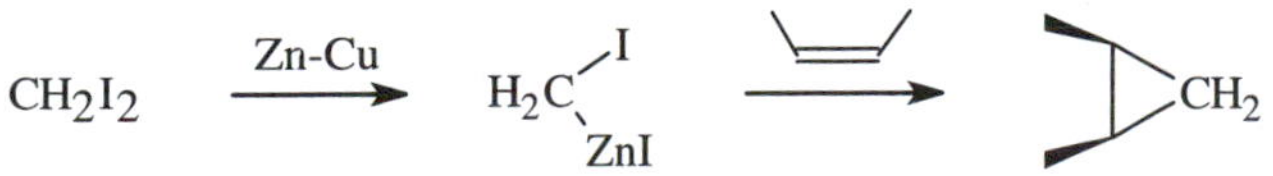

This addition is sensitive to steric biases in the olefin, and the methylene group will enter from the least hindered side of the molecule. Alcohol substituents in the olefin will facilitate the reaction and guide the methylene group *syn* to the alcohol.

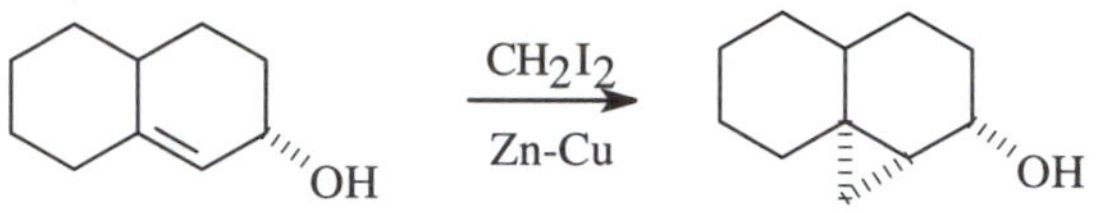

The base-promoted α eliminations of chloroform or bromoform provide a simple method for the production of dihalocarbenes. These add readily to olefinic double bonds, to give 1,1-dihalocyclopropanes. One or both halogens can be removed by reduction, the most common method being to use tri-*n*-butyl tin hydride.

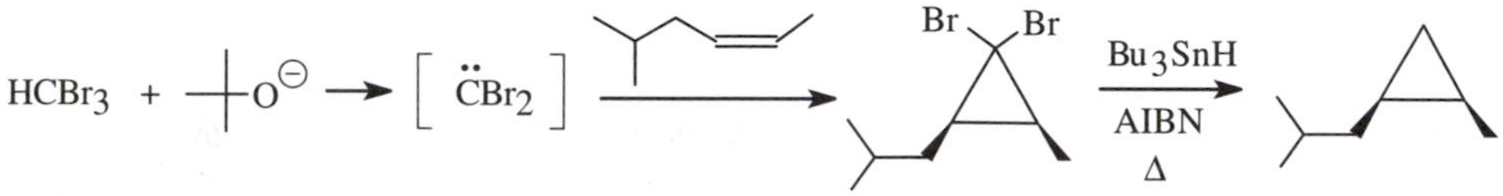

A third common method for forming cyclopropanes is to react α-diazoketones or esters with olefins under the influence of copper or, better yet, rhodium or ruthenium catalysis. Again, a metal carbenoid intermediate is produced which reacts with the olefin.

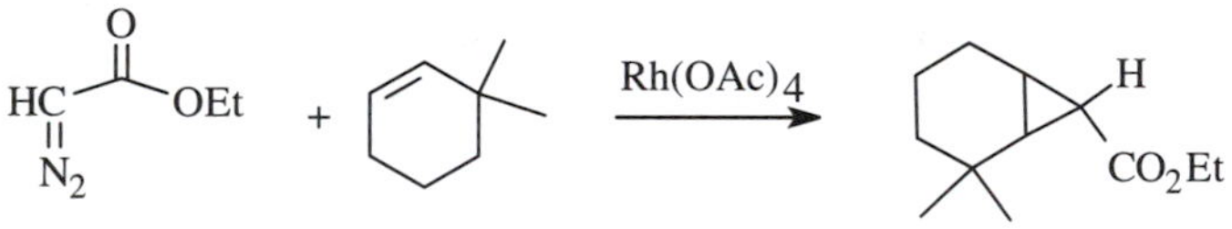

The importance of this strategy is that functionalized cyclopropanes are produced which can be further manipulated. The process can also be carried out intramolecularly with high efficiency.

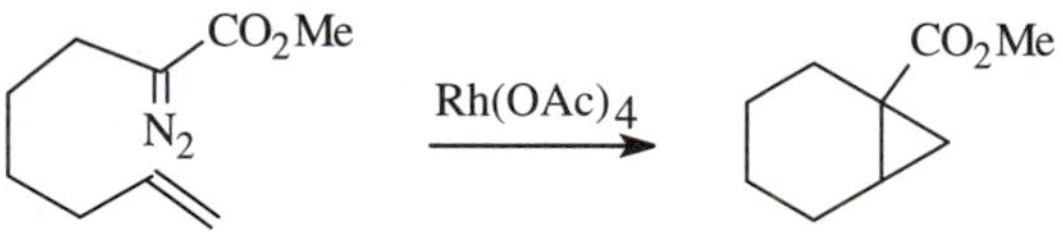

Bibliography

Undergraduate textbooks have discussions of the standard methods of carbon–carbon bond formation, including many of the named reactions such as the aldol, Claisen, and Reformatski. For example, see R. T. Morrison and R. N. Boyd, *Organic Chemistry,* 6th ed., Wiley, New York, 1993; M. A. Fox and J. K. Whitesell, *Organic Chemistry*, Jones & Bartlett, Boston, 1994; and J. McMurry, *Organic Chemistry, 3d ed.,* Brooks/Cole, Pacific Grove, CA, 1992.

For another discussion, see F. A. Carey and R. J. Sundberg, *Advanced Organic Chemistry, Part B: Reactions and Synthesis*, 3d ed., Plenum Press, New York, 1990, Chapters 1 and 2.

The index of J. A. March, *Advanced Organic Chemistry, Reactions Mechanism, and Structure,* 4th ed., Wiley Interscience, New York, 1992 can be used to look up individual reactions easily.

The compilation by B. P. Mundy and M. G. Ellerd, *Name Reactions and Reagents in Organic Synthesis,* Wiley, New York, 1988, has excellent descriptions of name reactions in a brief, yet informative, style.

For an excellent discussion of diastereoselectivity in the aldol and related reactions, see M. B. Smith, *Organic Synthesis,* McGraw-Hill, New York, 1994, Chapter 9, pp. 857–964.

Problems

1. Give the products of the following reactions. Indicate the new carbon–carbon bond that has been formed and identify the carbon nucleophile and electrophile.

(*a*)

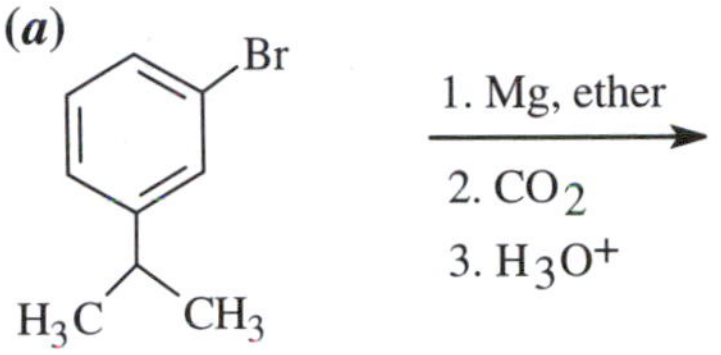
Br
H3C
CH3
1. Mg, ether
2. CO2
3. H3O+

(*b*)

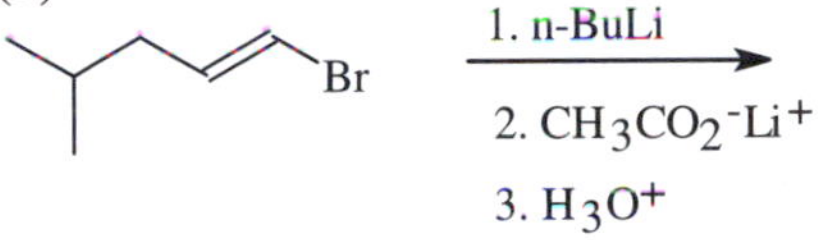
Br
1. n-BuLi
2. CH3CO2-Li+
3. H3O+

(*c*)

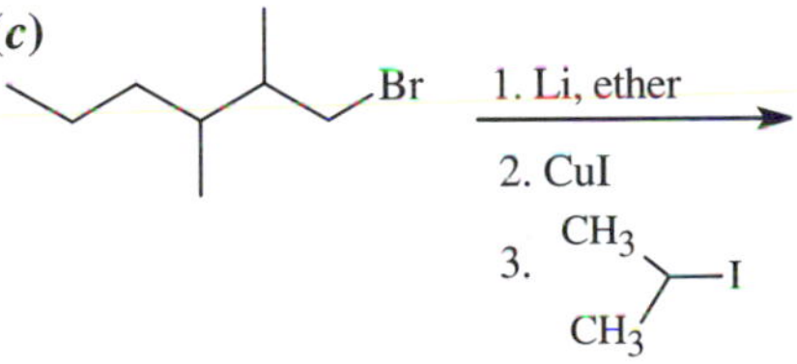
Br
1. Li, ether
2. CuI
3. CH3
I
CH3

(*d*)

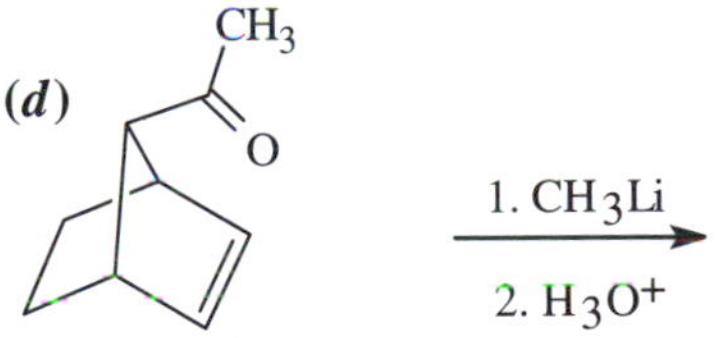
CH3
O
1. CH3Li
2. H3O+

(*e*)

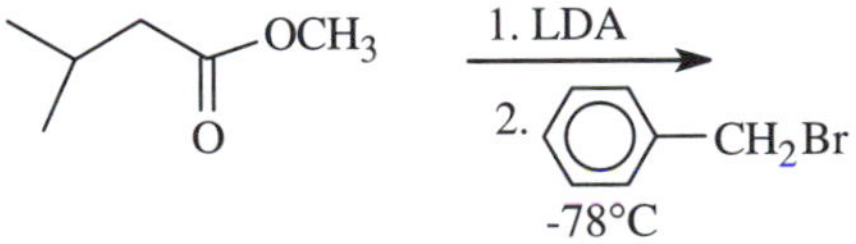
OCH3
O
1. LDA
2. CH2Br
-78°C

(*f*)

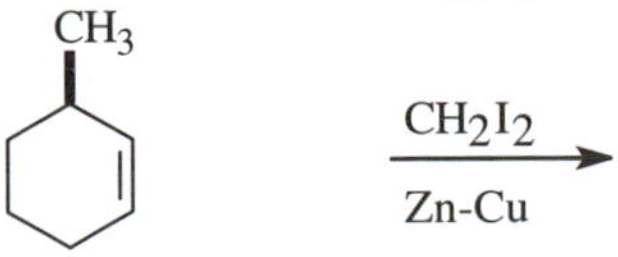
CH3
CH2I2
Zn-Cu

(*g*)

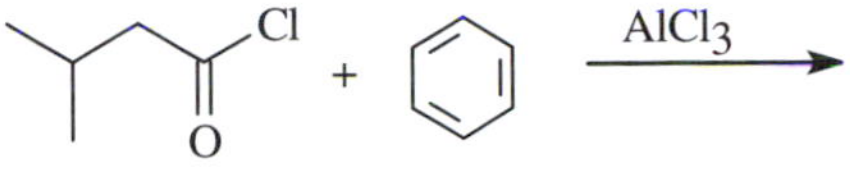
Cl
O
+
AlCl3

(*h*)

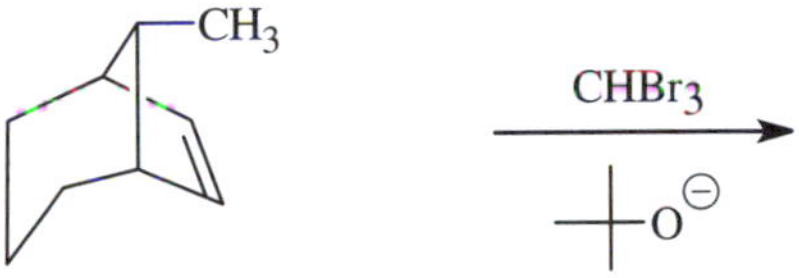
CH3
CHBr3
O

(*i*)

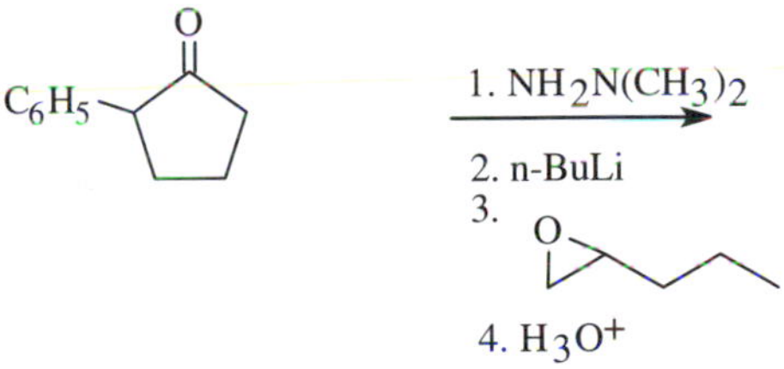
O
C6H5
1. NH2N(CH3)2
2. n-BuLi
3. O
4. H3O+

(*j*)

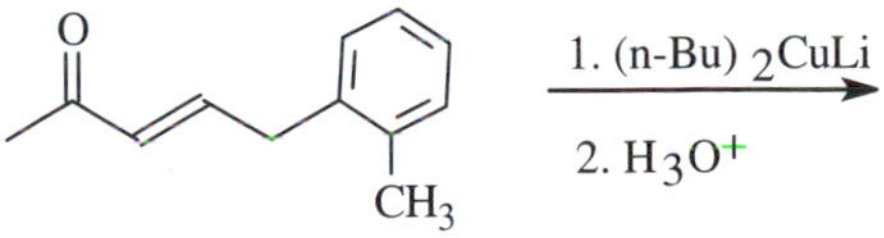
O
CH3
1. (n-Bu) 2CuLi
2. H3O+

(*k*)

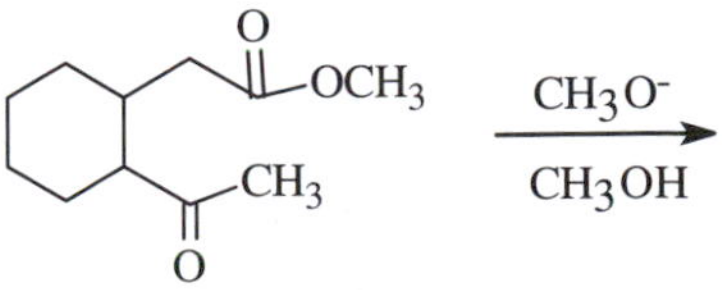
O
OCH3
CH3
O
CH3O-
CH3OH

(*l*)

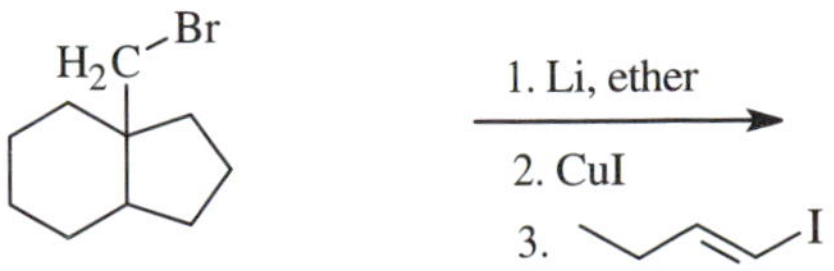
Br
H2C
1. Li, ether
2. CuI
3. I

(m)

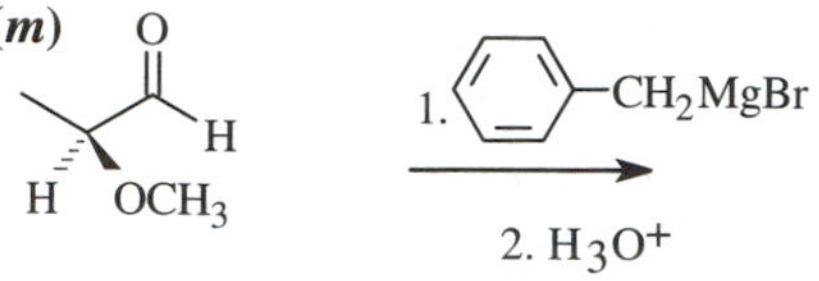

1. [benzyl]−CH$_2$MgBr

2. H$_3$O$^+$

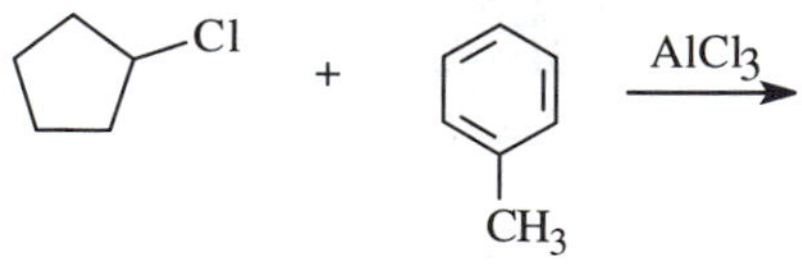

(t)

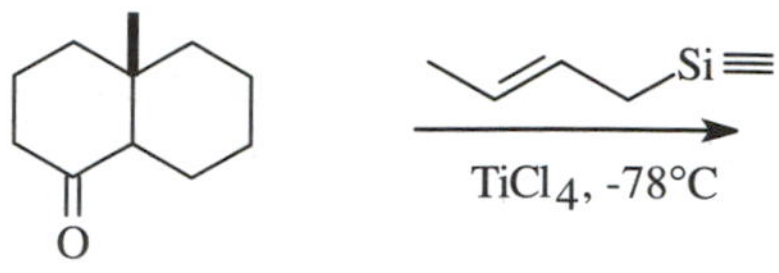

TiCl$_4$, -78°C

(n)

1. Cl−C(O)−C(O)−Cl

2. AlCl$_3$

(u)

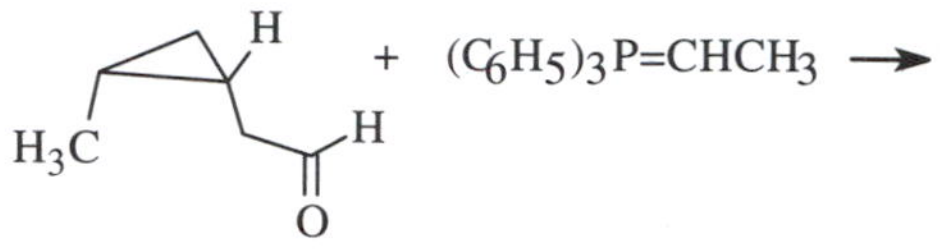

+ (C$_6$H$_5$)$_3$P=CHCH$_3$ →

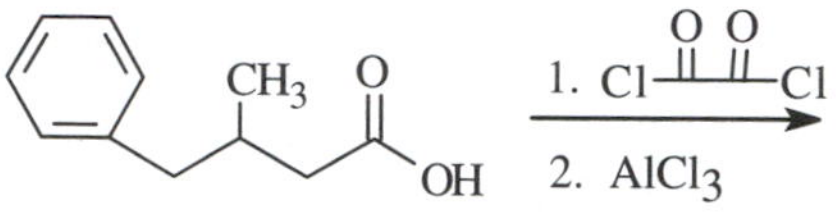

(o)

1. CH$_3$CH$_2$CH$_2$MgCl

2. H$_3$O$^+$

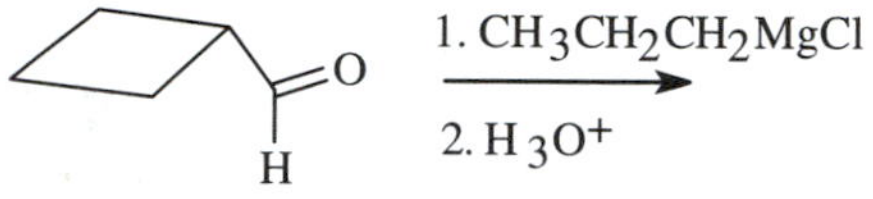

(v)

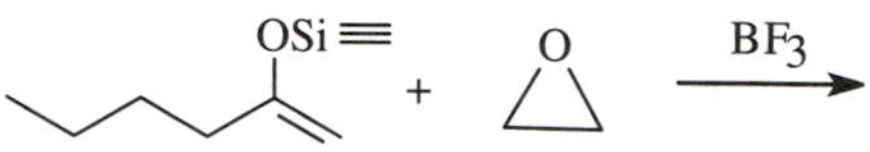

+ [epoxide]

BF$_3$

(p)

1. NaH

2. [bromomethylcyclohexane] Br

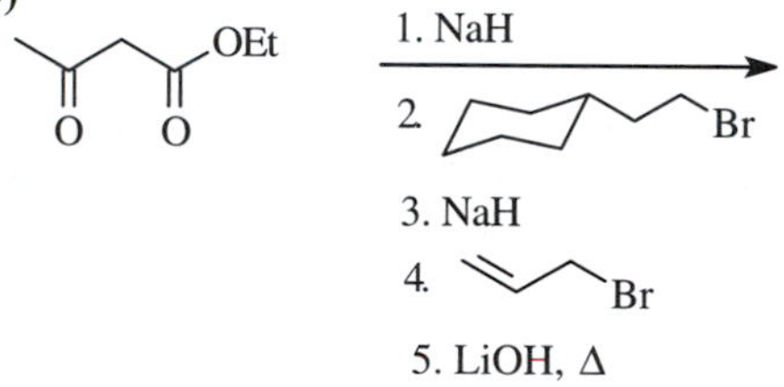

3. NaH

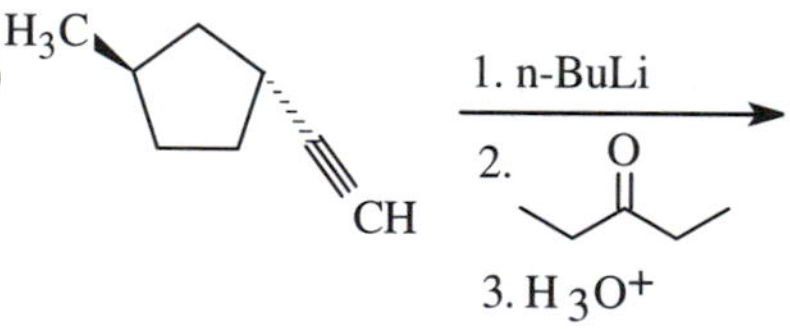

4. [allyl] Br

5. LiOH, Δ

(w)

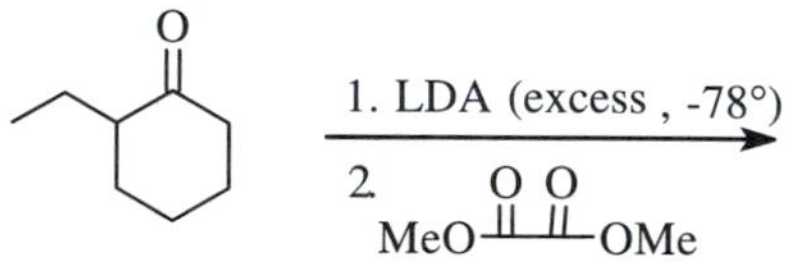

1. LDA (excess , -78°)

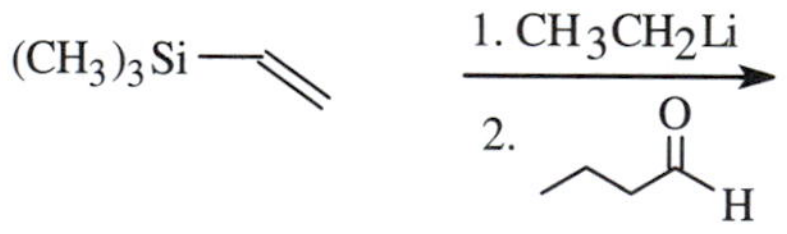

2. MeO−C(O)−C(O)−OMe

(q)

1. n-BuLi

2. [2-pentanone]

3. H$_3$O$^+$

(x)

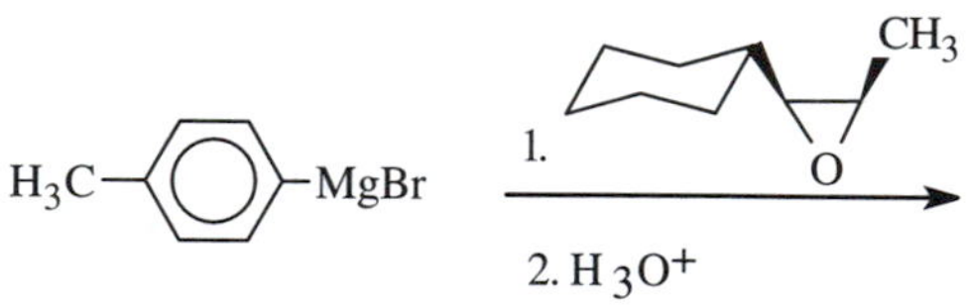

H$_3$C−[phenyl]−MgBr

1.

2. H$_3$O$^+$

(r)

(CH$_3$)$_3$Si−[allyl]

1. CH$_3$CH$_2$Li

2. [butanal]

(y)

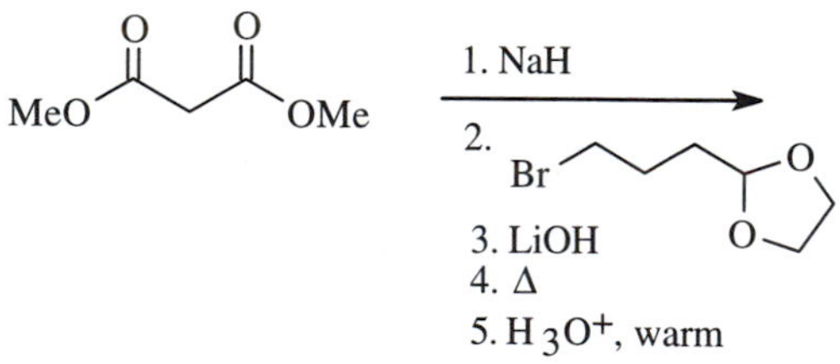

MeO−C(O)−CH$_2$−C(O)−OMe

1. NaH

2. [dioxolane bromide] Br

3. LiOH

4. Δ

5. H$_3$O$^+$, warm

(s)

[cyclopentyl]−Cl + [toluene] CH$_3$

AlCl$_3$

(z)

1. LDA (excess, -78)

2. C$_6$H$_5$CHO

3. H$_3$O$^+$

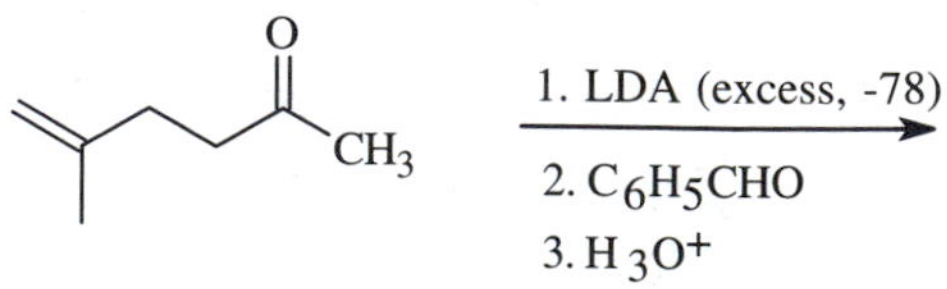

(aa)

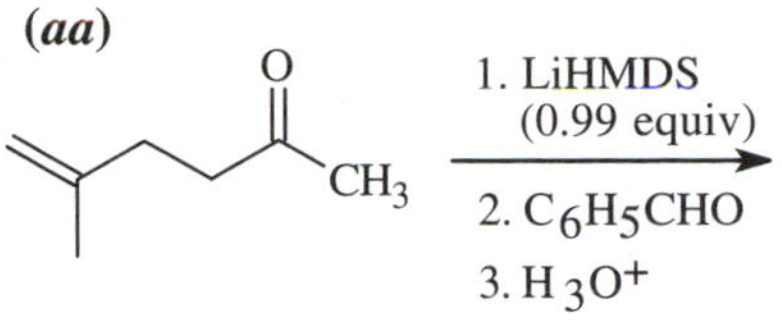

1. LiHMDS
   (0.99 equiv)
2. C₆H₅CHO
3. H₃O⁺

(bb)

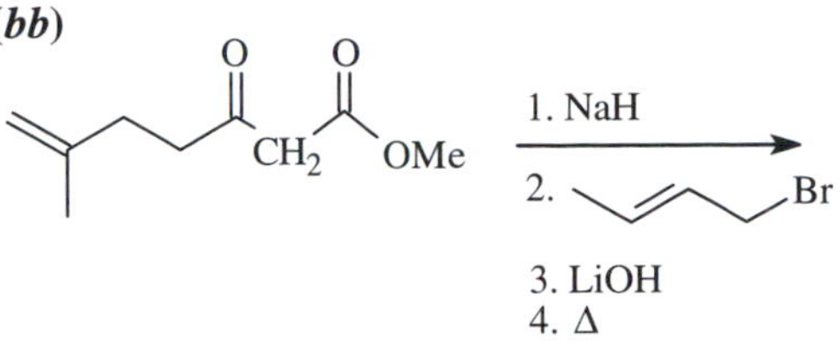

1. NaH
2. [allyl bromide structure] Br
3. LiOH
4. Δ

2. Give reactions which would produce the indicated bond in the following compounds. Give the reacting partners and tell which is the electrophile and nucleophile. Also tell how you would generate any reactants which are not stable compounds.

(a)

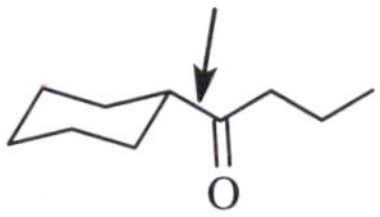

(b)

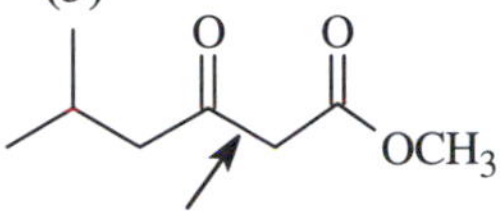

(c)

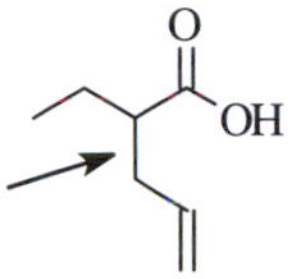

(d)

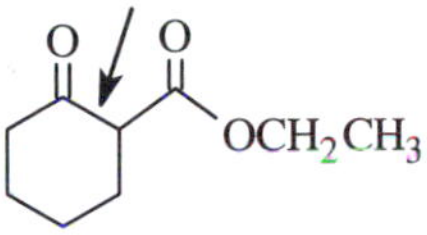

(e)

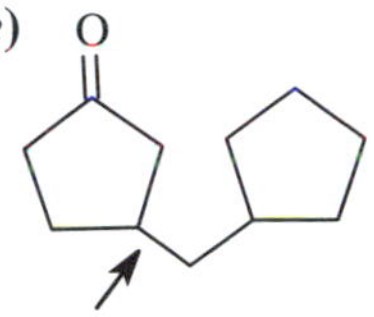

(f)

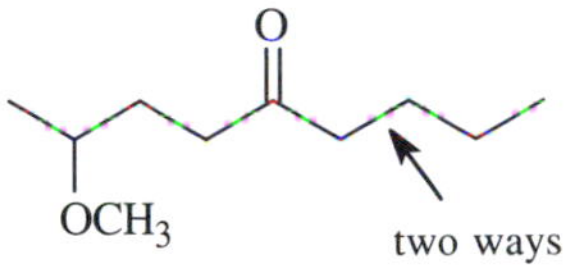

two ways

(g)

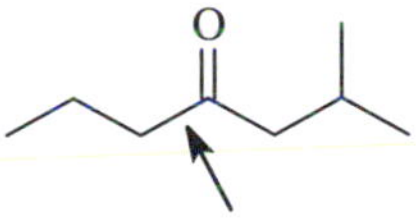

(h)

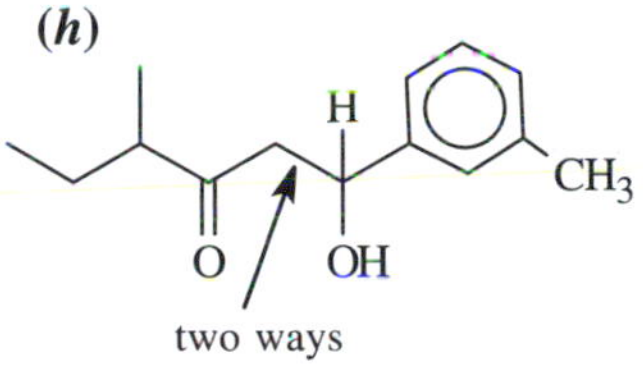

two ways

(i)

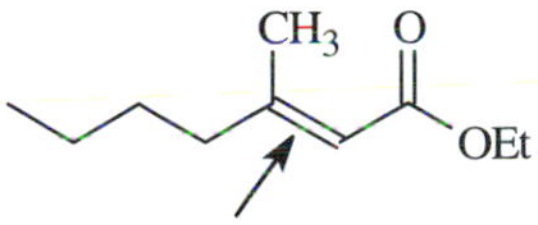

(j)

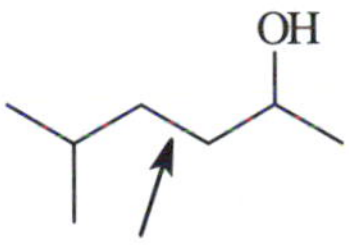

(k)

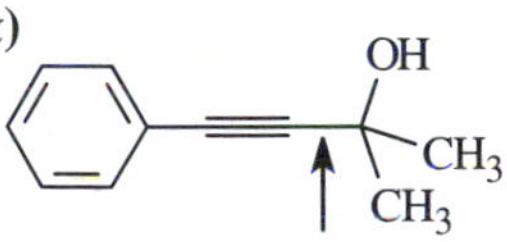

(l)

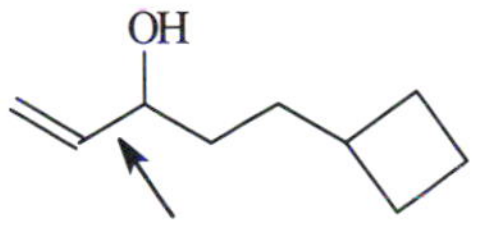

(m)

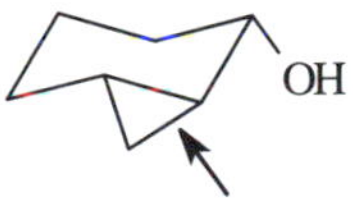

(n)

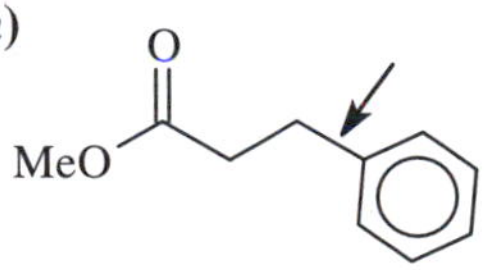

(o)

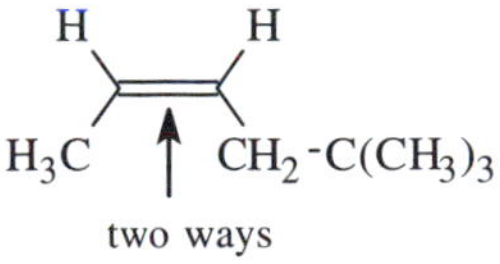

two ways

(p)

two ways

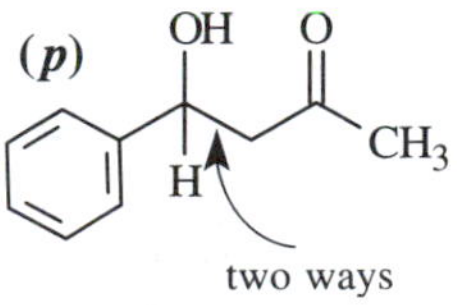

3. Show how you would prepare each of the following molecules from the indicated starting materials. Where more than one step is required, show each step clearly.

(a)

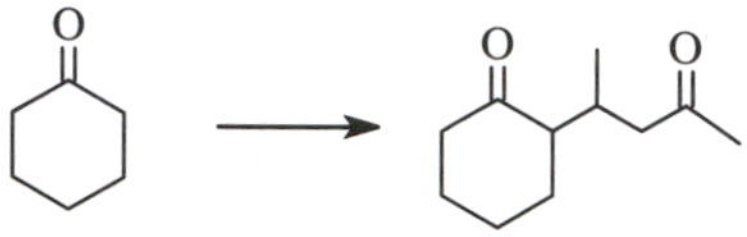

(b)

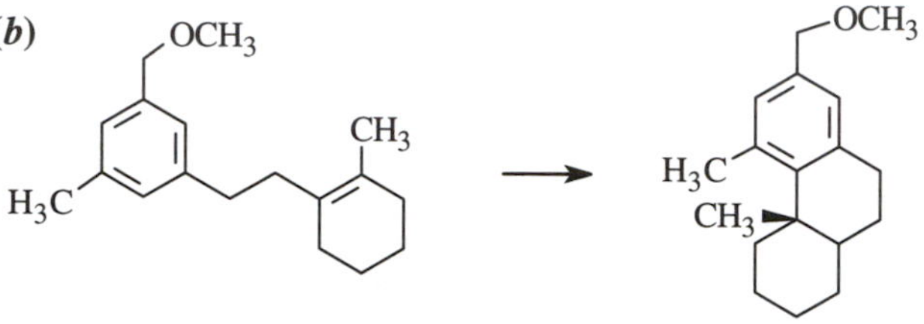

(c)

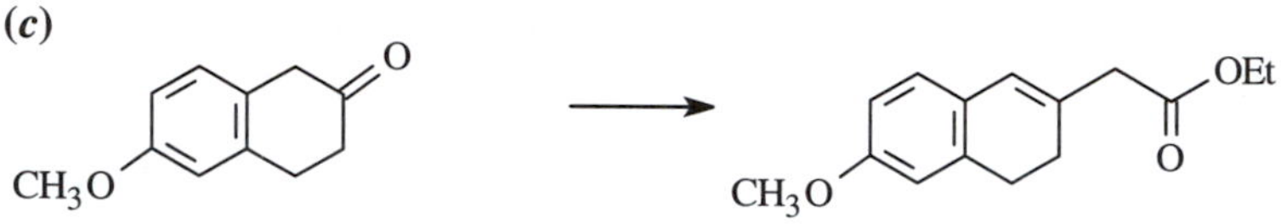

(d)

(e)

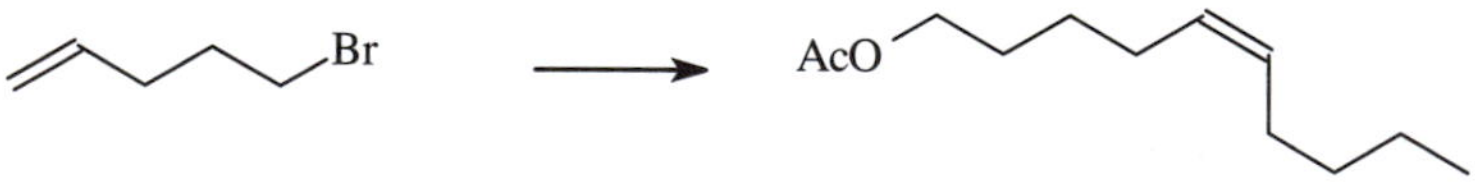

(f)

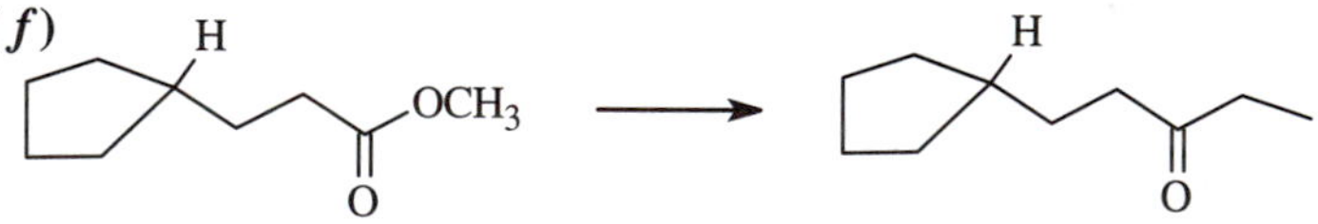

(g)

(h)

C H A P T E R 8

CARBON–CARBON BOND

FORMATION BY NONPOLAR

REACTIONS: FREE RADICALS AND

CYCLOADDITIONS

Free Radical Reactions

In the previous chapter, the formation of carbon–carbon bonds was discussed in terms of polar or two-electron processes. In such reactions, one carbon serves as an electron pair donor (nucleophile) and a second carbon serves as an electron pair acceptor (electrophile). The result of the donor–acceptor interaction of these two species is a new carbon–carbon bond in which the electron pair is shared by the donor and acceptor.

$$R_3C^{\oplus} \quad \text{and} \quad {}^{\ominus}CR'_3$$
$$\text{or} \qquad\qquad \longrightarrow \qquad R_3C-CR'_3$$
$$R_3C^{\ominus} \quad \text{and} \quad {}^{\oplus}CR'_3$$

Another way to form a bond between two carbons is for each carbon atom to supply one electron. In this case, interaction between two carbons which each have a single, unshared electron would result in formation of a carbon–carbon bond. Species with unshared electrons are called free radicals, and formation of carbon–carbon bonds by this strategy requires carbon-centered free radicals as reactants.

$$R_3C\cdot \quad \text{and} \quad \cdot CR'_3 \longrightarrow R_3C-CR'_3$$

Carbon-centered free radicals are carbon atoms which have three bonds and seven valence shell electrons, with the unshared electron occupying a valence orbital. They are generally thought to be planar (sp^2-hybridized), with the unshared electron in a 2p atomic orbital. They are not rigidly planar, and are easily deformed to a pyramidal geometry.

facile deformation

Because they have only seven valence shell electrons, free radicals are very reactive intermediates and rapidly undergo a variety of reactions. Because of their high reactivity, the formation of bonds by the combination of two free radicals is actually rare; the free radical species must survive long enough to encounter another free radical with which to react. Normally free radicals undergo other reaction processes before they encounter a second free radical with which they can combine.

The reactivity of carbon-centered free radicals results from their drive to achieve an octet electronic configuration, which they do by two principal reaction processes. The first process is atom transfer. In this case an atom with one electron is transferred from a closed-shell molecule (fully paired, valence octets) to the free radical (see Figure 8.1). Due to the conservation of spin, a new radical species is formed. If the atom that is transferred is a hydrogen, the process is called hydrogen abstraction. This is the most common atom transfer reaction, although other atoms can also be transferred to free radicals. The driving force for atom transfer (abstraction) reactions is usually the formation of a stronger bond, the formation of a more stable free radical, or both.

The second common free radical reaction is addition to π systems to give a new bond and a new free radical. In this process the π bond is broken.

This process is quite common for carbon-centered free radicals because the carbon–carbon σ bond which is formed is about 30 kcal stronger than the π bond which is broken. Other radical species are also known to undergo olefin additions. The addition of bromine atoms to olefins is the key step in the anti-Markovnikov addition of HBr to olefins.

Another common feature of free radical reactions is that they tend to be chain processes. Since any chemical reaction must exhibit conservation of spin, the reaction of a free radical with a closed-shell (fully electron paired) molecule must result in the production of a new free radical species which can participate in subsequent free radical reactions.

Free radical chain reactions can generally be divided into three phases:

1. *Initiation:* the phase in which free radicals that can start the chain reaction are produced.

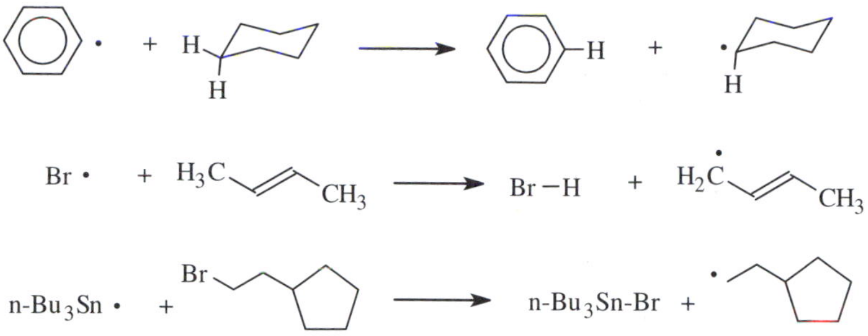

FIGURE 8.1

2. *Propagation:* the phase in which free radicals undergo reactions which form products and produce new free radicals which can continue the chain.
3. *Termination:* the phase in which free radicals are removed from the system by recombination or other reactions, thus interrupting the chain reaction.

A classic example of a chain reaction is the free radical addition of chloroform to olefins, initiated by benzoyl peroxide (see Figure 8.2).

Initiation normally requires that molecules with weak bonds undergo homolytic cleavage to produce free radicals. Since bond homolysis, of even weak bonds, is endothermic, energy in the form of heat (Δ) or light (hν) is usually required in the initiation phase. Some type of initiation is *required* to get any free radical reaction to proceed. That is, you must first produce free radicals from closed-shell molecules in order to get free radical reactions to occur. Benzoyl peroxide contains a weak O—O bond that undergoes thermal cleavage and decarboxylation (probably a concerted process) to produce phenyl radicals, and these can initiate free radical chain reactions.

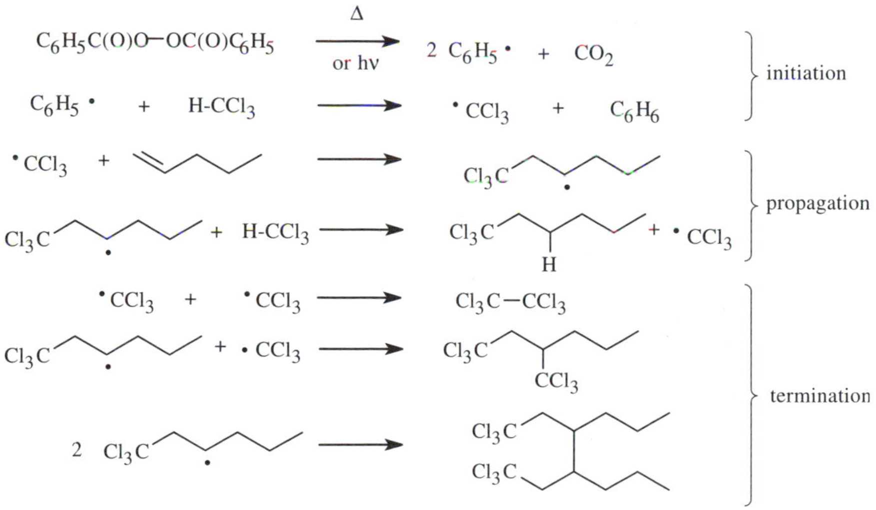

FIGURE 8.2

Azobisisobutyronitrile (AIBN) is perhaps the most widely used initiator. It undergoes either thermolytic (Δ) or photolytic cleavage ($h\nu$) to give isobutyronitrile radicals which can initiate free radical reactions.

AIBN

Hexa-*n*-butylditin can be photolyzed to two tri-*n*-butyl tin radicals which are initators for tin-based free radical reactions.

$$(n-Bu)_3 Sn - Sn(n-Bu)_3 \xrightarrow{h\nu} 2 \ (n-Bu)_3 Sn \bullet$$

Many other free radical initiators are available as well, and the choice of initiator is normally based on literature precedent and ease of use.

The propagation phase of a free radical chain reaction is usually a cyclic sequence in which a molecule of product is produced and the propagating radical is regenerated. In the above example, the trichloromethyl radical adds to the double bond to give a new carbon-centered radical. The radical abstracts a hydrogen from chloroform to produce a molecule of product and another trichloromethyl radical that continues the chain. Notice that this stage is cyclic and infinite. If a single $\cdot CCl_3$ were generated, it would continue to form one molecule of product and another $\cdot CCl_3$ until the olefin was converted completely to product, one molecule at a time. Clearly this would be very slow, but because a single initiation event can lead to many molecules of product, one needs very little initiation. If, for example, the "chain length" of the propagation cycle is 200 to 300 (a common value), one would need only a 1/200 to 1/300 ratio of initiator to olefin to convert the olefin completely to product. Thus initiation at 0.5 percent to 0.3 percent would suffice. If chain lengths are longer, less initiation is required; if they are shorter, more initiation is required.

Termination reactions, while rare, do occur; they serve to interrupt propagation cycles by removing propagating radicals from the system. Often these reactions are radical recombinations. However, termination reactions also include the reaction of propagating free radicals with other species in solution (called scavengers) to give radicals incapable of participating in the propagation cycle. In the above example, a scavenger could react with either the trichloromethyl radical or the trichloromethyl addition product to give an unreactive free radical and thus interrupt the chain process.

scavenger   unreactive radical

The more effective a termination step, the shorter the propagation cycle and the less product will be produced per initiation event. In the limiting case, if each initiation event is terminated, *no* product is produced. This is the role of "antioxidants" added to many products and most processed food. These additives scavenge free radicals and prevent them from participating in oxidation propagation cycles; thus oxidative degradation is stopped or markedly slowed.

It is the reactivity of free radicals which has made them difficult to understand and control. Because of their great reactivity, they are quite unselective and they tend to react with anything in solution. Hence multiple pathways and many products are often the rule. Moreover, many initiation methods fail to produce single free radical species in a controlled and efficient fashion. As a result, the use of free radicals in preparative organic chemistry has seen two distinct phases.

Free Radical Polymerization

The first phase was the use of free radicals in olefin polymerization. Polymerization reactions are amenable to free radical initiation for several reasons. First, the olefin is the only reagent present, so competing reactions are mimimized. Second, the initiator radical is produced by heat, light, or catalysis in the presence of a huge excess of the olefin. Under these conditions, free radical addition to the double bond is virtually the only process that occurs. The new radical species is also produced in the presence of a huge excess of olefin so that it adds to another olefin molecule, giving a larger free radical. The process continues. By controlling the purity of the starting olefin, and the reaction conditions (so that terminations are rare), chain lengths in the tens of thousands can be achieved. This leads to the formation of thousands of carbon–carbon bonds per polymer molecule and extremely long polymer chains (see Figure 8.3).

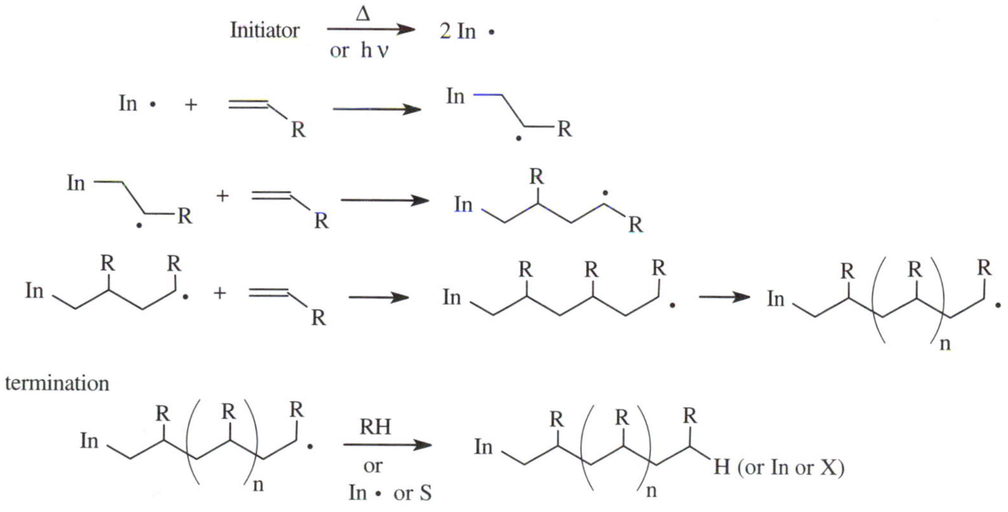

FIGURE 8.3

Since free radical reactions are chain processes, they are very well suited for the preparation of polymers rather than single products. That is, products are obtained whose size is determined by the number of propagation cycles that occur before a termination event. If the number of propagation cycles is between 200 and 300, the product mixture will contain molecules which contain between 200 and 300 monomers. It is more reasonable to describe the product mixture in terms of the "average molecular weight", rather than a single product with a discrete molecular weight. The physical properties reported for a polymer are those of a mixture of polymeric molecules rather than those of a single polymeric compound.

Free radical polymerization was a mainstay of the plastics industry for many years. Although new and better methods have been developed for the polymerization of many substrates, free radical polymerization is still used for the preparation of many plastics and composites. The success of these methods is based on an understanding of the process. Huge amounts of effort have been expended in finding controllable and reproducible initiation reactions that produce free radicals. The reaction environment has been studied intensively so that propagation reactions are maximized and termination events minimized. Finally, the rational control of termination reactions, which are necessary to control the chain length and thus the average size of the polymer molecules produced, has been successfully developed. It is important to emphasize that the properties of the product mixture were the gauge by which the understanding and control was measured.

Nonpolymerization Reactions

The use of free radical reactions for the preparation of single molecules requires greater control of the various steps in the process. Traditional free radical addition reactions carried out in solution (to minimize polymerization) often give low yields and mixtures of products and thus are not of real synthetic value. In the ideal case, one would want to be able to generate a specific free radical species, which would undergo a single reaction process and then be terminated. The termination step should give a single product plus the reactive radical to continue the chain. Thus very selective methods of initiation are needed, and a clear understanding of propagation and termination steps is required in order to control product formation more closely.

$$\text{Initiator} \xrightarrow[\text{or } h\nu]{\Delta} 2 \text{ In} \cdot$$

$$\text{In} \cdot + \text{ M} \longrightarrow \text{M} \cdot$$

$$\text{M} \cdot \longrightarrow \text{P} + \text{ In} \cdot$$

Based on these requirements for the controlled production of single molecule products from free radical reactions, the second phase of free radical chemistry began about 20 to 30 years ago. At that time, fast kinetic methods were developed to measure the rate constants for known free radical processes in solution. Thus it became possible to measure the comparative rates of hydrogen abstraction versus olefin addition and the relative rates of intermolecular versus intramolecular reactions. The results of these investigations led to the realization that intramolecular olefin additions (which produce rings) are often much faster than other reaction pathways, especially in dilute solutions. Moreover, the formation of five-membered rings by

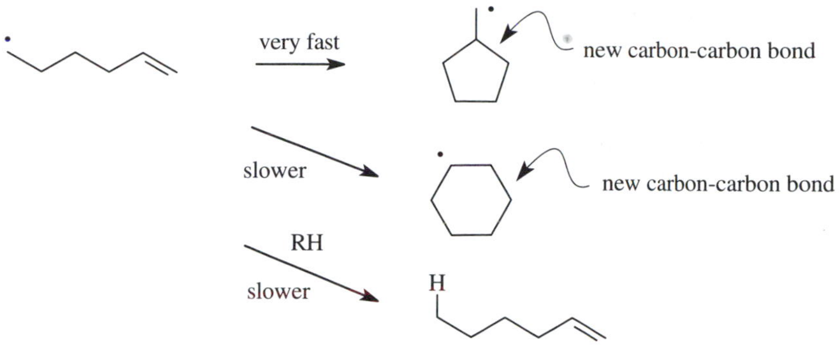

FIGURE 8.4

intramolecular olefin addition is much faster than the formation of rings of other sizes. It is this kinetic selectivity which can be used as the basis for efficient carbon–carbon bond-forming processes (Figure 8.4).

The identification of this single reaction process which occurs much faster than others can be used as a focal point for selective carbon–carbon bond construction. That is, if a radical can be produced on a carbon which is five carbons away from a double bond, the fastest reaction which occurs is cyclization to a five-membered ring. Because free radicals are uncharged, nonpolar entities, this type of cyclization is also found to be largely unaffected by polar substituents, solvents, or substitution patterns. Thus protecting groups are normally not needed for free radical cyclizations, and a wide range of reaction conditions are compatible with an efficient reaction.

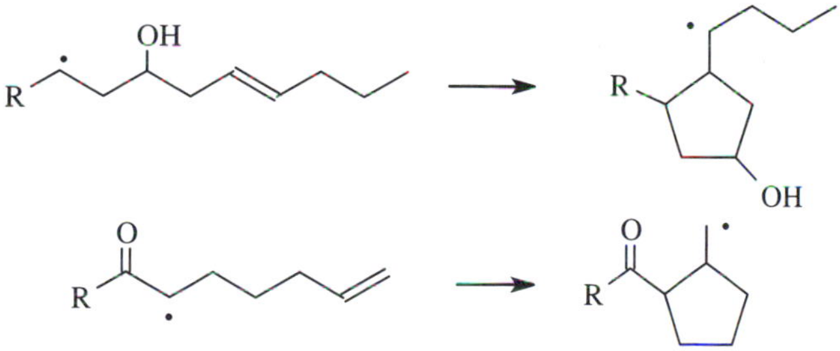

If this cyclization can be incorporated into a chain process, the cyclized radical will be trapped and will yield a single product.

Free Radical Initiation

In addition to new insights into the reactivity of free radicals, methods for the production of carbon-centered free radicals have also seen major improvements in the last several years. One common new method is to use tin-based reagents as radical chain carriers. Trialkyl tin radicals readily abstract bromine or iodine from carbon to produce a carbon-centered free radical. Placement of a bromide or iodide substituent on a substrate thus permits formation of a carbon-centered free radical at that position using tin-based methodology. This process was initially developed for the reduction of alkyl halides, and it remains an excellent synthetic method for that purpose.

One equivalent of tributyl tin hydride is required to reduce one equivalent of alkyl halide. Each step in the sequence is energetically favored, so side reactions are minimized. That is, the tributyl tin radical does not abstract hydrogen from the alkyl, it only attacks the halide (Br or I). Likewise, the carbon radical abstracts the hydrogen from tin much more readily than it does other C—H hydrogens.

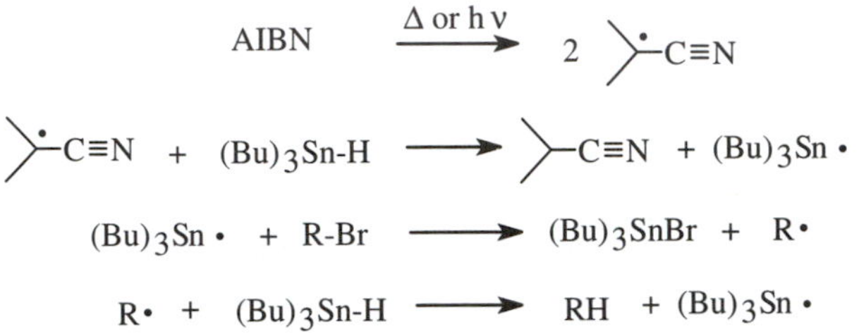

The process is clean from a mechanistic point of view and consequently results in clean reduction products.

If this method of free radical generation is applied to an unsaturated alkyl halide, the free radical has two pathways available. It can abstract a hydrogen from tributyl tin hydride and give a reduced product, as in the reduction process described above. Alternatively, it can cyclize to give a new cyclic radical. This radical can abstract hydrogen from tributyl tin hydride to give a cyclized product. These competing propagation steps are shown in the following:

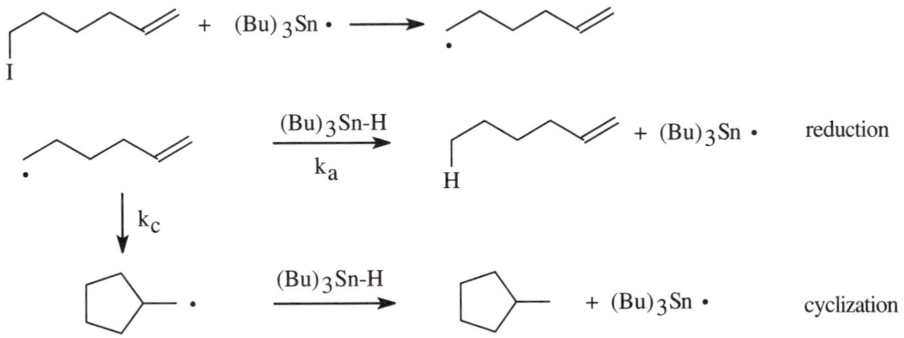

The ratio of products obtained from these two processes depends on their competing rates. The rate of reduction of the radical is given by $k_a[R\bullet][Bu_3SnH]$ while the rate of cyclization is given by $k_c[R\bullet]$. As was mentioned previously, it has been found that k_c is sufficiently large that cyclization is often the major process. However, reduction is always possible. One way to control reduction is to note that it is a second-order process which depends on the concentration of both the radical R• and tributyl tin hydride. Thus the rate of reduction can be lowered by simply running the reaction under dilute conditions. As the concentration goes down, the rate of reduction goes down much faster than the rate of cyclization, and cyclization is favored.

A second common way to produce free radicals for use in carbon–carbon bond-making reactions is to use esters of N-hydroxypyridine-2-thione.

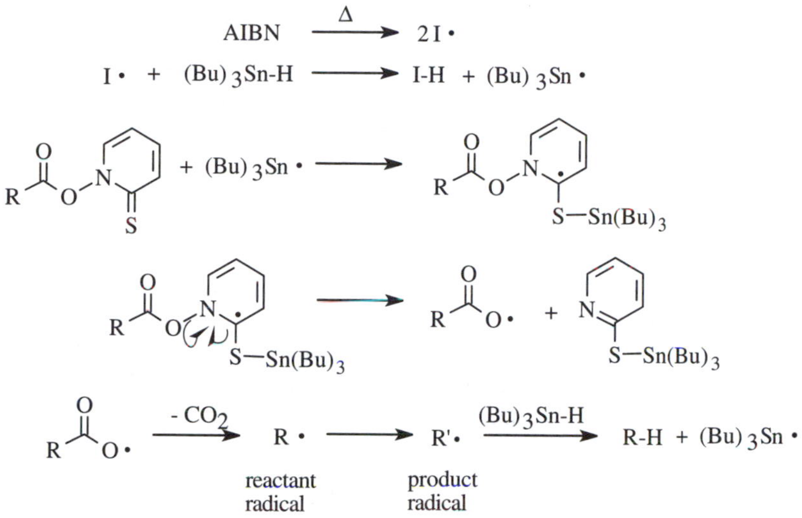

This method is also tin-based and relies on the propensity for tin radicals to add to carbon–sulfur double bonds. Subsequent fragmentation reactions lead to free radicals that are ultimately used in the propagation steps.

The free radicals R• participate in a chain process that regenerates Bu₃Sn• and thus continues the chain. While there are many steps in this overall sequence, it is a chain process. If each step is exothermic and kinetically favored, each step will proceed selectively. Thus the formation of product is efficient and reproducible, regardless of the number of the steps in the chain. This method is a clean and reliable sequence for initiating free radical reactions.

It is also known that, once the tributyl tin radical adds to the sulfur of N-hydroxypyridine-2-thione, both the breakage of the N—O bond and decarboxylation are very fast and probably concerted. They are driven energetically by formation of the very stable carbon dioxide molecule. It comes as no surprise that if atoms other than carbon are attached to the carboxyl group, they will also end up as free radicals after decarboxylation. This is shown for a urethane analog as a source of nitrogen-centered free radicals. A wide variety of other free radical species can be produced by this strategy, and it is thus quite useful.

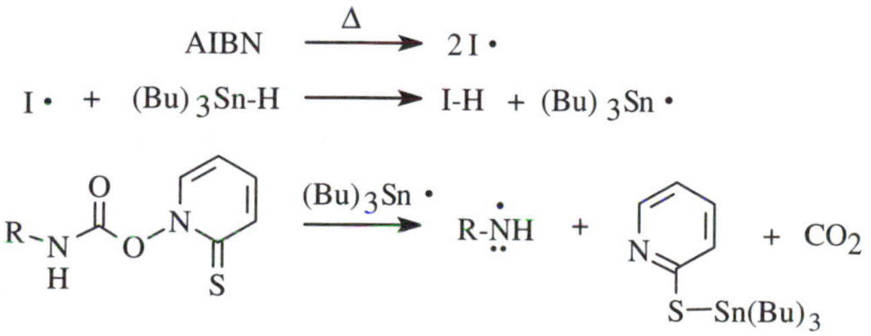

A third tin-based method of free radical production also utilizes tin radical addition to a carbon–sulfur double bond as a key reaction.

$$\text{AIBN} \xrightarrow{\Delta} 2\,\text{I}^\bullet$$

$$\text{I}^\bullet + (\text{Bu})_3\text{Sn-H} \longrightarrow \text{I-H} + (\text{Bu})_3\text{Sn}^\bullet$$

In this case, a thiono ester (usually a thiono carbonate) is the reactant. As in the previous method, addition of tin to the sulfur atom is followed by fragmentation to a carbon-centered radical.

The driving force for the fragmentation is formation of the C—O double bond. If R• reacts with Bu₃SnH, a tributyl tin radical is produced which continues the chain. This reaction is called the Barton deoxygenation of alcohols, since alcohols are precursors for the thiono esters.

If, on the other hand, R• is unsaturated and can undergo cyclization rapidly, it will do so. This competition between reduction of the first-formed radical R• and its cyclization to a new cyclic radical R′• is the same as the formation of free radicals from alkyl halides and tributyl tin radicals. The only difference is in the way in which the carbon-centered radical is produced.

A number of other strategies have been developed for the production of free radical intermediates for carbon–carbon bond construction. However, the tin-based methods described above are by far the most common.

Thus alkyl halides, carboxylic acids, and alcohols are all excellent precursors for free radicals by these methods. The choice of method can be made by selecting the substrate that fits into the synthetic scheme most efficiently, or by selecting the most readily available precursor.

Free Radical Cyclization

With good methods available for producing carbon-centered free radicals, the cyclization process can be examined in greater detail. Cyclization involves the intramolecular addition of a
free radical to a double bond. This requires that the two reacting parts of the molecule, the free
radical center and the π bond, come within bonding distance of one another.

It is easy for open-chain systems to undergo intramolecular cyclization because of their many
rotational degrees of freedom. More rigid systems undergo efficient cyclization only if the free
radical center and the π system are held in close proximity, as in the first example below. If
the molecular geometry is fixed in such a way as to prevent effective interaction between the
free radical center and the π system, cyclization is inefficient and reduction predominates.
Cyclization in the second example is an obvious impossibility.

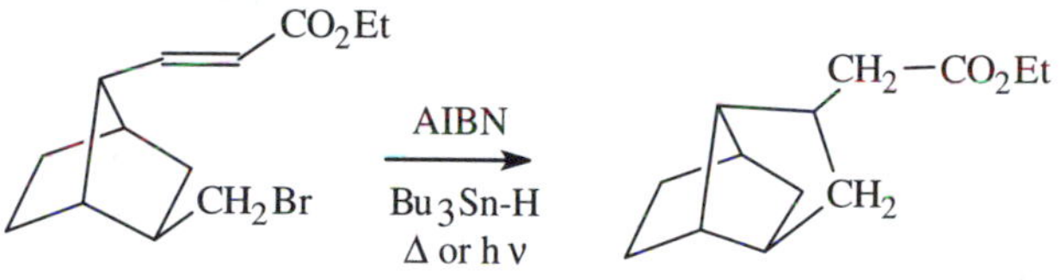

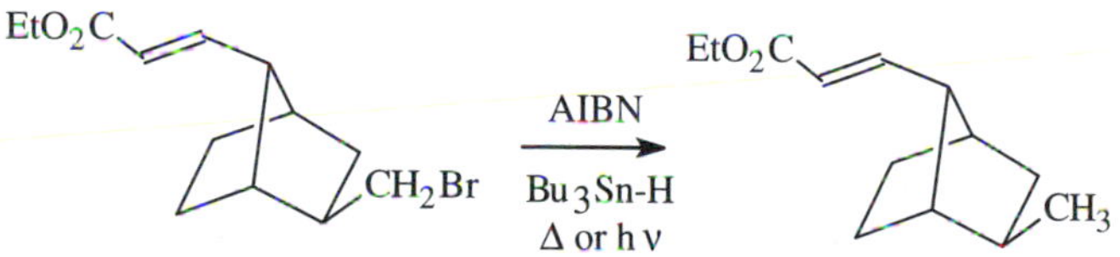

Other cases are not always so obvious, yet any structural or steric feature which influences
the close approach of the π bond and the free radical center will influence the rate of cyclization
and hence the yield of cyclized product. For example, *trans*-fused cyclopentyl systems are
much higher in energy than *cis*-fused ones, so the *trans*-fused cyclopentyl compound does not
cyclize effectively.

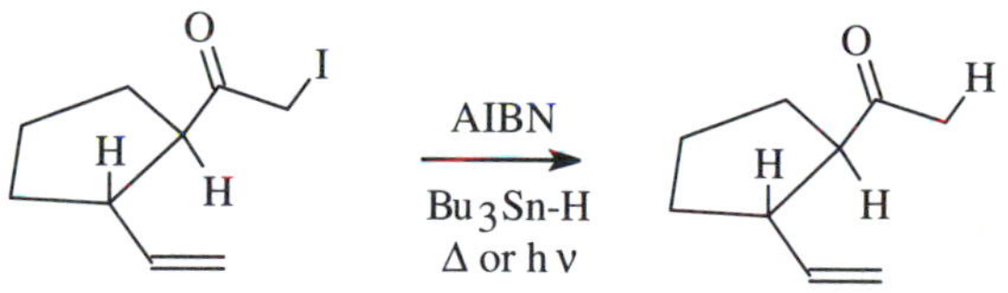

It gives only reduction, whereas the *cis*-fused cyclizes efficiently with little reduction.

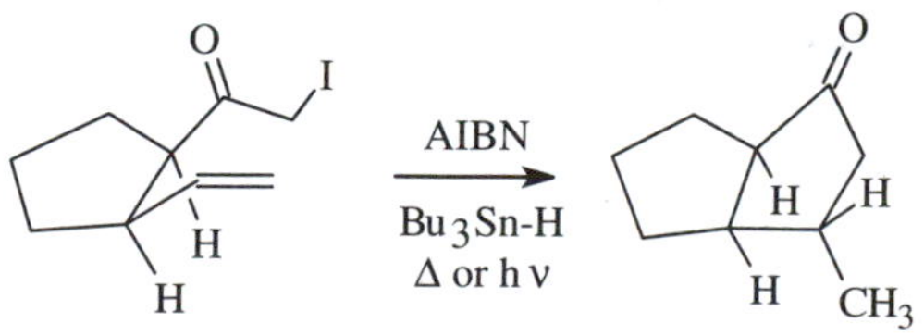

The cyclization itself can produce two different ring sizes, depending on which carbon of the double bond is attacked. Of the two possibilities, it is seen that one mode of cyclization gives a secondary radical while the other mode produces a primary free radical.

Reaction A

Reaction B

Since the order of free radical stabilities falls in the order $3° > 2° > 1°$, product stability would dictate that cyclization should occur to give the more stable secondary radical. This results in a six-membered ring in reaction *A* (path *a*) and a seven-membered ring in reaction *B* (path *a*).

In contrast, it is known that the rates of ring-forming free radical cyclizations are $5 > 6 > 7$. Experimentally, it was found that reaction *A* gives the five-membered ring product (path *b*) exclusively, and reaction *B* gives the six-membered ring product (path *b*). Thus the regioselectivity of ring formation is not controlled by thermodynamic considerations, but rather by kinetic control of the cyclization. It turns out that bond formation between a radical and a π system stereoelectronically requires an approach angle of about 110° between the free radical center and the olefinic plane. This is due to the fact that free radical addition results from donation of the unpaired electron on the radical into the $\pi*$ antibonding orbital of the olefin, which coincidentally makes an angle of about 110° with the olefinic plane.

In an intramolecular cyclization, attack on the end of the double bond closest to the radical center (an exocyclic cyclization) achieves the proper approach angle. Attack on the other olefinic carbon requires that the radical reach across the double bond to achieve the proper approach angle. This is a higher energy path and is kinetically disfavored. The same arguments hold for cyclizations which can produce six- or seven-membered rings (see Figure 8.5).

A final feature of radical cyclizations is that they are mainly influenced by steric factors and are practically insensitive to electronic effects. Since free radicals are charge neutral, their

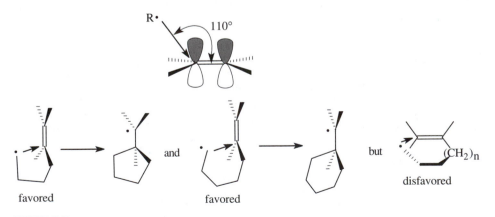

FIGURE 8.5

reactivity is not greatly influenced by either electron-donating or electron-withdrawing groups. For instance the following cyclizations occur with similar efficiencies, even though the electronic character of the cyclizing radicals are vastly different:

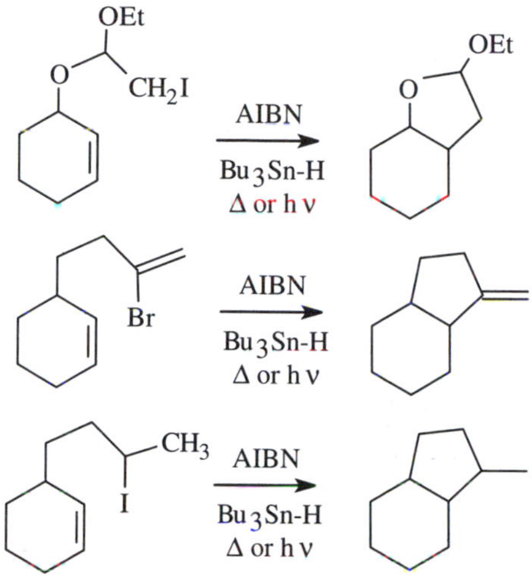

It has also been shown that the electronic character of the olefin has little influence on the efficiency of the intramolecular cyclization. Intramolecular competition between an electron-rich enol ether and a simple double bond gives a one-to-one ratio of products. This demonstrates that free radical cyclizations have a remarkable insensitivity to electronic effects.

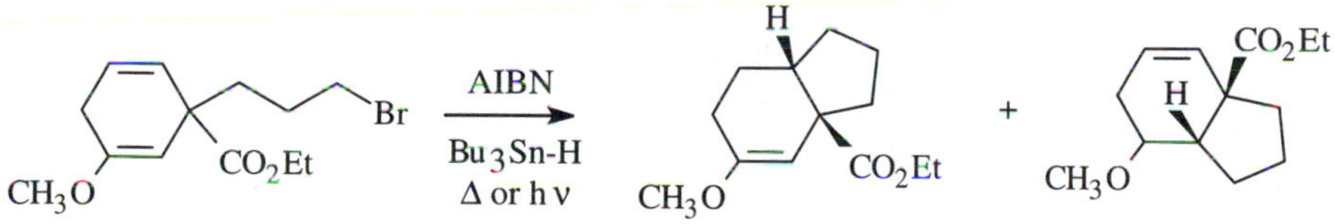

Steric effects, however, can markedly influence the cyclization process. Bulky substituents, which hinder the approach of the free radical to the π system, can prevent cyclization altogether and give only reduced product.

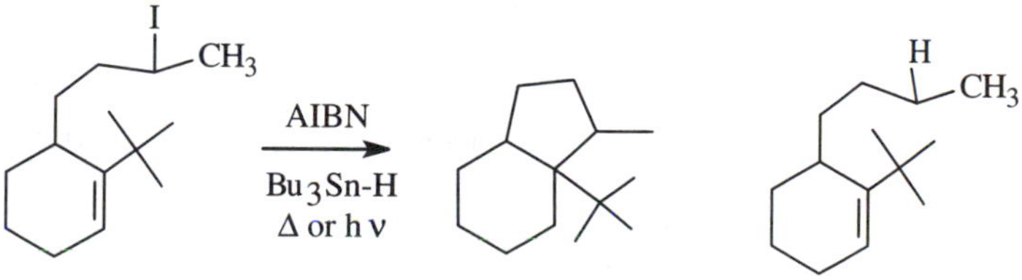

In the following examples of the types of complex structures that can be assembled by intramolecular free radical cyclization, note the presence of a great many polar functional groups in the cyclization substrates which are compatible with the process. While functional groups in the examples shown do not require protecting groups, a great number of other free radical cyclizations are known which have unprotected alcohols, carbonyl groups, and carboxylic acids in the cyclization precursor.

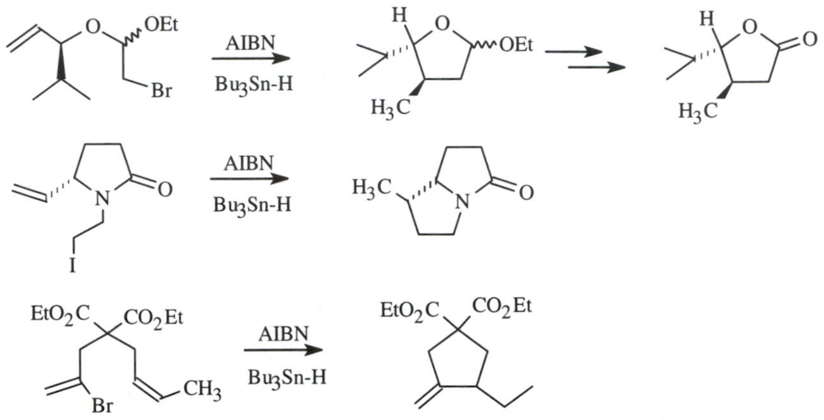

The advances made in using free radicals as synthetic intermediates in the last 10–20 years has been extraordinary. This is due both to new methods to effectively generate free radicals and to new insights into reactivity patterns which allow them to be controlled.

Diels–Alder Reaction

A widely used reaction for the formation of six-membered rings is the Diels–Alder reaction. This is a reaction between a diene and an olefin to give a new six-membered ring. It is also termed a 4+2 *cycloaddition* because one partner (the diene) containing four π electrons reacts with a two-electron fragment (the olefin) containing two π electrons, in a cycloaddition process, to yield a ring.

The Diels–Alder reaction has six different bonds which are being either made or broken, all at the same time. The three π bonds in the reactants are all broken, and two new carbon–carbon σ-bonds and one new carbon–carbon π bond are formed. Curved arrows can be used for keeping track of electrons but they are not of mechanistic significance. Alternatively, dotted lines are sometimes used to describe the transition state; they imply that all bonds are made and broken simultaneously. No matter which mechanistic formalism is used to describe the process, the product contains a six-membered ring which has a double bond.

The Diels–Alder reaction is one type of a large class of ring-forming reactions termed *cycloaddition reactions.* In most instances, new carbon–carbon bonds are formed during cycloaddition processes. Such reactions are neither polar reactions (although the reactants might be polar molecules), nor are they free radical reactions. Rather, cycloadditions are usually concerted reactions in which all bond making and bond breaking is taking place at the same time. Such reactions have been described as having a "no-mechanism mechanism." As we have seen for the Diels–Alder process, however, it is possible to use curved-arrow notation to keep track of the electron changes that occur during the cycloaddition.

Concerted cycloaddition reactions result from the interaction between π systems in two different molecules. The π system in one molecule reacts with the π system in a second molecule to produce new bonds. Since both the diene and the olefin are closed-shell, ground state molecules, the first question that comes to mind is "How do such systems react?"

The formation of new bonds in the Diels–Alder reaction requires that the π electrons in the individual diene and olefin π systems become reorganized and shared in the new bonding pattern of the cyclic product. For this bonding change to occur, the two π systems must overlap so that electrons can move into new orbitals. The most straightforward way the necessary orbital overlap can occur is for one π system to function as an electron donor and the other π system to function as an electron acceptor. Therefore, the bonding changes in the Diels–Alder reaction result from a donor–acceptor interaction between the diene and olefin π systems.

It is well-known that π electrons are less tightly bound than σ electrons; consequently, they can readily be donated to a variety of electrophiles (e.g., bromine, protons, mercuric ion). Therefore, it is easy to imagine a π system acting as an electron donor. In contrast, π systems, being electron-rich, are not often thought of as electron acceptors (which would make them even more electron-rich).

In order for a π system to function as an electron acceptor, it must have unfilled orbitals available to accept electrons. In the case of olefins or dienes those are $\pi*$ antibonding MOs. Thus interaction of the HOMO of one π system with the LUMO of a second π system produces a donor–acceptor pair (HOMO donating to LUMO). This allows electrons to be transferred from one π system to another with resulting bond formation.

In picture form, this can be drawn as

HOMO–LUMO Interactions

One requirement for a successful HOMO–LUMO interaction is that the symmetry of the HOMO must match the symmetry of the LUMO (either both symmetric or both antisymmetric). If so, the interaction is symmetry "allowed" and will lead to productive cycloaddition. If the symmetries do not match, the HOMO–LUMO overlap is symmetry "forbidden" and

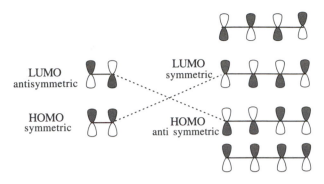

FIGURE 8.6

cycloaddition will not proceed. Molecular orbitals can be classified by their phase symmetry, with respect to a plane normal to the π system. The symmetry is related to the number of nodal planes which occur in each molecular orbital. For the olefin component, the π orbital (HOMO) is symmetric with respect to this plane, and the $\pi*$ orbital (LUMO) is antisymmetric with respect to this plane. For the diene component, the HOMO is antisymmetric and the LUMO symmetric. Based on these symmetries, the HOMO–LUMO interaction between butadiene and ethylene is symmetry "allowed" and can thus proceed to product. It turns out that the orbitals for any diene and any olefin have the same symmetry properties, so all Diels–Alder reactions are symmetry-allowed (see Figure 8.6).

While symmetry requirements dictate whether a cycloaddition can occur, they do not determine the strength of the HOMO–LUMO interaction. Both the strength of the donor–acceptor interaction and the rate of cycloaddition are inversely related to the difference in energy between the interacting HOMO and LUMO. If the HOMO–LUMO energy gap is small, the interaction is strong and the reaction is rapid, whereas if the HOMO–LUMO energy gap is large, the interaction is weak and the reaction is slow. The Diels–Alder reaction between butadiene and ethylene is very slow; the donor–acceptor interaction is very weak because the HOMO–LUMO energy gap is large.

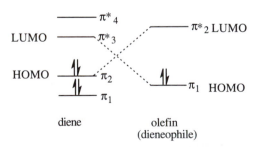

It should be recognized that there are two HOMO–LUMO interactions possible between butadiene and ethylene. In one, the HOMO of the diene acts as the electron donor; in the other, the HOMO of the olefin would be the electron donor. A "normal" Diels–Alder reaction is one in which the diene is electron-rich and acts as the electron donor, while the olefin (dienophile) is electron-poor and acts as the electron acceptor. In such a case, the diene HOMO and the dienophile LUMO are closer in energy, the donor–acceptor interaction between them is strong, and the reaction takes place at a more rapid rate.

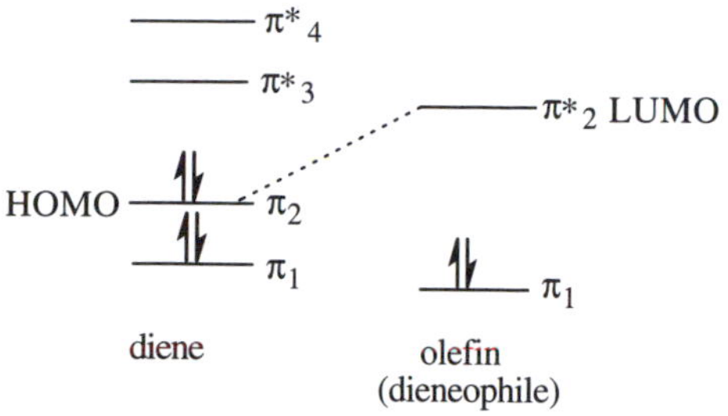

Examples of such partners are dienes, substituted with alkyl groups or other electron-donating groups (which makes them electron-rich), and dienophiles with carbonyl groups or other electron-withdrawing groups attached (which makes them electron-deficient). The dienophile need only be a two-electron π system. Olefin, acetylene, and azo π systems can all serve effectively as dienophiles as long as they have electron-deficient π bonds. Furthermore, the dienophile may be symmetrically or unsymmetrically substituted with electron-withdrawing groups.

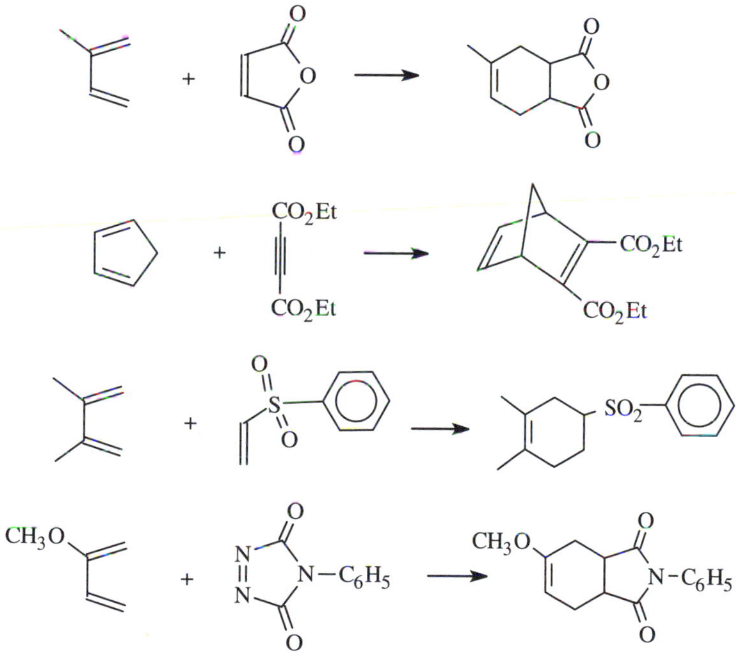

Stereoelectronic Factors

The interaction between the diene HOMO and the dienophile LUMO takes place when the ends of the two π systems overlap, permitting the transfer of electrons from the HOMO into the LUMO. This requirement of overlap imposes stereoelectronic constraints on the two reaction partners. First the diene must be able to adopt an s-*cis* conformation so the ends of the diene can contact and overlap with the ends of the dienophile π system. For acyclic dienes, even though the s-*trans* conformer is favored, rotation about the central carbon–carbon bond is rapid, and there will be a steady state population of the required s-*cis* form present so that cycloaddition can occur.

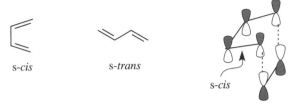

s-*cis*   s-*trans*   s-*cis*

However, when the diene system is constrained to the s-*cis* conformation by a cyclic frame-work, the effective concentration of the s-*cis* diene is much higher than for acyclic dienes, which have the s-*cis* conformer as a minor component of the rotomeric equilibrium. Such con-formationally constrained dienes react more easily and are excellent Diels–Alder dienes. Ex-amples include cyclopentadienes, 1,3-cyclohexadienes, and furans.

s-*cis* Diels-Alder dienes

Second, substituents on the dienophile (olefinic or azo) can also adopt a position in the transition state, either *exo* or *endo* to the diene system. It has been found that the *endo* transition state is favored significantly over the *exo* transition state. This preference has been attributed to secondary orbital interactions (attraction) between the diene π system and polar substituents on the dienophile.

exo transition state   endo transition state

This distinction is important because *exo* and *endo* transition states lead to different diastereomers. Control of diasteroselection is extremely important to the utility of the Diels–Alder reaction, since mixtures of diastereomers are avoided and control of multiple stereogenic centers is achieved.

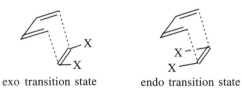

exo transition state   *anti*

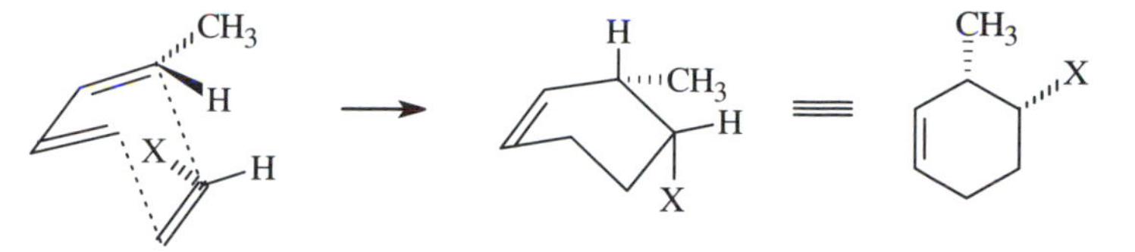

endo transition state   *syn*

For example, reaction of (E,E)-2,4 hexadiene with methyl crotonate gives a single product. The relative stereochemistry of four contiguous stereogenic centers in the product is explicitly defined by the geometry of the starting materials and the endo transition.

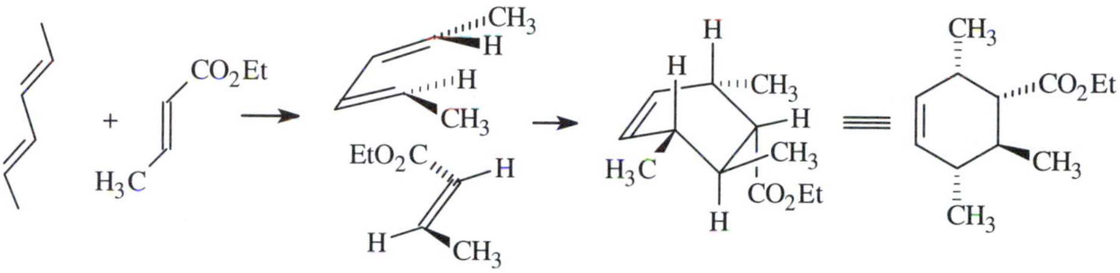

A third aspect of the Diels–Alder reaction is the regiochemistry of the products. If either the diene or the dienophile is symmetric, only a single regioisomer is possible. If both the diene and the dienophile are unsymmetric, however, two regioisomers are possible. The regioisomer produced depends on the relative orientation of the substituents at the transition state. Usually, one of these regioisomers is favored over the other.

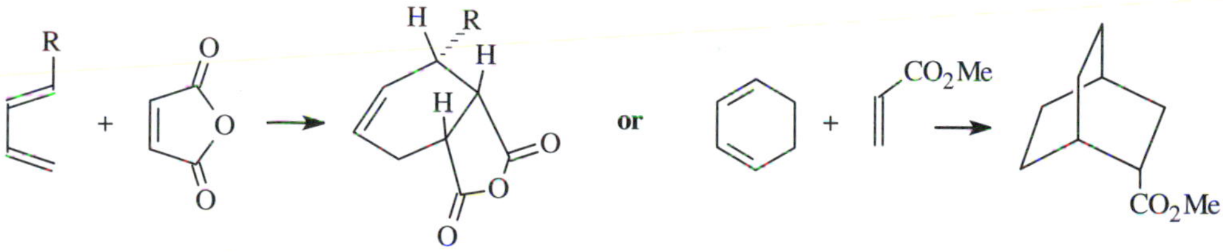

symmetric
dieneophile

single product

or

symmetric
diene

single product

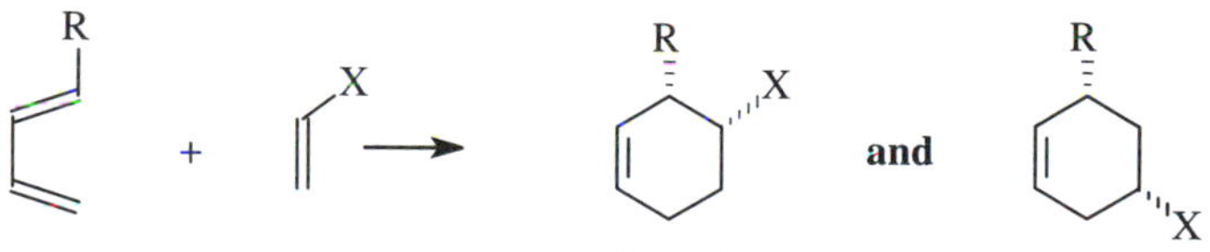

unsymmetric   unsymmetric
diene   dieneophile

favored

and

regioisomers

The control of regiochemistry has been rationalized on the basis of the orbital coefficients of the HOMOs and LUMOs but, in fact, it is not well understood. In most cases such cycloadditions are not regiospecific and isomeric mixtures are formed, although one regioisomer usually predominates. Qualitative estimation of the electron distributions in the diene and dienophile can often be used to predict the major product. For example, C-1 of siloxy diene (A) should be much more electron-rich than C-4. In addition, C-3 of acrylate (B) should be more electron-deficient than C-2.

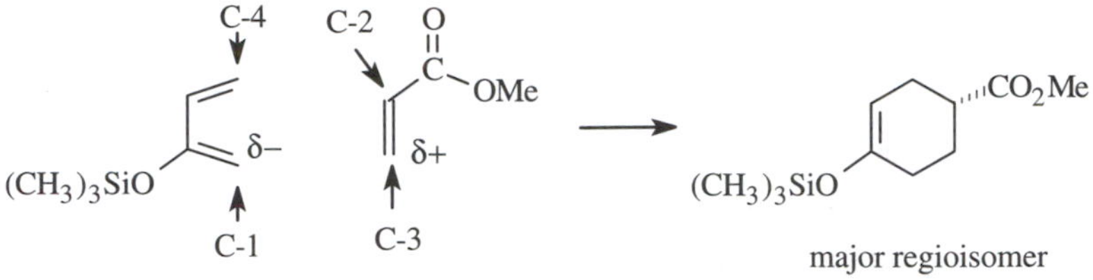

Thus the most favorable donor–acceptor interaction should occur between C-1 of the diene and C-3 of the dienophile. This interaction would favor 1,4-orientation of the substituents in the Diels–Alder product, as is observed.

In spite of the fact that the major product is often predictable, these systems are rarely regiospecific. Because mixtures of regioisomers which must be separated are the rule, either a symmetric diene or a symmetric dienophile is usually employed to avoid such regiochemical issues.

1,3-Dipolar Cycloaddition Reactions

The essential features of the Diels–Alder reaction are a four-electron π system and a two-electron π system which interact by a HOMO–LUMO interaction. The Diels–Alder reaction uses a conjugated diene as the four-electron π system and a π bond between two elements as the two-electron component. However, other four-electron π systems could potentially interact in a similar fashion to give cycloaddition products. For example, an allyl anion is a four-electron π system whose orbital diagram is shown in Figure 8.7. The symmetry of the allyl anion non-bonding HOMO matches that of the olefin LUMO (as does the olefin HOMO and the allyl anion LUMO). Thus effective overlap is possible and cycloaddition is allowed. The HOMO–LUMO energy gap determines the rate of reaction, which happens to be relatively slow in this case.

Molecules isoelectronic with the allyl anion, but which are neutral and have at least one resonance form with formal positive and negative charges in a 1,3-relationship, are called 1,3-dipoles (see Table 8.1). All these molecules have an orbital diagram analogous to the allyl anion, in which three interacting p orbitals give rise to three MOs containing a total of four π electrons. For example, a nitrone is seen to have a C—N π bond interacting with a filled

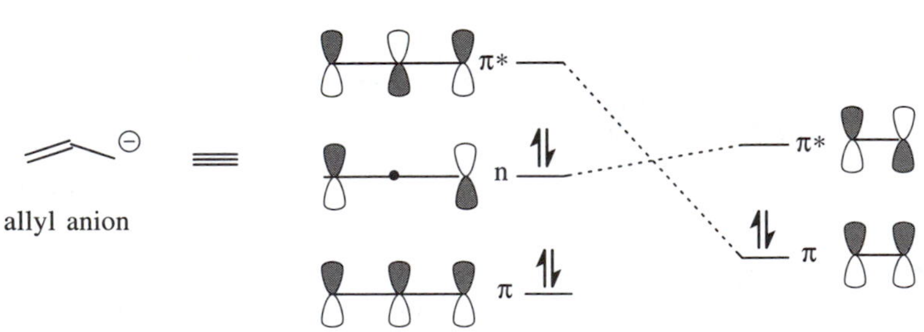

FIGURE 8.7

TABLE 8.I Representative 1,3-Dipoles

Azides	$R-\overset{\cdot\cdot}{N}=\overset{\oplus}{N}=\overset{\cdot\cdot}{N}\ominus$	$\longleftrightarrow$	$R-\overset{\cdot\cdot}{N}-\overset{\oplus}{N}=\overset{\cdot\cdot}{N}\ominus$
Diazoalkanes	$R_1-\overset{\overset{\oplus}{C}}{\underset{R_2}{C}}=\overset{\cdot\cdot}{N}=\overset{\cdot\cdot}{N}\ominus$	$\longleftrightarrow$	$R_1-\overset{\overset{\oplus}{C}}{\underset{R_2}{C}}-\overset{\cdot\cdot}{N}=\overset{\cdot\cdot}{N}\ominus$
Nitrile oxides	$R-C\equiv\overset{\oplus}{N}-\overset{\cdot\cdot}{\underset{\cdot\cdot}{O}}\colon^{\ominus}$	$\longleftrightarrow$	$R-\overset{\oplus}{C}=\overset{\cdot\cdot}{N}-\overset{\cdot\cdot}{\underset{\cdot\cdot}{O}}\colon^{\ominus}$
Nitrones	$R_1-\overset{\overset{\oplus}{C}}{\underset{R_2\ \ R_3}{C}}=\overset{\cdot\cdot}{N}-\overset{\cdot\cdot}{\underset{\cdot\cdot}{O}}\colon^{\ominus}$	$\longleftrightarrow$	$R_1-\overset{\overset{\oplus}{C}}{\underset{R_2\ \ R_3}{C}}-\overset{\cdot\cdot}{N}-\overset{\cdot\cdot}{\underset{\cdot\cdot}{O}}\colon^{\ominus}$

orbital on the oxygen atom to define a new π system containing four π electrons (Figure 8.8). Interaction of the HOMO of the 1,3 dipole with the LUMO of a simple π bond (called a dipolarophile in this process) leads to bond formation between the ends of the 1,3-dipole and the olefin, producing a new five-membered ring.

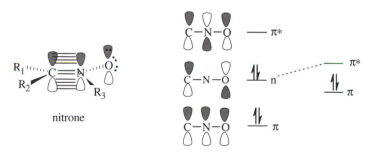

The process ia a concerted $4_\pi + 2_\pi$ cycloaddition and is related electronically to the Diels–Alder reaction. The formal charges are destroyed during the cyclization, and a wide variety of heteroatom components are possible in the 1,3 dipole. Moreover, other π bonds besides alkenes and alkynes can be used as dipolarophiles. As a result, 1,3-dipolar cycloadditions have been used to make a large number of heterocyclic compounds.

Since the 1,3-dipolar cycloaddition is concerted, the reaction is stereospecific and the geometry of the olefin is maintained in the cyclic product.

FIGURE 8.8

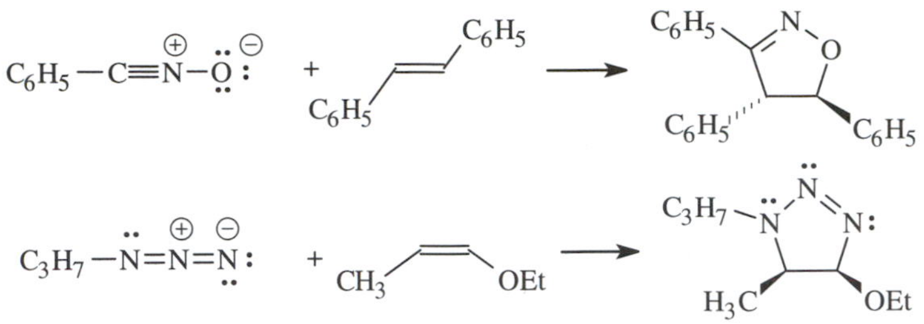

If a symmetric dipolarophile is used, there are no regioisomers possible. If, however, the dipolarophile is unsymmetric, regioisomers are possible. As in the case of the Diels–Alder reaction, the regioselectivity can be understood in terms of the electron distribution in the 1,3-dipole and the dipolarophile.

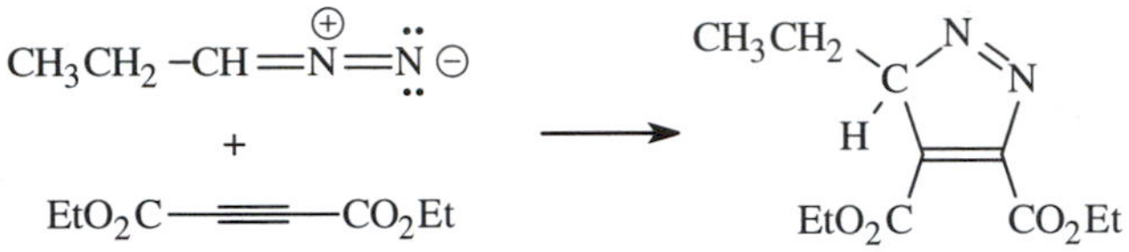

For example, a nitrile oxide should have a relatively electron-deficient carbon and a relatively electron-rich oxygen. Reaction with propene, which has greatest electron density at C-2, because of the inductive effect of the methyl group, gives the regioisomer **C**. By matching the polarity of the dipole and the dipolarophile, one can predict this product. Conversely, reaction with methyl acrylate, which because of conjugation has electron deficiency at C-3, gives regioisomer **D** as the major product.

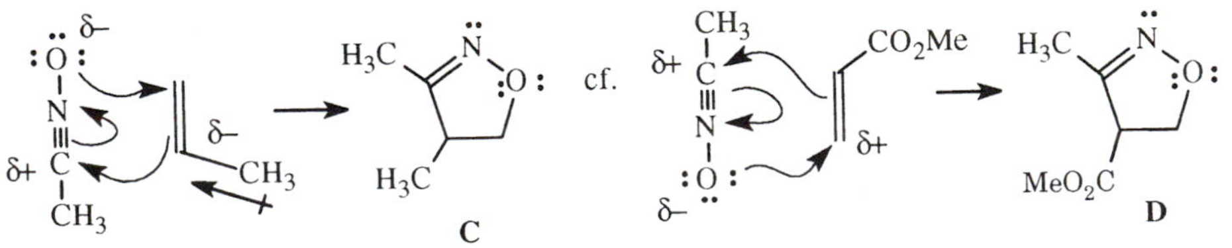

Polarity matching, to predict the major product of 1,3-dipolar cycloadditions, is qualitative only and frequently fails to predict the major product correctly, because each 1,3-dipole tends to exhibit a characteristic regioselectivity toward particular dipolarophiles that may be modified by steric or strain effects. In fact, there is still some uncertainty as to just what factors do influence the regioselectivity in these systems.

Nevertheless, 1,3-dipolar cycloadditions are an important method for the synthesis of a wide variety of heterocyclic compounds. Furthermore, they illustrate the generality of 4+2 cycloaddition reactions as a means to prepare cyclic products efficiently from acyclic precursors.

Bibliography

For a discussion of the general properties and reactivity of free radicals see F. A. Carey and R. J. Sundberg, *Advanced Organic Chemistry, Part A: Structure and Mechanisms,* 3d ed., Plenum, New York, 1990, Chapter 12, pp. 651–708.

For an advanced discussion of intramolecular free radical cyclizations in synthesis see M. B. Smith, *Organic Synthesis,* McGraw-Hill, New York, 1994, pp. 1423–32.

For an excellent literature review of free radical synthetic methods see D. P. Curran, *Synthesis,* 1988.

For an expanded discussion of the Diels–Alder and related 4+2 cycloadditions see Carey and Sundberg, *Advanced Organic Chemistry, Part B: Reactions and Synthesis,* 3d ed., Plenum, New York, 1990, Chapter 6, pp. 283–307.

An excellent comprehensive discussion of the use of the Diels–Alder reaction and its variants in synthesis can be found in M. B. Smith, *Organic Synthesis,* McGraw-Hill, New York, 1994, pp. 1113–82. In the same volume, 1,3-dipolar cycloadditions are treated on pp. 1208–25.

For an excellent review of the early work on 1,3-dipolar cycloaddition reactions, see R. Huisgen, *Angew. Chem. Int. Ed., English,* 1963, 2, 533 and 633.

Problems

1. Give balanced chemical equations for each mechanistic step for the following transformations.

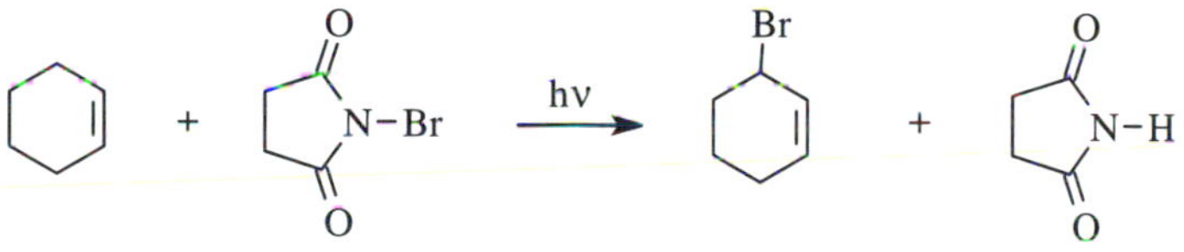

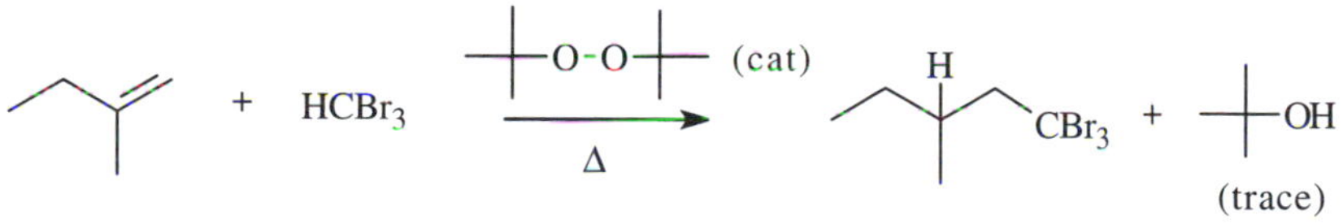

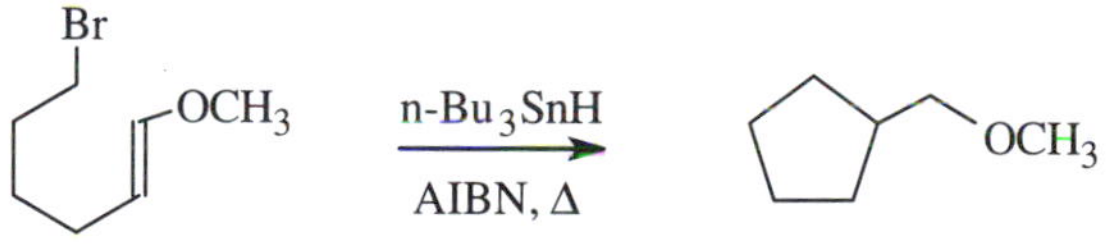

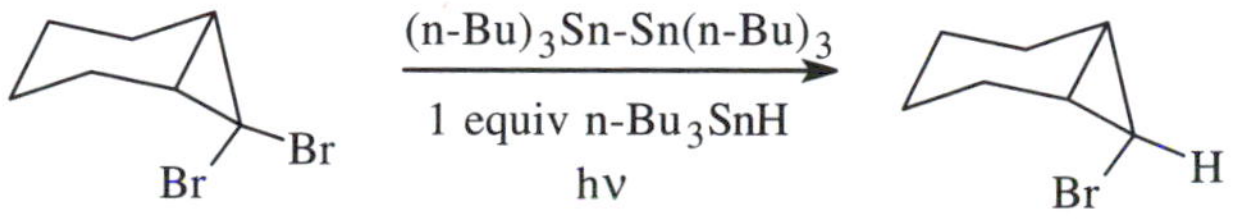

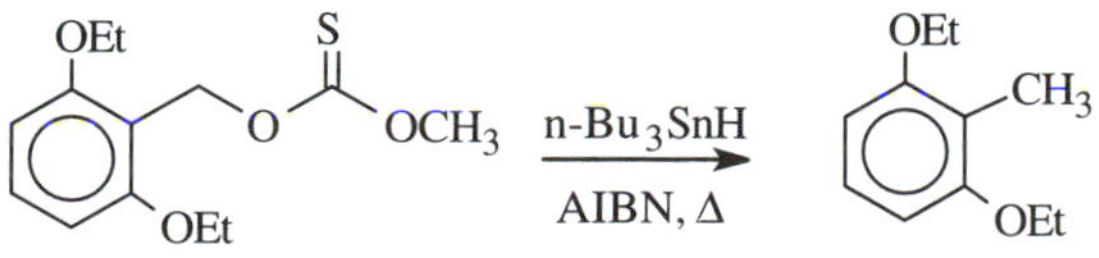

(*f*)

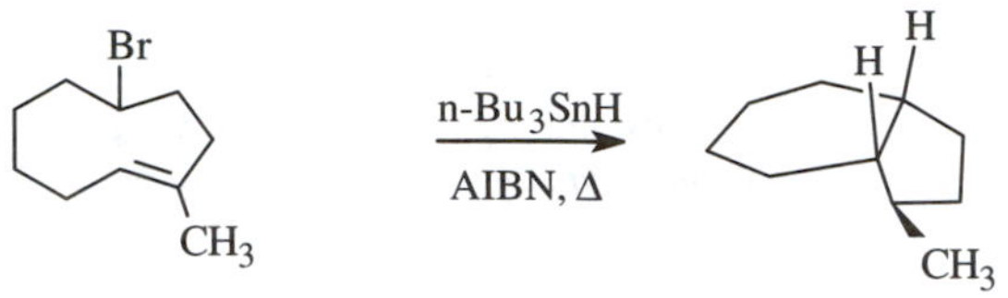

(*g*)

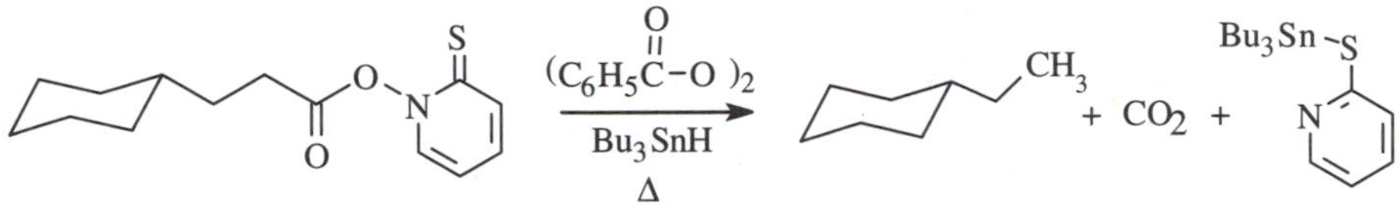

(*h*)

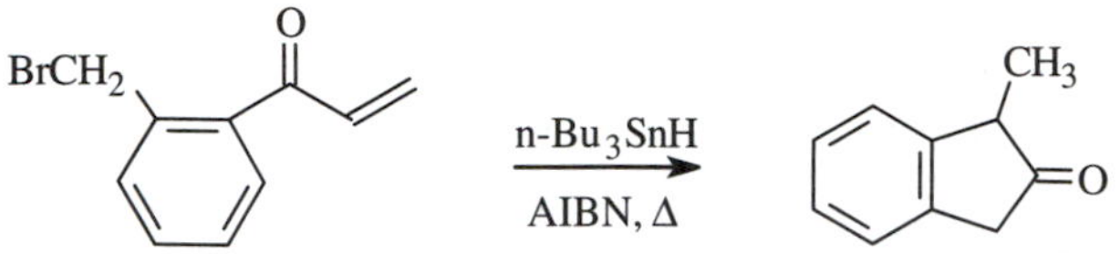

(*i*)

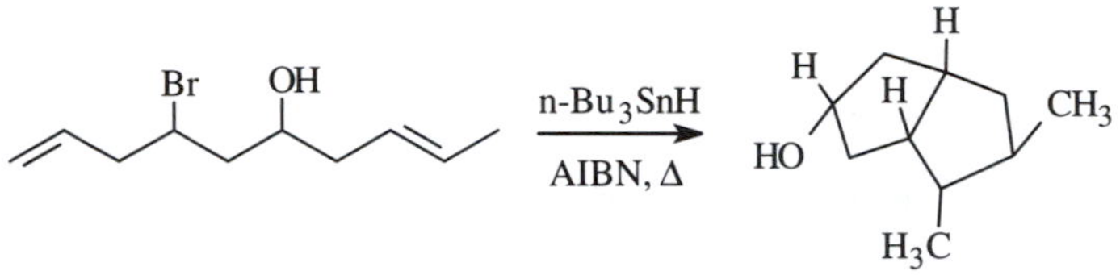

(*j*)

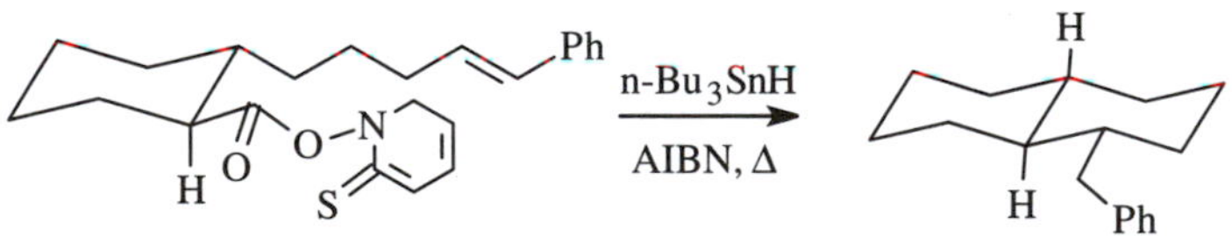

2. For each of the following reactions, the structures of possible isomeric products are shown. Predict which product will be favored and give the reasons for your prediction.

(*a*)

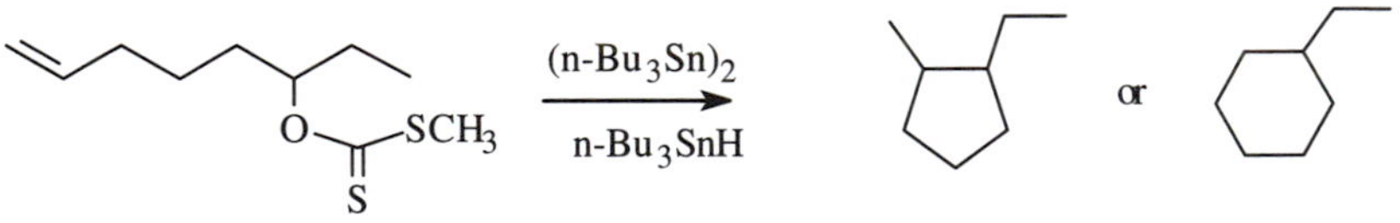

(*b*)

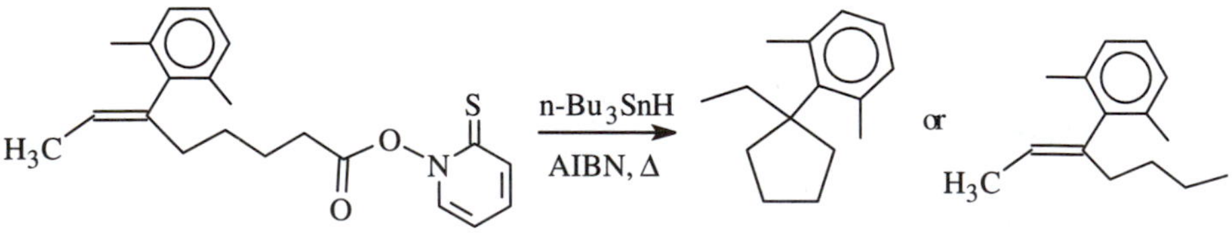

(*c*)

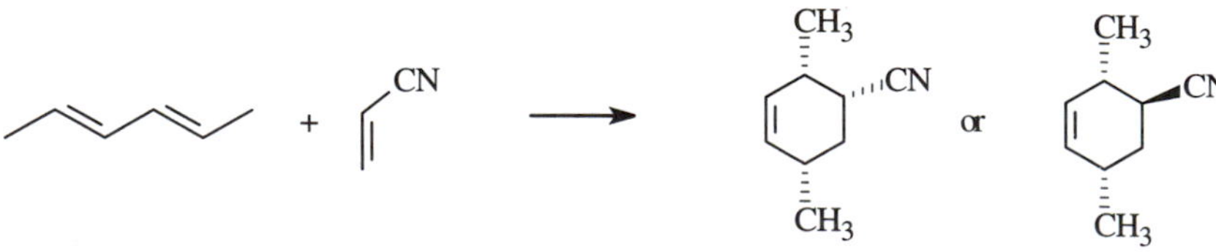

(d)

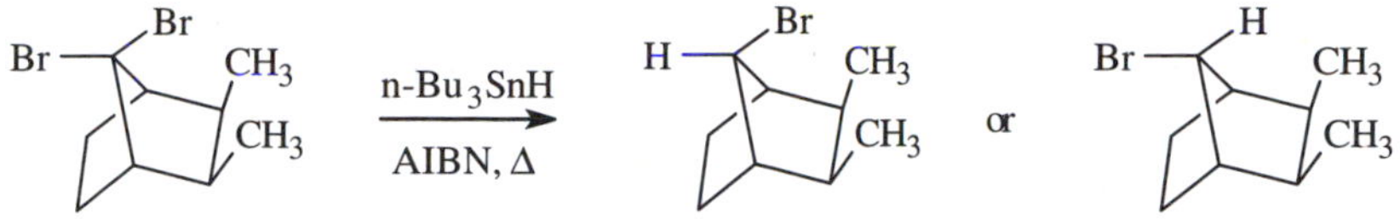

n-Bu$_3$SnH
AIBN, Δ

or

(e)

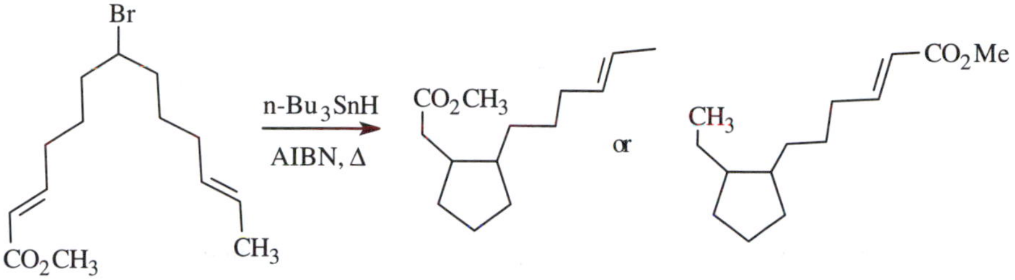

n-Bu$_3$SnH
AIBN, Δ

CO$_2$CH$_3$

or

CO$_2$Me

CH$_3$

(f)

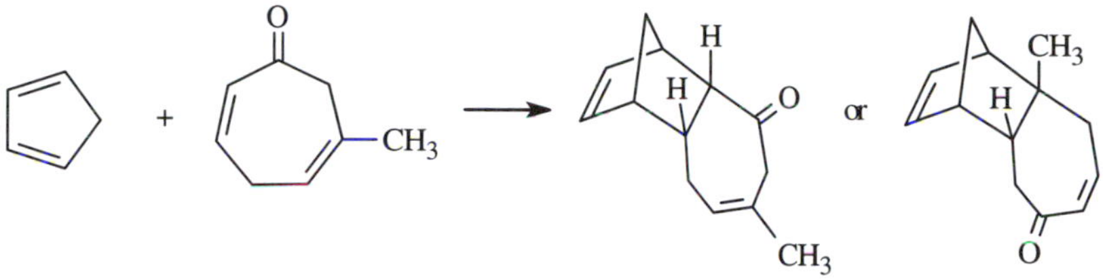

+

→

or

(g)

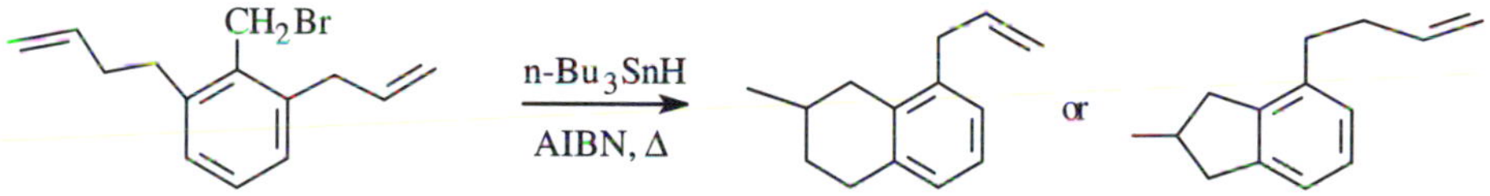

CH$_2$Br

n-Bu$_3$SnH
AIBN, Δ

or

(h)

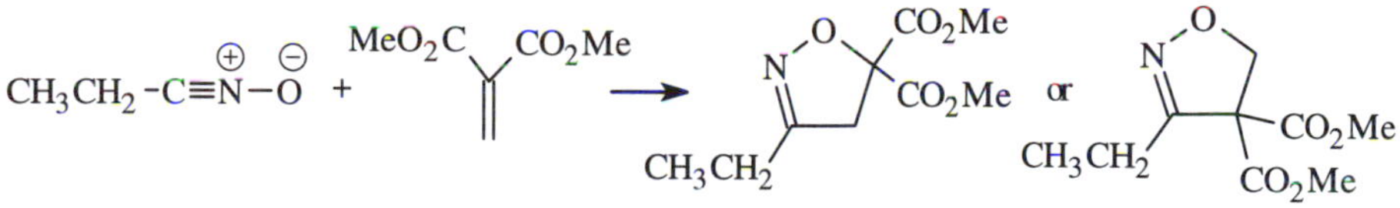

CH$_3$CH$_2$–C≡N–O +

MeO$_2$C   CO$_2$Me

→

CO$_2$Me
CO$_2$Me

CH$_3$CH$_2$

or

CO$_2$Me
CO$_2$Me

CH$_3$CH$_2$

(i)

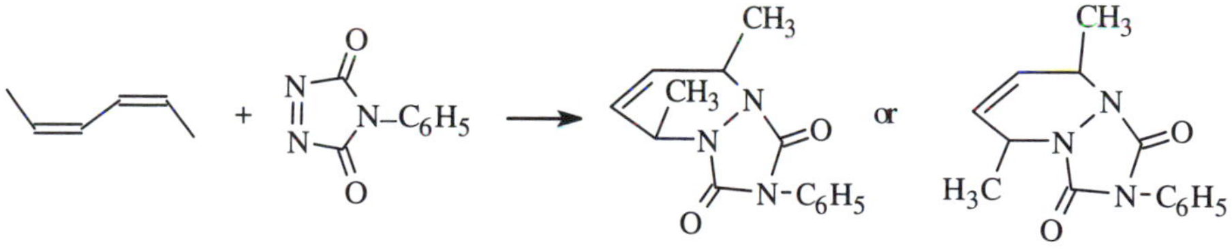

+

N–C$_6$H$_5$

→

CH$_3$
CH$_3$

N–C$_6$H$_5$

or

CH$_3$

H$_3$C

N–C$_6$H$_5$

(j)

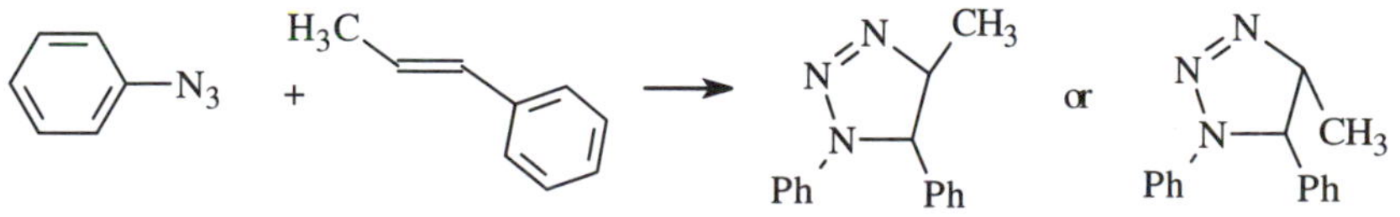

N$_3$

+

H$_3$C

→

CH$_3$

Ph   Ph

or

CH$_3$

Ph   Ph

3. Cyclopentadiene is purchased as a dimer which has the structure shown below. Heating the dimer at about 150°C causes cyclopentadiene to distill from the flask until no dimer remains. Upon standing at room temperature, cyclopentadiene dimer is rapidly formed. Explain the significance of these observations.

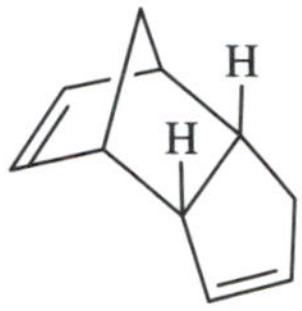

cyclopentadiene dimer

4. Give the products of the following reactions. Where more than one product is likely to be formed in significant yield, indicate which will be the major product.

(a)

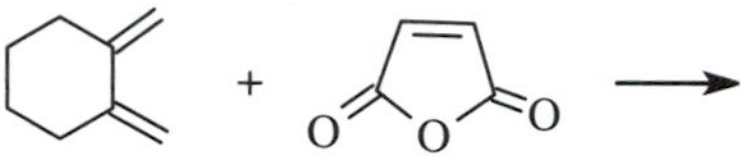

(b)

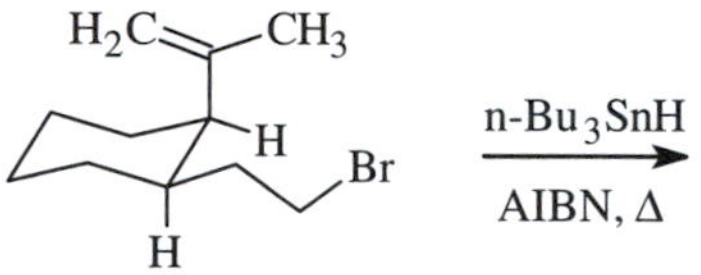

$\text{n-Bu}_3\text{SnH}$

AIBN, Δ

(c)

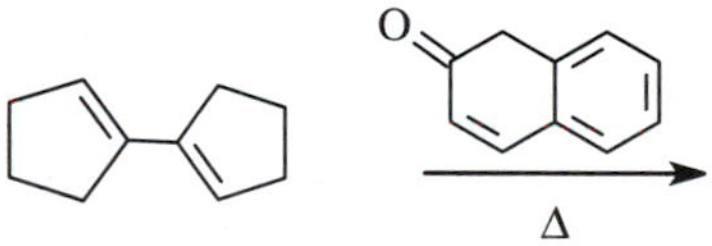

Δ

(d)

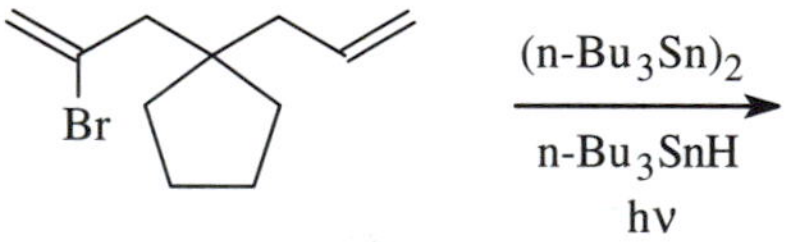

$(\text{n-Bu}_3\text{Sn})_2$

$\text{n-Bu}_3\text{SnH}$

hv

(e)

1. NaH

2. CS_2

3. CH_3I

4. $\text{n-Bu}_3\text{SnH, AIBN,}\Delta$

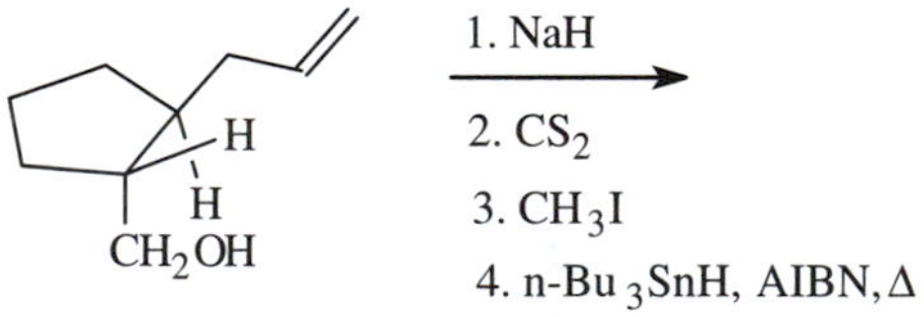

(f)

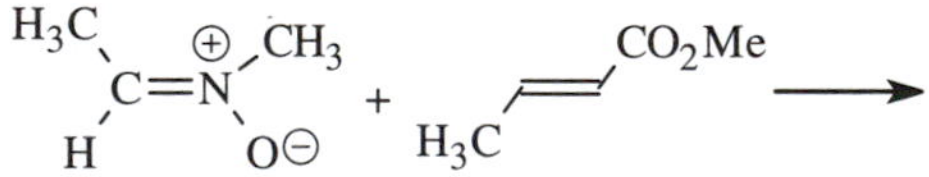

(g)

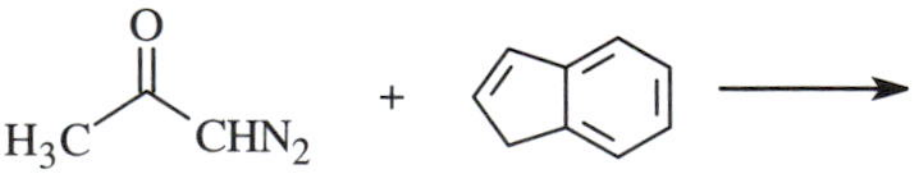

(h)

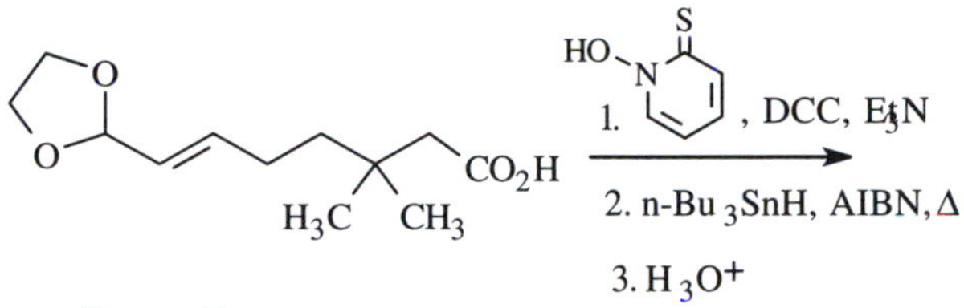

1. HO-N, S (with DCC, Et₃N structure) , DCC, Et₃N

2. n-Bu₃SnH, AIBN, Δ

3. H₃O⁺

(i)

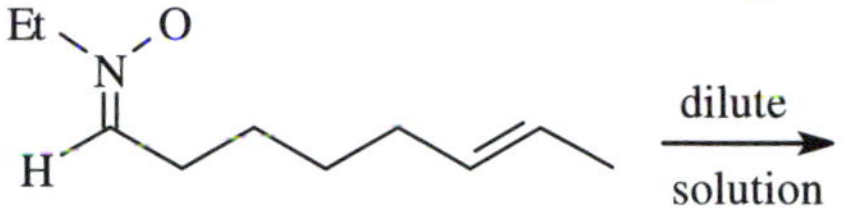

$\xrightarrow{\text{dilute solution}}$

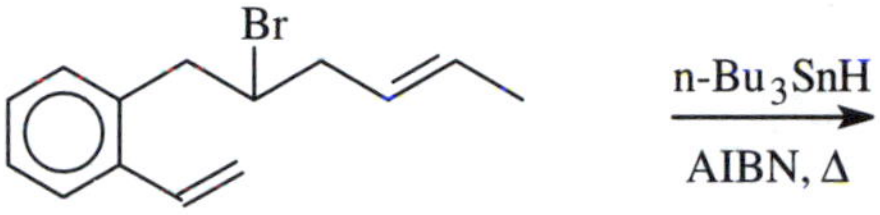

(j)

$\xrightarrow[\text{AIBN, } \Delta]{\text{n-Bu}_3\text{SnH}}$

PLANNING ORGANIC SYNTHESES

Retrosynthetic Analysis

The last several chapters have been concerned with learning how to manipulate functional groups, how to make carbon–carbon bonds, and how to put pieces of molecules together. Next, all of these ideas must be integrated into the general idea of synthetic planning. If a particular molecule is to be synthesized (the target), we must be able to plan the actual chemical route to be used in the preparation of the target. The task is to devise a strategy whereby a particular starting material is converted, by a series of steps (reactions), to the desired target.

One fact must be recognized at the outset: the target is the compound which must be produced from the starting material. This rather obvious statement of the problem is often overlooked by students, but it lies at the heart of synthetic planning. Targets are chosen in order to achieve some purpose. A particular target might be chosen as a drug candidate, as a potential insecticide, or as a motor oil additive. Whatever its purpose, a target will have particular structural features that must be produced by the synthetic sequence. Getting close won't do.

Starting materials can be chosen by a variety of criteria. For example, a particular starting material might have the same carbon skeleton as the target, or, it might be a by-product of a chemical plant and therefore readily available. It might be very cheap and easily obtained from a specialty chemical company, or it might contain a chiral center necessary in the target. Whatever the reason, a particular starting material imposes certain requirements on the synthetic route that must be taken in producing the target.

In order to transform the starting material into the target, reactions must be chosen which will accomplish the conversion efficiently and with the correct selectivity. Reactions which do not or cannot deliver the target are without merit for the synthesis of that particular target. This relatively simple idea is also often forgotten. It is common to force a particular reaction to give a product even though the oxidation level, reactivity, regioselectivity, and/or stereochemistry of the reaction is wrong for the target being considered.

Herein lies one of the difficulties of synthetic planning. One tends to learn organic reactions in the forward direction; that is, reactants A and B give product C. This type of information handling is a convergent process, in that a set of conditions is imposed which leads to a limited number of potential outcomes—in most cases a single product:

$$A + B \rightarrow C$$

Yet synthetic planning asks the opposite question: What reactants are necessary to give product C? This requires thinking backward from products to reactants. This type of problem solving is a divergent process, in that a great many potential reactants are possible and many possibilities must be explored before any one set of reactants is chosen as the solution.

Here is a very simple one-step synthesis to illustrate this difference. Given the following set of reactants, methyl cyclohexanol and chromic acid in acetone, it is very easy to write the product as 4-methylcyclohexanone because secondary alcohols are oxidized efficiently to ketones by chromic acid (this is a Jones oxidation).

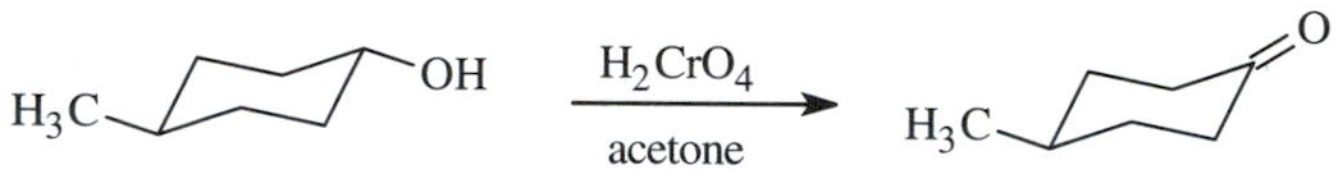

If 4-methylcyclohexanol is recognized as a secondary alcohol, it must give a ketone with Jones reagent.

In contrast, if asked to synthesize the target bicyclic ketone **K** (Figure 9.1), we must work backward. With an alcohol **A** of the same carbon skeleton, we could oxidize it to the ketone by a Jones oxidation. This step is indicated by a double arrow $\Longrightarrow$ and is called a retrosynthetic step. It shows how to work backward ("retro") from the target **K** to a given starting material **A**. (Of course, the actual synthesis would be carried out in the opposite direction—from the starting material **A** to the product **K**.) Alternatively, oxidative cleavage (O_3) of olefin **O** would furnish the target, as would the periodate cleavage of exocyclic diol **x-D**. Pinacol rearrangement of endocyclic diol **n-D** and hydrolysis of dibromide **B** would also furnish the target **K**. Thus,

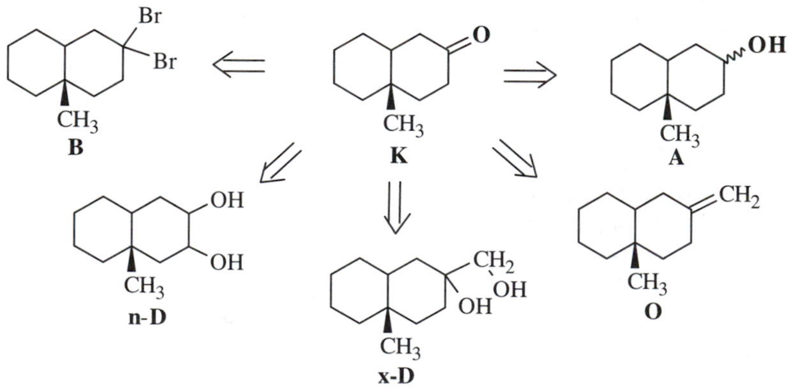

FIGURE 9.1

each retrosynthetic step in the backward analysis corresponds to a synthetic step to the target in the forward direction.

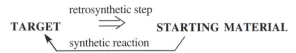

Although it is possible to generate a variety of potential solutions to the synthetic task at hand, the problem is not yet solved. The validity of each step must be checked. In order for each retrosynthetic step to be valid, there must be a reaction or reagent capable of effecting the transformation (with needed chemospecificity, regiospecificity, and stereospecificity) to give the target compound as the major product. In the previous example, oxidation of an alcohol **A** to a ketone **K** is a very facile step, and because there are many ways to do this conversion, it is a valid retrosynthetic step. Likewise, the olefin cleavage **O** or the diol cleavage **x-D** is also well known to be clean and would yield the target. Conversely, the pinacol rearrangement of **n-D** would probably not be regiospecific, and the hydrolysis of dibromide **B** is a very messy reaction which often gives complex mixtures rather than single products. These latter two retrosynthetic solutions are therefore invalid solutions to the synthetic problem.

Moreover, the following retrosynthetic analysis for the preparation of **K** is not valid either, because there is no good reaction to directly convert an olefin to a ketone. The olefin could be converted to an alcohol and then the alcohol could be oxidized to the target ketone, but this is actually two sequential reactions, not one.

![Retrosynthetic step showing ketone K derived from a methyl-substituted decalin with a double bond]

K

They must be written as shown:

![Retrosynthetic analysis showing ketone from alcohol OH from alkene, each bearing a CH3 group]

Here we see that the first retrosynthetic step, the preparation of a ketone from an alcohol, is valid, as discussed above. However, the second retrosynthetic step is not. Although the addition of water across a double bond is a straightforward, common reaction, its use in the present example requires that hydration take place regiospecifically, even though there is no significant control element present to ensure this. Thus, a mixture of alcohols is expected from the hydration reaction, rather than the single alcohol regioisomer needed for oxidation to the target ketone.

![Reaction showing alkene + H2O (any method) giving two regioisomeric alcohols with OH groups, CH3 substituents]

no regiochemical control

For each retrosynthetic step to be a valid solution, the reactants must give the appropriate product, with needed structural features and control in the forward direction (i.e., the direction taken in the actual synthesis). Thus, each retrosynthetic step must be checked in both the forward and backward directions to effectively plan workable synthetic routes to new molecules.

In order to carry out an effective retrosynthetic analysis, one must keep in mind certain basic features which must be dealt with during the synthesis. These include functional groups, oxidation levels, and stereochemistry. In addition, it is often necessary to construct the carbon skeleton so carbon–carbon bond-forming processes must be integrated into the sequence at the best juncture, all the while keeping in mind the above considerations.

Carbon Skeleton Synthesis

In general, it is most efficient to construct the carbon skeleton first and then adjust the functionality to give the target. Thus, in retrosynthetic analysis one must often move backward from the target, to compounds which contain functional groups important in carbon–carbon bond-making reactions. As a consequence, carbonyl groups play a very important role in retrosynthetic analysis. They are very useful sources of both electrophilic and nucleophilic carbon that can be used in making carbon–carbon bonds. Based on earlier discussions, a carbonyl group can be seen to influence the polarity of nearby carbons as seen in Figure 9.2.

By bond polarity and resonance, the carbonyl carbon and a carbon β to the carbonyl carbon can be utilized as electrophilic centers. The carbonyl group can react by direct nucleophilic addition to the carbonyl carbon and the β carbon by Michael addition to an α,β-unsaturated ketone. By resonance interaction, the α position in carbonyl compounds and γ positions in α,β-unsaturated carbonyl compounds can be converted to nucleophilic centers by proton removal. These "normal" polarities are used frequently in retrosynthetic planning, as points of disconnection which are used to establish potential bond-forming steps using carbonyl groups.

As an initial exercise, consider the synthesis of **C** from cyclohexyl bromide. One should note relevant facts about the target. First, it contains an ester; second, it has two more carbons than the starting material, so a two-carbon fragment will have to be attached by a new carbon–carbon bond.

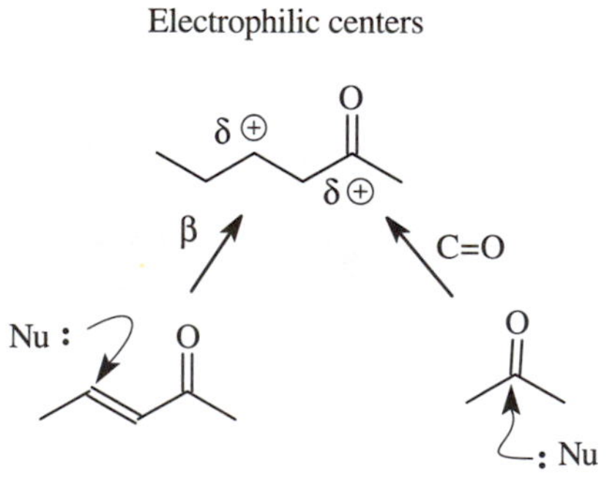

FIGURE 9.2

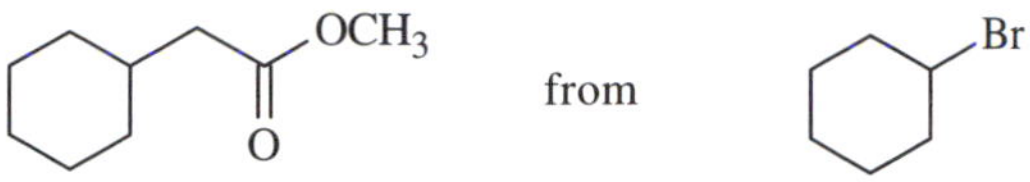

C from

Looking at the starting material, note that it contains an electrophilic carbon; thus, the needed carbon–carbon bond could be formed by reaction of a 2-carbon nucleophile with the electrophilic center of bromocyclohexane. Because the ester group needed in the product acidifies the α position, it could be used to make the nucleophilic carbon required for carbon–carbon bond formation. Retrosynthetically this can be written as

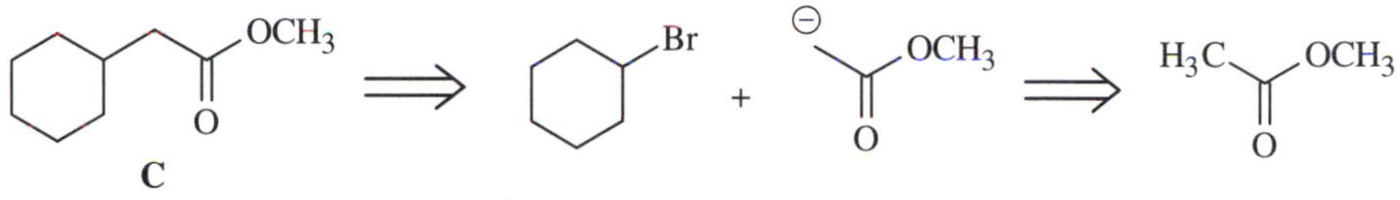

Therefore, the synthesis could be done in one step, by making the anion of methyl acetate and reacting it with bromocyclohexane. The polarities of the reaction partners match nicely, but alkylations of secondary bromides with enolates often give poor yields. The enolate is a strong base which promotes elimination in the secondary bromide, rather than giving the substitution product which is needed in the synthesis. Thus, elimination from cyclohexyl bromide to cyclohexene would be a major process if the reaction were attempted. While the retrosynthetic step seems reasonable, the synthetic step has known difficulties. It is important to work backward in the retrosynthetic analysis and then check each forward step for validity.

What is needed in the synthesis of **C** is a two-carbon nucleophile (or its equivalent) which is less basic than an enolate so that elimination is not competitive. If product **C** is recognized as an acetic acid derivative, the following analysis can be made:

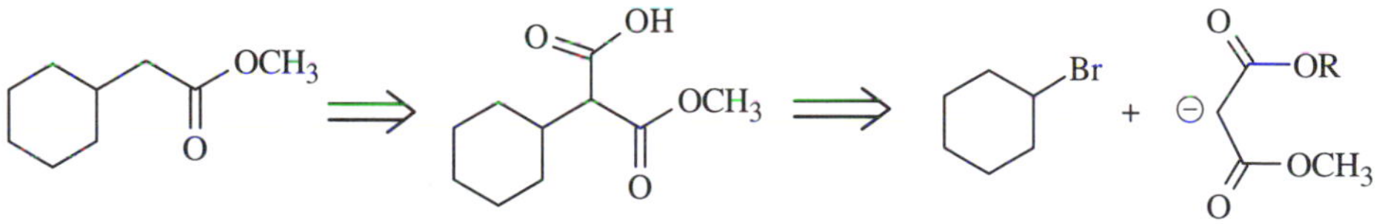

A malonate ion used as the carbon nucleophile is much less basic than a simple ester enolate and hence undergoes substitution readily, but does not promote elimination effectively, particularly in secondary systems.

Clearly, there will have to be some functional group adjustment in the synthesis, because a hydrolysis of the alkylated malonate must be carried out to give decarboxylation to the acetic ester derivative. Either an unsymmetric malonate must be used that can be differentially hydrolyzed, or both ester functions of the malonate could be hydrolyzed, and after decarboxylation the acid could be reesterified. The actual synthesis could be planned as shown in A or B:

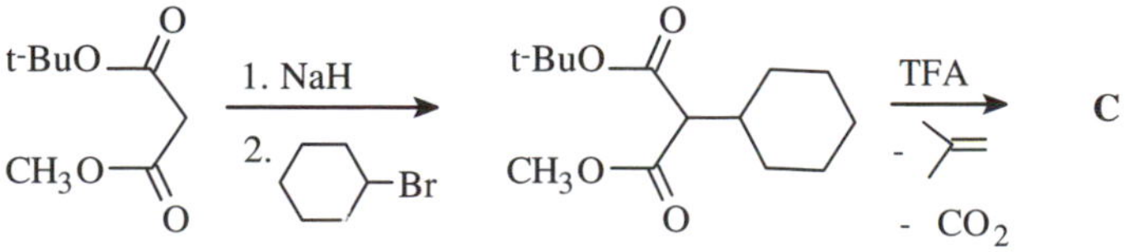

B.

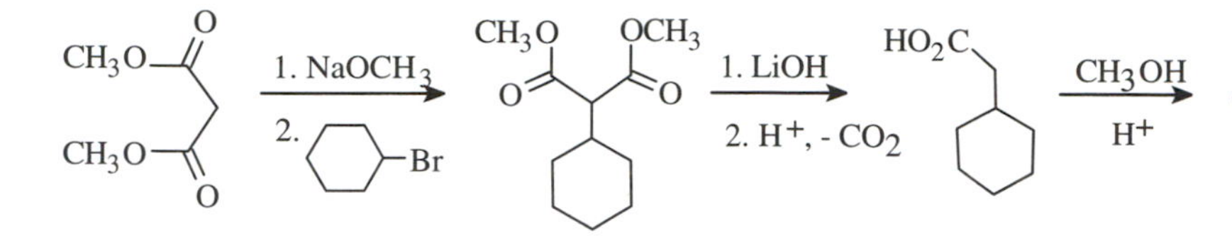

Synthesis B contains an extra step, but uses very cheap and available starting materials. Synthesis A is shorter and goes in high yield, but requires anhydrous conditions for the alkylation and a more expensive malonate starting material.

Next, consider the synthesis of **M** from "readily available" starting materials. The relevant facts about **M** are (a) it contains an aromatic ring, an acetate ester, and a vinyl group, (b) it has a straight chain attached to the ring, and (c) all the functional groups are isolated.

To begin the retrosynthetic analysis, note that the acetate ester is easily produced from the corresponding alcohol **A**. Therefore, conversion of **A** to **M** using acetic anhydride/pyridine could be used in the synthetic step. (Remember: for each retrosynthetic step, a reaction must be available to accomplish the synthetic step.)

The alcohol functional group in **A** is a natural point for bond disconnection to take place, since alcohols are the products of carbon nucleophiles and carbonyl groups. If one considers bond **a** in the retrosynthetic analysis, the next retrosynthetic step would be as shown:

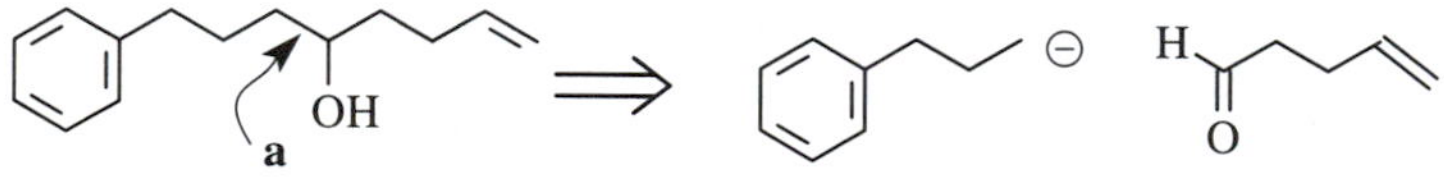

Since the anion **N** is a nonstabilized carbanion, an organometallic nucleophile, such as an organolithium or a Grignard reagent, could be prepared from the corresponding bromide:

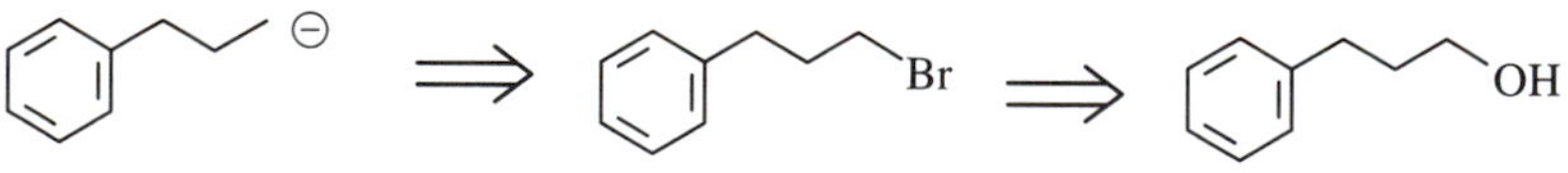

The bromide could be prepared from 3-phenyl-1-propanol ($59/kg). The unsaturated aldehyde **O** can be made by oxidation ($CrO_3 \cdot py$) of 4-penten-1-ol ($41.80/10 g). A cheaper way is to make ethyl 4-pentenoate from ethyl acetate ($15/gal) and allyl bromide ($19/100 g) and reduce it to the aldehyde **O** with DIBAH ($19/0.1 mol).

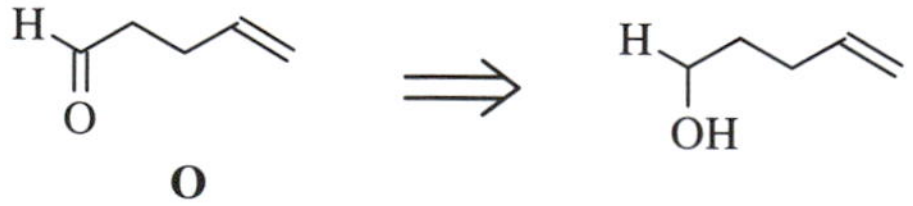

or

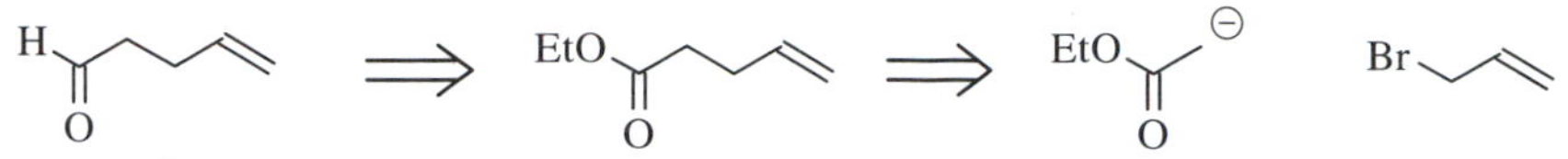

A synthesis of **M**, based on this retrosynthetic analysis would start with ethyl acetate, allyl bromide, and 3-phenyl-1-propanol (see Figure 9.3).

To develop an alternate synthesis, go back to **A** and consider disconnection at a different bond.

Recognizing that alcohol **A** could easily come from reduction of ketone **K** and considering the polarities possible, a great number of disconnections can be envisioned. Choosing bond *b* means that polarity (with respect to the carbonyl group) would be

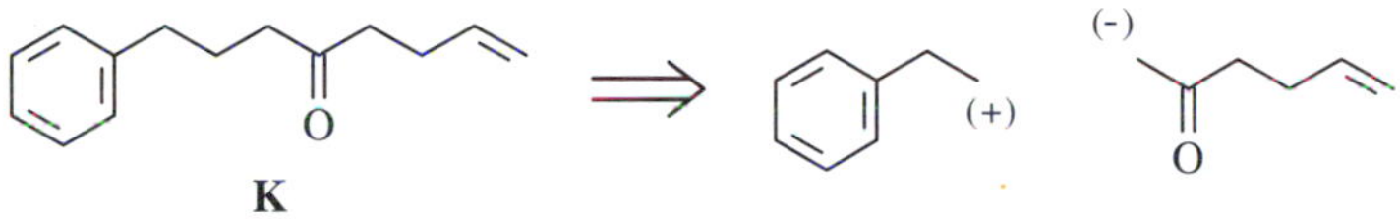

Thus, an enolate reacting with a carbon electrophile would be appropriate. A valid retrosynthetic step would be

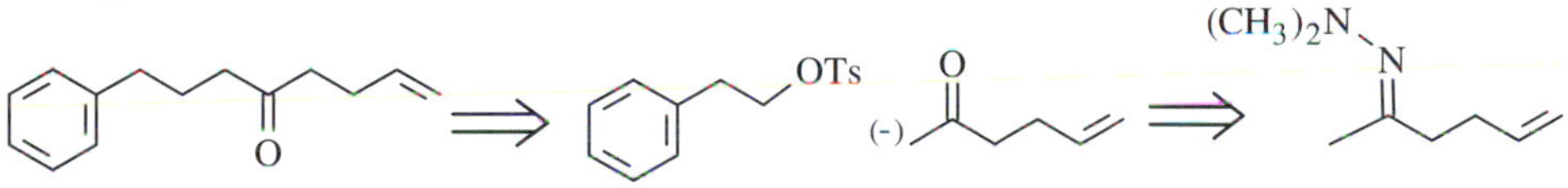

Because the tosylate is primary, substitution should be the major pathway (although, in this case, elimination could be problematic because of conjugation with the phenyl ring). Note, however, that the enolate needed is the kinetic enolate of 5-hexen-2-one. This poses a

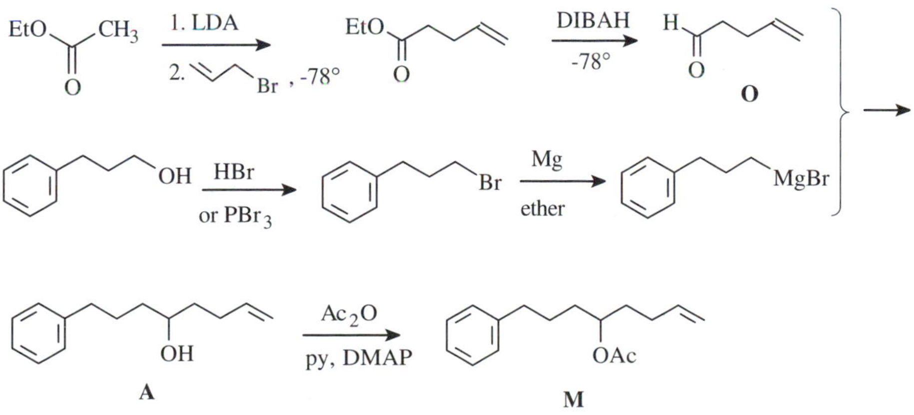

FIGURE 9.3

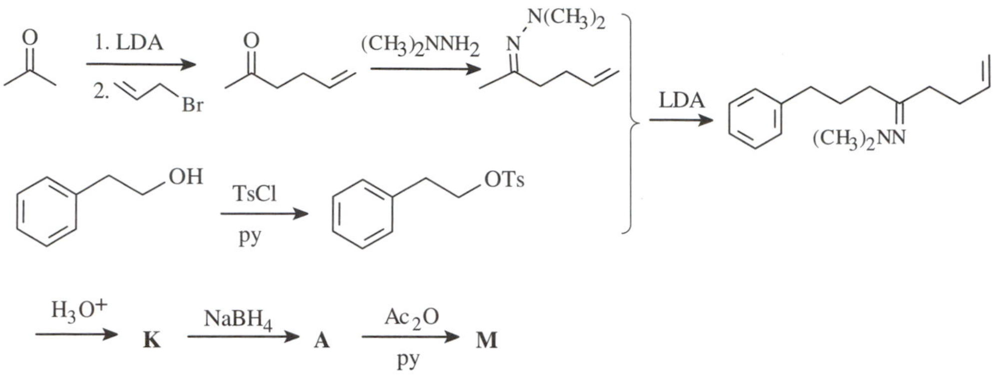

FIGURE 9.4

regiochemical control problem which can be solved by making the *N,N*-dimethylhydrazone of the ketone. The ketone 5-hexene-2-one is available ($46.20/25 g) or can be made by allylation of acetone.

Thus a synthesis based on this retrosynthetic analysis starts with β-phenylethanol ($35.60/1 kg), acetone, and allyl bromide. This route is comparable to the first, in both number of steps and cost. It differs in that regiochemical control of enolate formation is a crucial feature. Several other syntheses of **K** can be devised by other disconnections, suggested by the natural polarities engendered by the ketone group (see Figure 9.4).

Next, consider compound **R.** When the relevant facts are considered, **R** is seen to be merely an olefin with a saturated ring present. Because of the five-membered ring, and because the target has 12 carbon atoms, it is unlikely that compounds with the carbon skeleton of **R** will be available commercially. Hence, carbon–carbon bond-forming reactions will be needed to assemble the carbon skeleton.

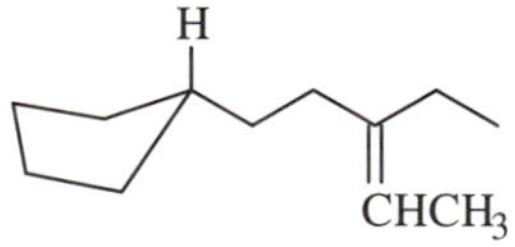

R

Moreover, the carbon–carbon double bond is a natural starting point for bond disconnection. A logical retrosynthetic step would be disconnection to a ketone because a Wittig reaction could be used to convert the ketone to the ethylidene product. (Note: dehydration of an alcohol to the olefin is not a viable synthetic step because dehydration would lead to a mixture of trisubstituted olefins.)

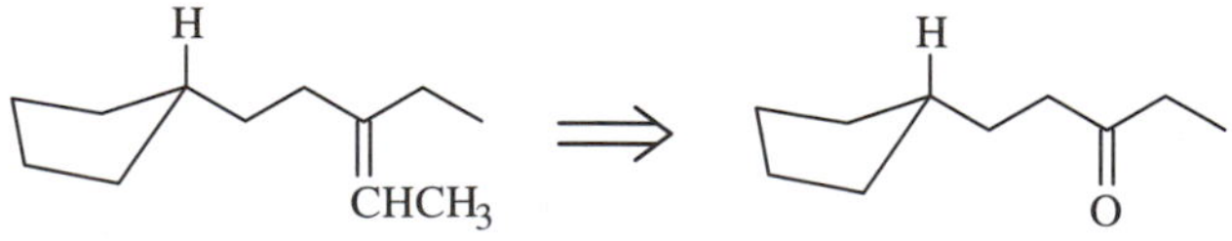

R

Once the ketone is recognized as a useful intermediate, normal polarities can be used to disconnect it retrosynthetically. A good disconnection could be as shown, where Michael addition to ethyl vinyl ketone by a cyclopentyl anion would give the needed ketone:

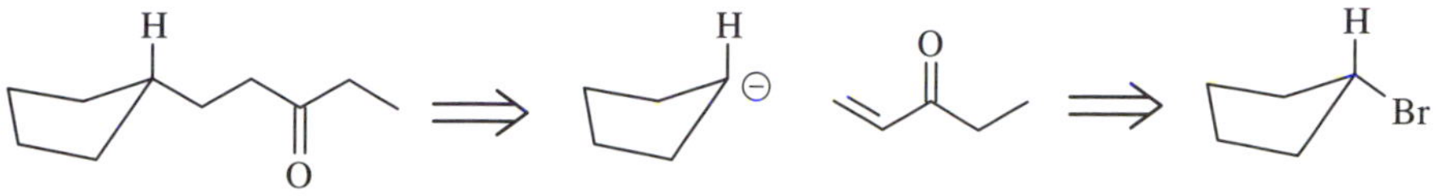

The cyclopentyl nucleophile (which should be an organocuprate, to ensure Michael addition) could be produced from cyclopentyl bromide. The synthetic sequence, consistent with the retrosynthetic analysis, turns out to be a rather simple synthesis of what is at first sight a more difficult molecule.

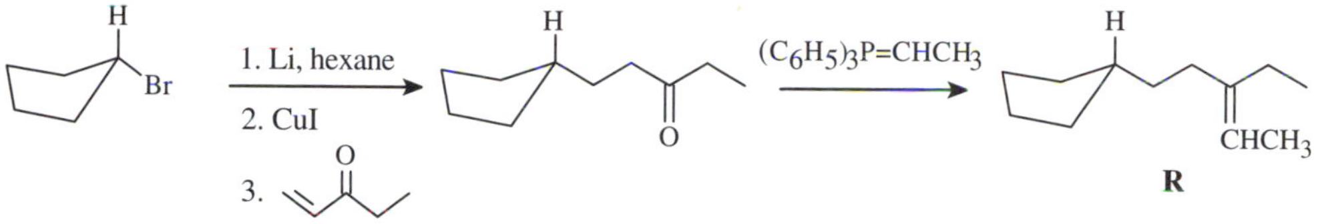

There are a variety of other ways to disconnect **R** in the retrosynthetic analysis. As long as each synthetic step is valid, and the target can be produced by the proposed synthetic route, each is a correct solution. There can be many correct synthetic solutions for a given target, and the "best" one may depend on factors other than those related strictly to the synthetic viability. Availability of starting materials, disposal of reaction by-products, number of steps, reagent sensitivity, expected yields, number of purifications, and the stereochemistry (among others) all contribute to the evaluation of a synthetic route.

Umpolung Synthons

Because of the polarities associated with carbonyl groups, some difunctional compounds are much easier to produce than others. For example, 1,3-dicarbonyl compounds and 1,5-dicarbonyl compounds are easy to produce using standard retrosynthetic steps, with normal polarities induced by the carbonyl group.

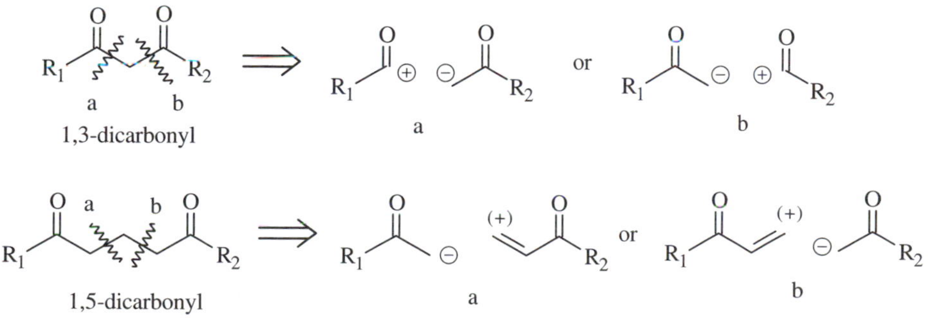

In contrast, 1,2-dicarbonyl compounds or 1,4-dicarbonyl compounds are more difficult to disconnect by valid retrosynthetic steps. Consider a 1,2-diketone. Disconnection of the bond between the carbonyl groups requires that one of the carbonyl groups have the normal electrophilic character. However, the other carbonyl carbon must have nucleophilic character (an acyl anion or its equivalent), which is not the normal polarity of a carbonyl group.

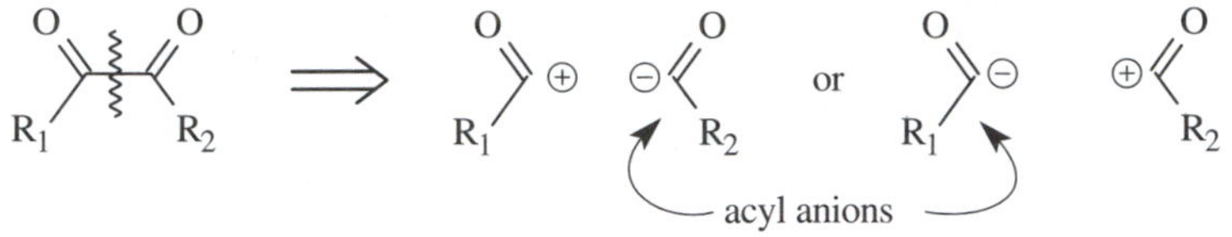

acyl anions

In the same way, disconnection of a 1,4-diketone requires either an acyl anion equivalent reacting with a normal β-carbonyl electrophile or a normal α-carbonyl nucleophile reacting with an abnormal α-carbonyl electrophile. These abnormal or reversed polarity reagents are said to have *umpolung* reactivity.

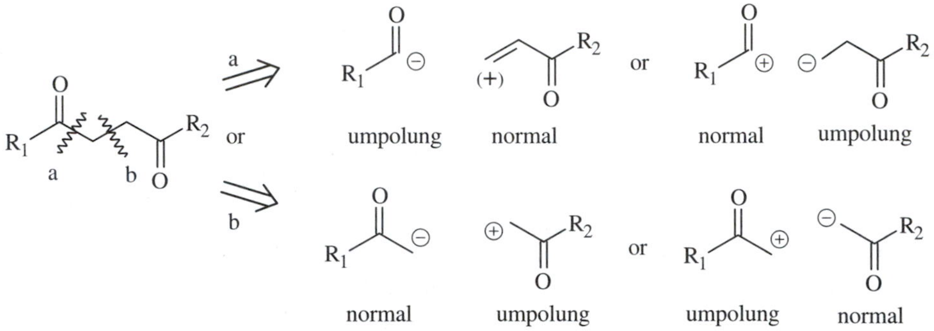

| umpolung | normal | | normal | umpolung |
| normal | umpolung | | umpolung | normal |

Consequently, there is a need for synthetic equivalents (synthons) of these reversed polarity (umpolung) reagents. The development of reagents with umpolung reactivity has been an important addition to modern synthetic methodology. Acyl anion equivalents, among the most common umpolung synthons, can be produced by many strategies. For instance, nitroalkanes can be used as nucleophiles and the nitro function can be cleaved to the carbonyl group. Thus, nitronates can be thought of as acyl anion equivalents.

$pK_a = 9\text{-}10$

acyl anion equivalent

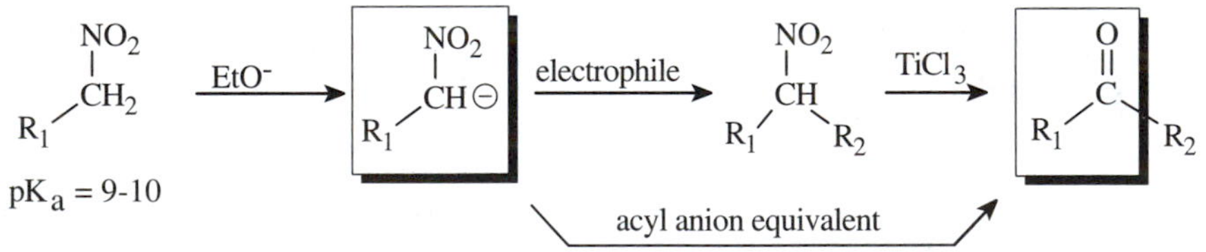

Likewise, 1,3-dithianes can be deprotonated by alkyl lithium bases, and the resulting anions are strong nucleophiles. The dithane group can be hydrolyzed back to the carbonyl group. Thus, the dithiane serves as a synthon for the acyl anion.

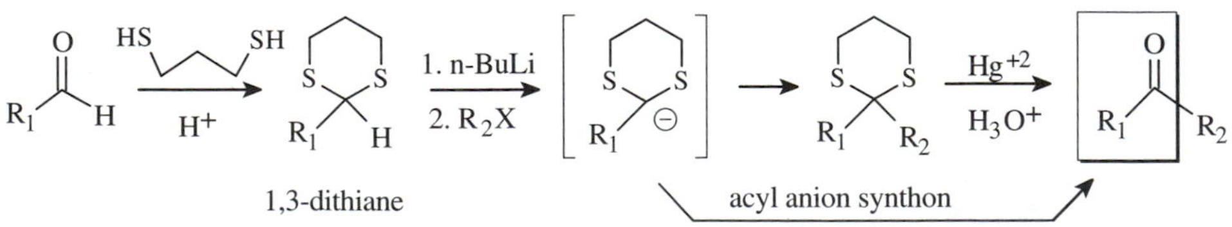

1,3-dithiane

acyl anion synthon

Cyanohydrin derivatives have also been widely used as acyl anion synthons. They are prepared from carbonyl compounds by addition of hydrogen cyanide. A very useful variant is to use trimethylsilyl cyanide to produce a trimethylsilyloxy cyanide. The cyano group acidifies the α position (p$K_a \approx 25$) and the α proton can be removed by a strong base. Alkylation of the anion and unmasking of the hydroxy group causes elimination of cyanide and reformation of the carbonyl group.

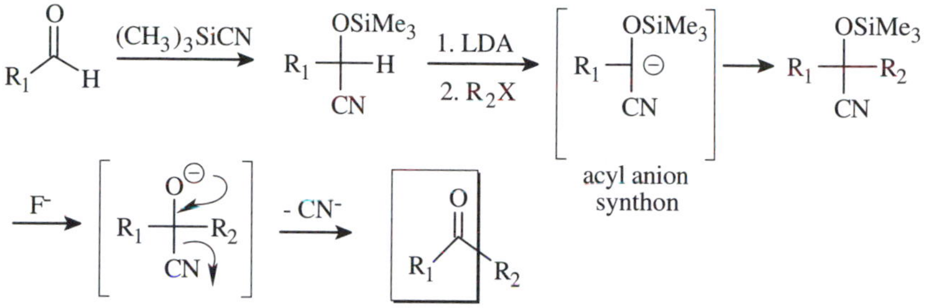

These are only three of many ways that have been reported for the formation of acyl anion equivalents, which are among the most common umpolung synthons to be found in the literature. All are prepared by a similar strategy, in that they contain functional groups which can sustain a negative charge on an adjacent carbon *and* can be converted back to a carbonyl group.

Another common umpolung synthon is a homoenolate. Normally, the β position of a carbonyl compound is an electrophilic center (by Michael addition to an α,β-unsaturated carbonyl derivative). In order to make it a nucleophilic center, an organometallic is needed, since it is unactivated and non-conjugated. A common way to do this is to use a β-bromo acetal (see Figure 9.5). The bromine substituent can be metallated to give a carbanion equivalent β to the acetal group. Since the acetal is easily hydrolyzed to the ketone, it is a synthon for a β-carbonyl anion—an umpolung reagent. It is thus important to recognize normal and reversed polarities when doing retrosynthetic analysis of a target, in order to use umpolung synthons when they are needed.

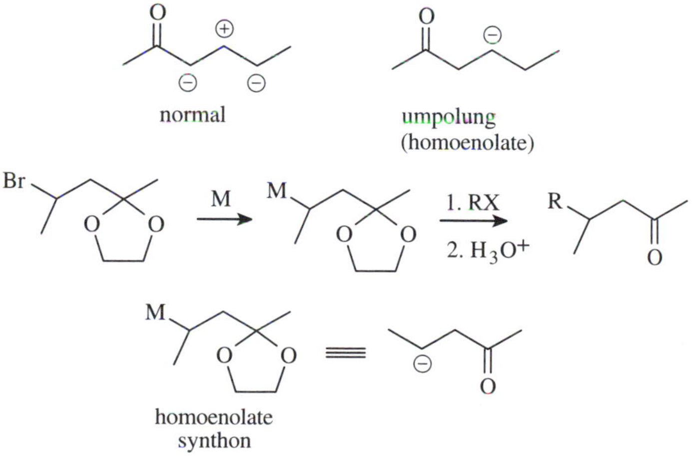

FIGURE 9.5

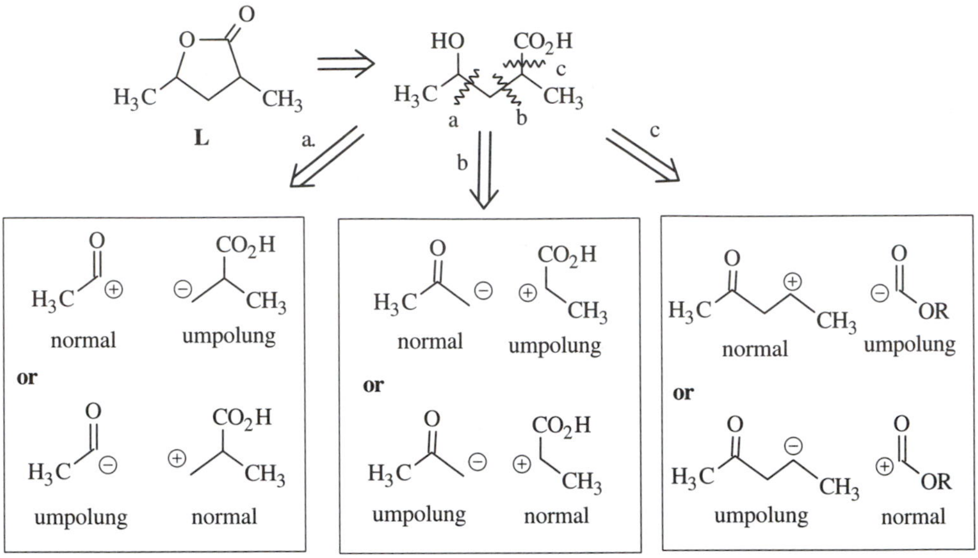

FIGURE 9.6

For the lactone target **L** (see Figure 9.6), cleavage of the lactone ring gives a γ-hydroxy acid. This can be disconnected at any one of the three intervening bonds between the hydroxyl group and the carbonyl group (*a, b, c*). If each of these is considered independently, it is easily seen that none of the disconnections has normal carbonyl polarities. (The same conclusion could be reached by merely noting that the hydroxy group and the carboxyl group have a 1,4-relationship. It was noted above that normal carbonyl-based polarities are not suited to the formation of 1,4-difunctional systems.) Thus, one has to either use an umpolung synthon or go to functional groups other than carbonyl groups to guide the reactivity. For example, one could use an epoxide electrophile rather than a carbonyl electrophile for the bond-forming reaction:

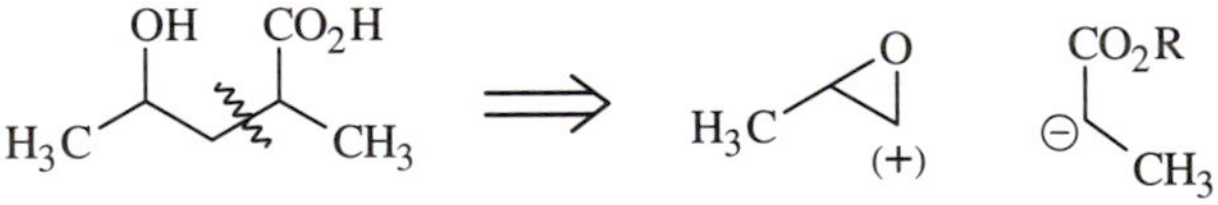

Acetylide Nucleophiles

Although the carbonyl group is a very common starting point for bond disconnections in retrosynthetic analysis, an olefinic or acetylenic unit is also a useful reference point in many instances. This is true because a terminal acetylene can be used as an effective nucleophile to install the triple bond into molecules, and it can be reduced stereospecifically to either the *cis* or *trans* olefin. Thus, for the *cis* olefin target **T**, the following retrosynthetic analysis leads to an efficient synthetic pathway which uses a nucleophilic displacement by an acetylide anion as a key carbon–carbon bond-forming step:

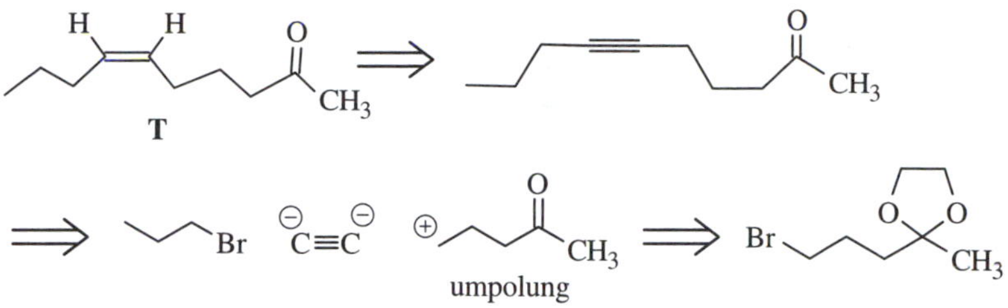

The synthetic sequence would be

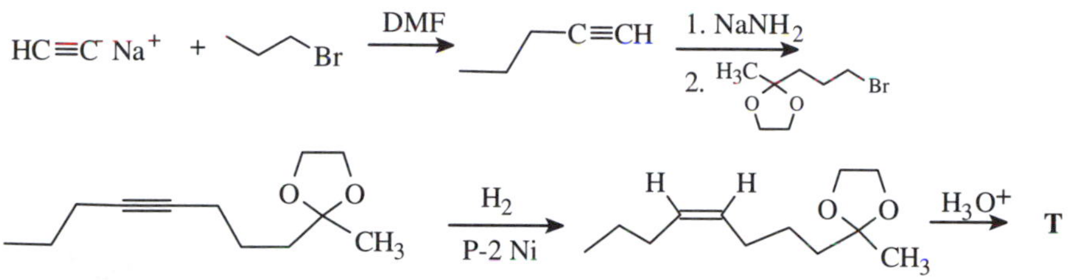

Ring Construction

The use of carbonyl groups to set the polarity of bond disconnections in retrosynthetic analysis is also useful for the construction of rings. If a carbon electrophile and a carbon nucleophile are connected by a carbon chain, they can react with each other to form a carbon–carbon bond. This is an absolutely normal type of carbon–carbon bond-forming process, but the fact that the carbon nucleophile and carbon electrophile are connected by a chain means the new carbon–carbon bond closes up the ends of the chain, forming a ring.

For example, the Claisen reaction (presented earlier) is a reaction of an ester enolate with an ester to produce a β-ketoester.

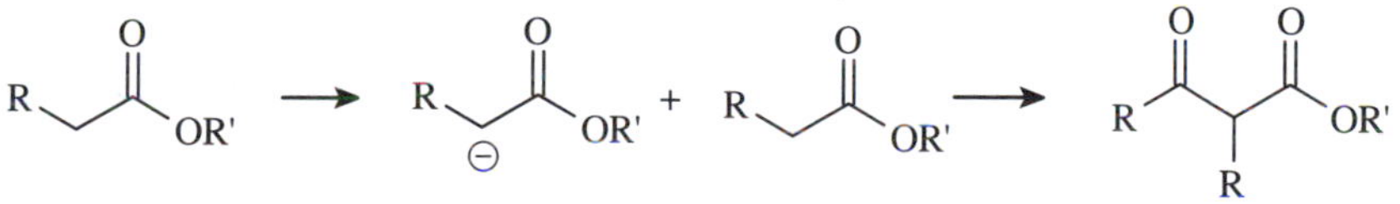

If both ester groups are in the same molecule and are connected by a chain, a Claisen type reaction between the α position of one ester and the carbonyl group of the other gives a new carbon–carbon bond and closes the ring. (This reaction is actually called the Dieckmann condensation, but it is nothing more than an intramolecular Claisen reaction.)

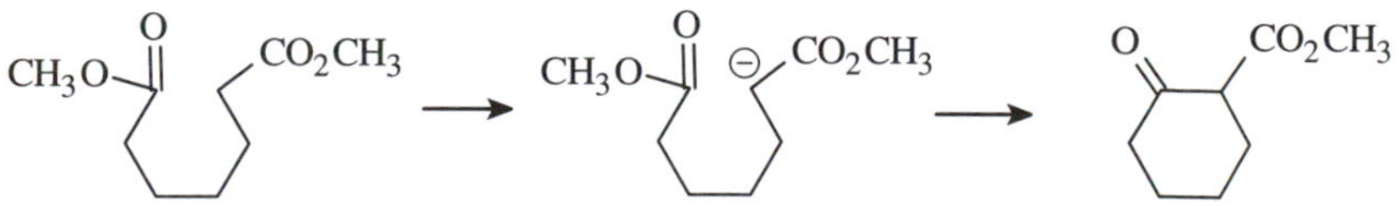

acyclic diester   cyclic β-ketoester

Ring-forming reactions are very important in retrosynthetic analysis, because many interesting targets are cyclic compounds and rings must often be installed, rather than being present in the starting materials. From a retrosynthetic point of view, there is really no difference between ring-forming reactions and other carbon–carbon bond-forming reactions. One looks for the same polarities and functional group features as in acyclic systems.

The only difference is that some rings are more easily formed than others. The rule of thumb is that rings of three, five, and six, members are routinely formed, whereas rings of four or more than six members are formed with difficulty. For example, reaction of diethyl malonate with 1,2-dibromoethane and two equivalents of base gives diethyl cyclopropane-1,1-dicarboxylate in high yield. Ring formation occurs by a double displacement sequence:

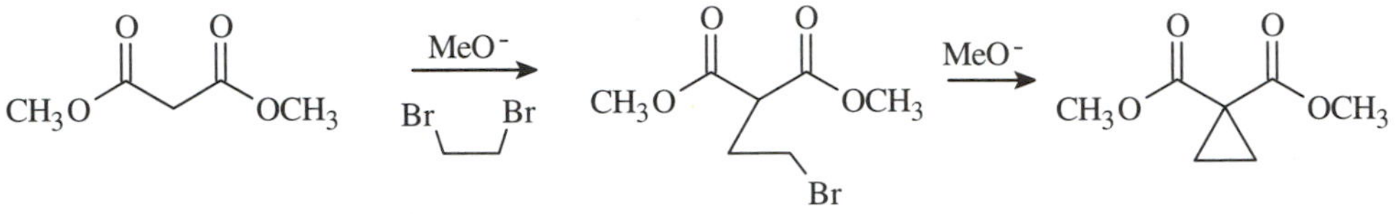

Likewise, reaction with 1,3-dibromopropane, 1,4-dibromobutane, or 1,5-dibromopentane gives the corresponding cyclobutyl-, cyclopentyl-, and cyclohexyl-1,1-dicarboxylates.

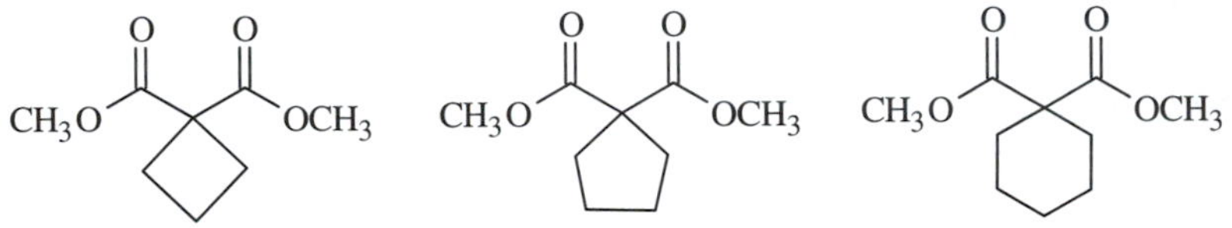

The yields of these reactions are not the same, however, and reactions which produce three-, five-, and six-membered rings are generally more effective. Use of 1,6- dibromohexane fails to give the cycloheptyl product.

Ring closure requires that a reactive center at one end of the chain encounter a reactive center at the other end *of the same chain* in the bond-forming process. The alternative is for the reactive center on one chain to react with a reactive center of a different chain. The first case produces a ring; the second case produces a polymer. An analogy is a chain with complementary hooks at each end, representing electrophilic and nucleophilic carbons at the ends of the chain. If two hooks on the same chain link up, a ring is formed, whereas if hooks from different chains link up, a larger molecule is formed, with hooks remaining on each end. These can link up further to form progressively larger molecules, as seen in Figure 9.7.

For short chains (three- to six-membered rings upon ring closure), there is a higher probability that one end of the chain will encounter the other end of the same chain and react *intra*molecularly, before it will encounter the end of another chain and react *inter*molecularly. Thus, ring closure is normally favored over oligomerization, for smaller rings of three to six members. On the other hand, as the chains become longer, it becomes less likely that the end of a chain will encounter the other end of the same chain. More likely, it will encounter and react with the end of another chain. The breakpoint is between six-membered rings, which are formed readily, and seven-membered rings, which are not easily formed. While this reasoning is a great simplification, it suffices to provide a good working model to predict the success for ring-forming reactions.

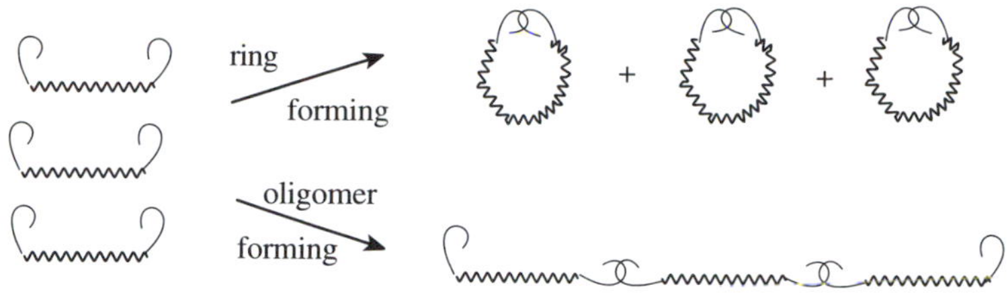

FIGURE 9.7

We can use this model in retrosynthetic analysis quite successfully. Suppose one were asked to produce cyclohexanone **C** from acyclic starting materials:

$$\text{acyclic starting materials}$$

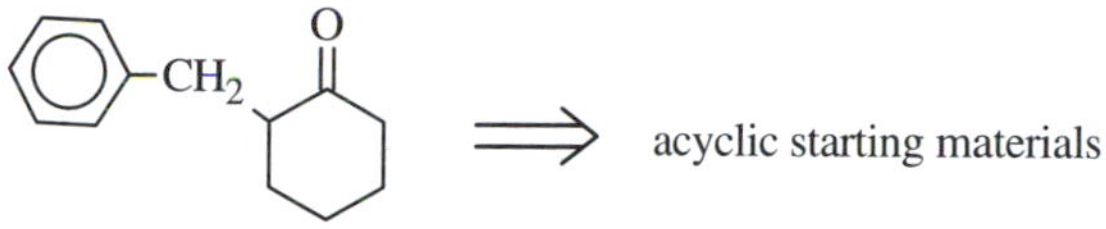

In this monofunctional compound, the ketone could serve as an electrophilic center in a cyclization step. Disconnection at the indicated bond leads to the polarity shown. However, it is immediately obvious that the carbon nucleophile occurs at an unactivated position and there is no good way to produce it there without a control element at that position.

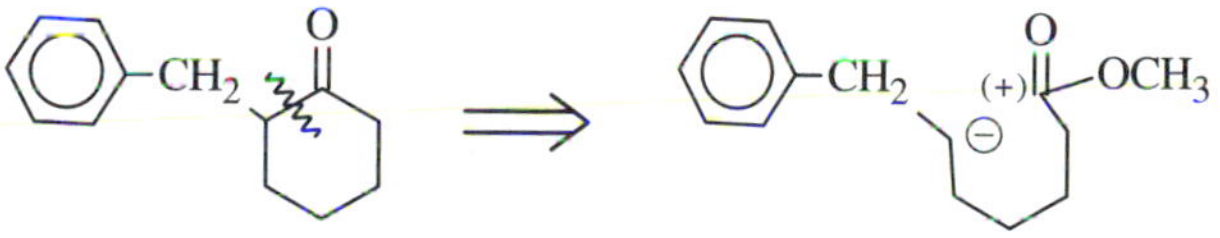

Use of an ester group could activate this position towards anion formation, as shown:

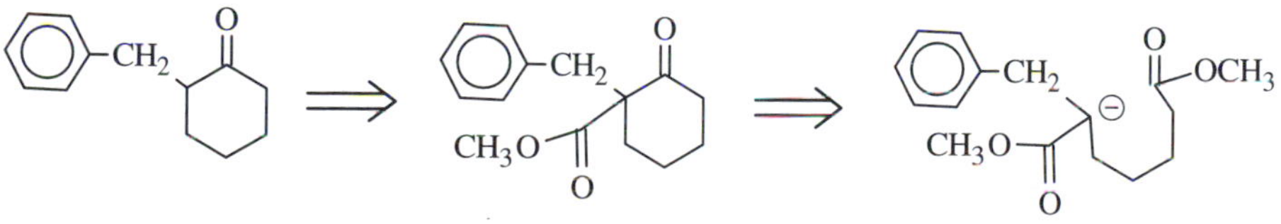

Now all is well in terms of polarity and we recognize this as a Deickmann reaction, followed by hydrolysis and decarboxylation of the β ketoester product. Proceeding backward, we write

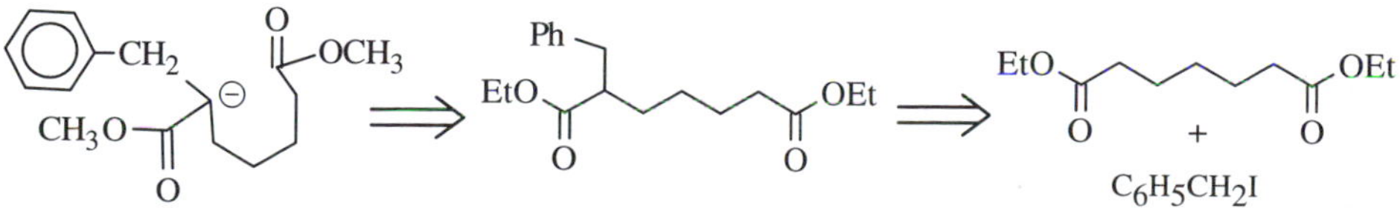

Alkylation of diethyl suberate with benzyl iodide would produce the α-benzylated product which would cyclize in the presence of base. This particular target does not present a

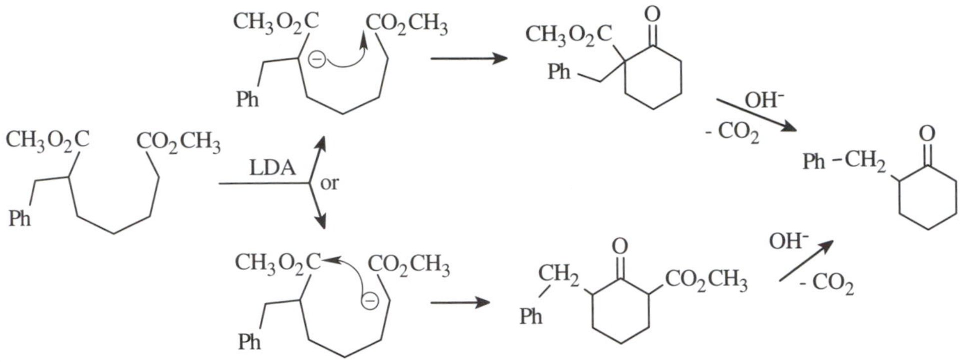

FIGURE 9.8

regiochemical difficulty. Base could pull off either α proton, and two different enolates would be produced. However, both enolates cyclize to give the same product after decarboxylation (see Figure 9.8).

    If molecular symmetry were included in the cyclization precursor, we would not have to worry about regiochemistry in the ring closure. Noting that the product of ring formation is a β-ketoester, which itself is a good carbon nucleophile, an alternate retrosynthesis (which actually is much better) is the following, in which the benzyl group is added after the ring is formed (in the forward synthesis):

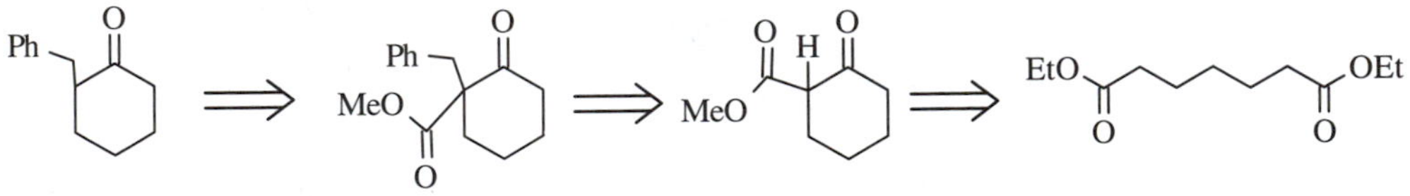

Robinson Annulation

Another example of a very common ring-forming sequence is the Robinson annulation. This sequence allows a six-membered ring to be appended to an existing carbonyl group:

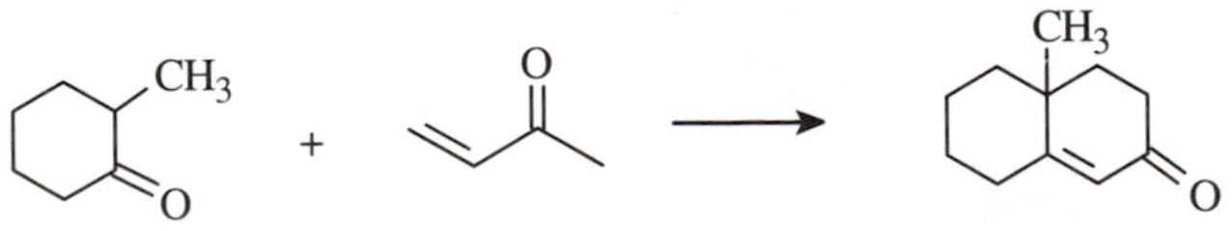

The strategy of the sequence is a Michael addition to an α,β-unsaturated ketone, followed by an intramolecular aldol reaction. Treatment of a ketone enolate with a Michael acceptor gives a diketone intermediate which is poised to produce a six-membered ring, if an enolate is produced. This intramolecularly adds to the carbonyl group.

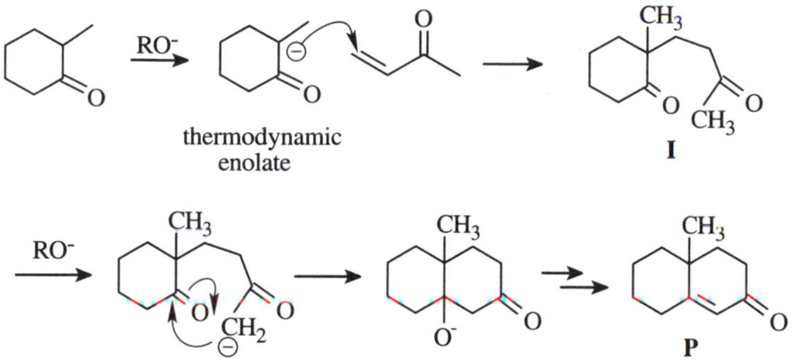

This process nicely accounts for formation of product, but considering intermediate **I**, there are several different α protons that could be removed by base: H_a, H_b, and H_c (see Figure 9.9). Furthermore, the acidities of these various α protons should be comparable, so all should be removed to similar extents under the reaction conditions. Sequentially removing each proton and writing the product from an intramolecular carbonyl addition gives the products shown in Figure 9.9. The facts are that only **P** is produced to any extent. This is due to the preference of six-membered ring formation over the formation of the more strained four-membered ring product or the more strained bridged product. Thus, the enolate formed by the removal of H_a closes faster than the enolates formed by removal of either H_b or H_c. Furthermore, since the aldol reaction is reversible, if any of these higher energy products were formed, they could open again under the reaction conditions. The exclusive formation of **P** is an example of kinetic as well as thermodynamic control, as the more stable product is formed fastest.

For the purposes of retrosynthetic analysis, a six-membered ring in a target can be related to a Robinson annulation of an existing ketone with an α,β-unsaturated ketone. Normally α,β-unsaturated methyl ketones are used to facilitate the ring closure, but this is not an

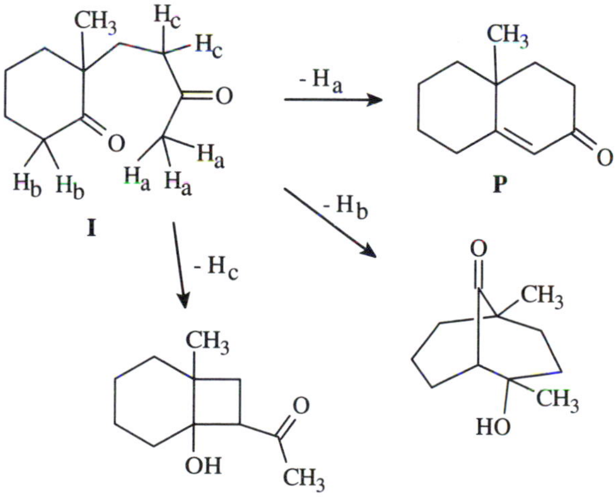

FIGURE 9.9

absolute requirement. The target steroid **S** could potentially be constructed by a series of Robinson annulations, as shown. The last retrosynthetic step (the first synthetic step) could be problematic, as a mixture of regioisomers would be formed.

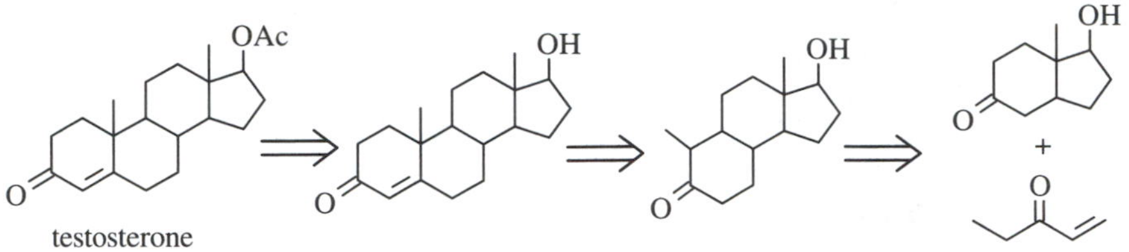

testosterone
acetate

Furthermore, the bicyclic starting material could also be constructed by a Robinson annulation on a cyclopentanedione. In this case, the final functionality must be achieved by selective reductions of the olefin and ketone functions, at appropriate stages in the synthesis.

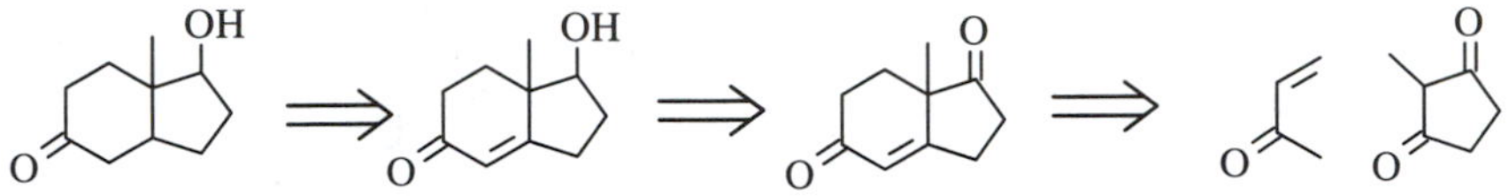

A great many complex multicyclic targets have been synthesized by the Robinson annulation, attesting to its generality and versatility.

Diels–Alder Reaction

Another approach for the construction of rings is to use cycloadditions reactions which start with two acyclic compounds and produce cyclic products. There are many of these processes, but the most used and most useful is the Diels–Alder reaction. In order for the Diels–Alder reaction to be considered as a viable retrosynthetic step, its requirements must be accommodated in the reactants if the reaction is to be successful. First, the ends of the diene must contact the ends of the olefin. Therefore, the diene must be able to adopt a cisoid conformation. Dienes whose structure enforces the s-*cis* conformation are particularly reactive.

Second, there is an electronic bias which favors electron-donating substituents on the diene and electron-withdrawing groups on the olefin (the dienophile). When these polarities match, Diels–Alder cycloaddition proceeds readily.

Third, the stereochemical outcome of the Diels–Alder reaction results from *endo* addition, in which polar substituents on the dienophile are on the underside of the diene system. This sets the stereochemical relationship of the substituents on the diene to the substituents on the dienophile. For example, in the reaction of maleic anhydride with cyclopentadiene, the *endo* transition state dictates that the anhydride group will be on the side opposite the carbon bridge.

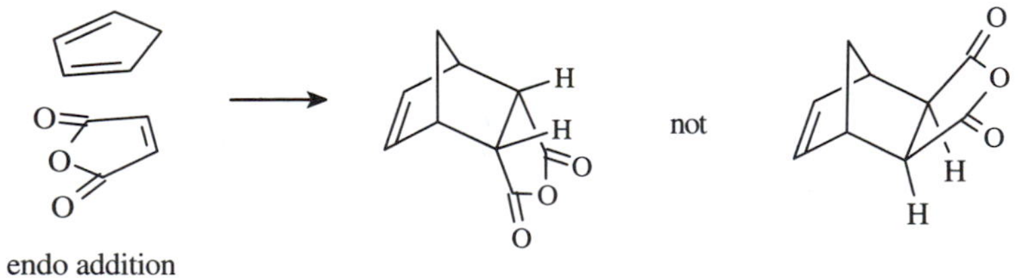

endo addition   not

With these considerations in mind, the Diels–Alder reaction can be used to create rings in many situations that would be difficult to accomplish by ring-closing approaches. Consider product **P**. This product can be made by a Diels–Alder reaction between diene **D** and acrylate **A**.

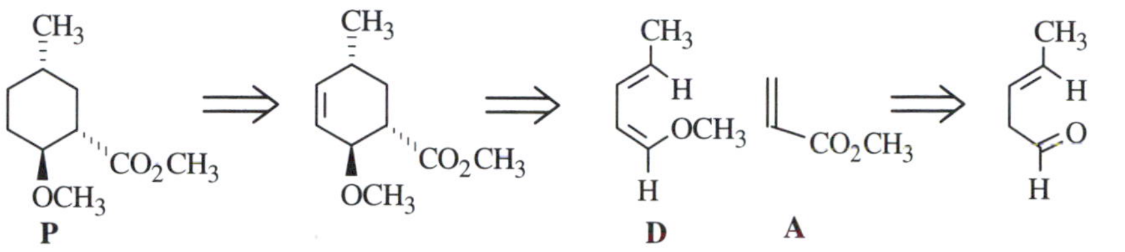

Because the diene is acyclic, it can achieve the required s-*cis* conformation by rotation. The polarity is correct, because the diene is electron-rich and the dienophile is electron-poor. The stereochemistry is fine because the *endo*-transition state gives the correct stereochemical relationship of the groups around the cyclohexyl ring. The stereochemistry can be seen more clearly by a drawing of the transition state.

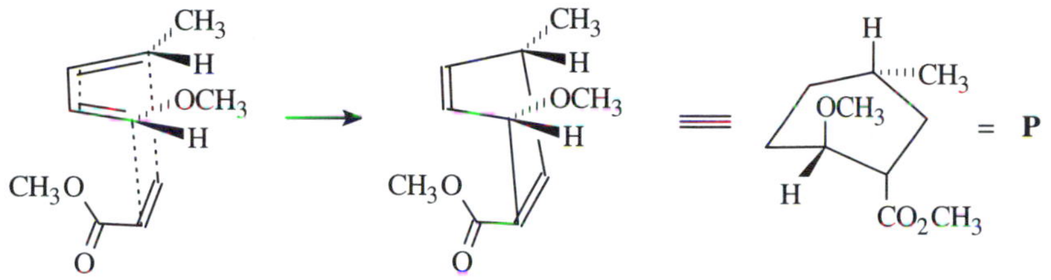

These are but a few examples of how retrosynthetic analysis can be used to develop one or more synthetic routes to a target. Developing synthetic strategies is one of the most creative activities that organic chemists perform. It requires that many different inputs and conditions be cohesively merged into a single thematic development that contains elements of texture and beauty, proportion and balance, and risk and reward. The process is every bit as creative as painting, sculpting, or writing the great American novel!

Bibliography

For a good overview of the process of synthetic analysis see M. A. Fox and J. K. Whitesell, *Organic Chemistry*, Jones & Bartlett, Boston, 1994, Chapter 15.

For an alternative introductory discussion, see A. Streitwieser, C. H. Heathcock, and E. M. Kosower, *Introduction to Organic Chemistry*, 4th ed., Macmillan, New York, 1992, Chapter 16.

An excellent approach can be found in S. Warren, *Designing Organic Syntheses, A Programmed Introduction to the Synthon Approach*, Wiley, New York, 1978.

For an alternate discussion, see M. B. Smith, *Organic Synthesis*, McGraw-Hill, New York, 1994, pp. 1–86. There is also a nice discussion of synthetic strategies in Chapter 10, pp. 980–1098.

Far and away, the best discussion of retrosynthetic analysis is found in E. J. Corey and X.-M. Cheng, *The Logic of Chemical Synthesis*, Wiley Interscience, New York, 1989, Part One, pp. 1–98.

Problems

1. The following group of syntheses are, fundamentally, functional group manipulations. Give a retrosynthetic pathway from the target on the left to the starting material on the right. Then provide a synthetic pathway, with proper reagents and conditions for each step.

(a)   from

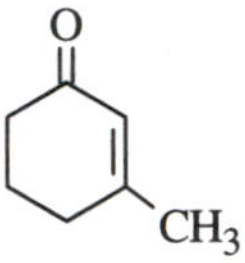

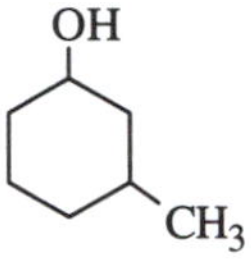

(b) 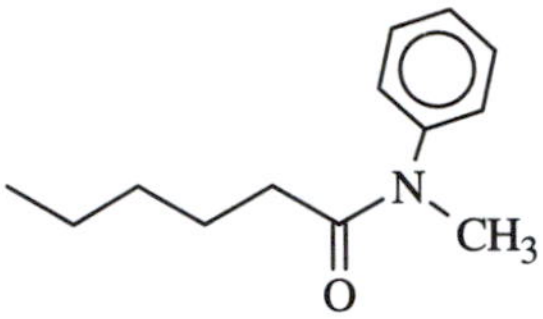   from

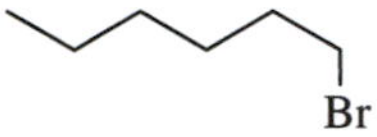

(c) 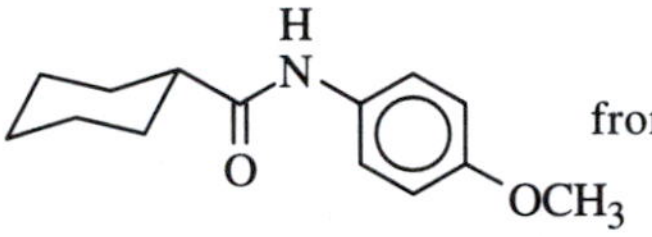   from

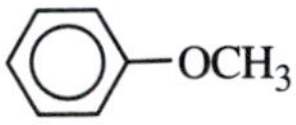

(d) 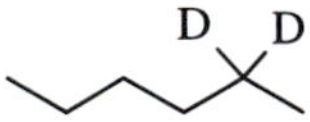   from

(e) 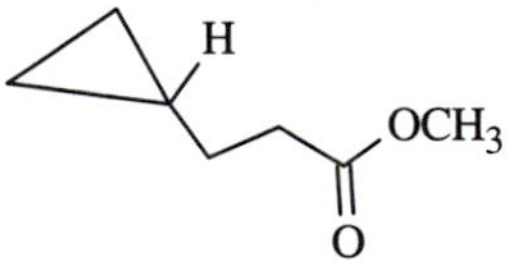   from

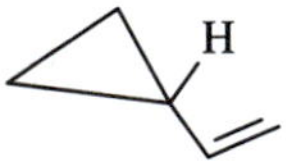

(f) 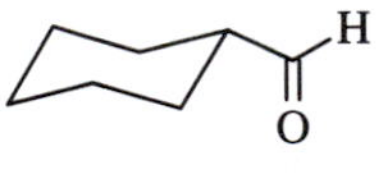   from

(g) 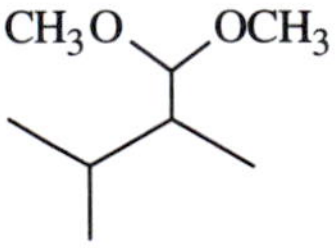   from

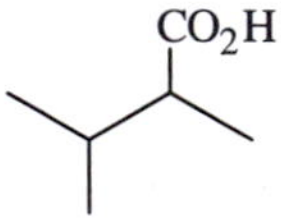

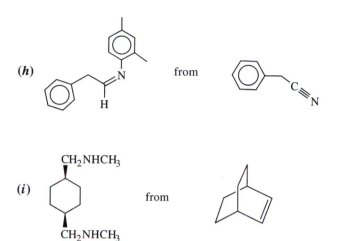

(h) from

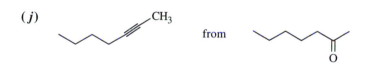

(i) from

(j) from

2. Show a retrosynthetic pathway to the following targets which involves the formation of the bond or bonds indicated by an arrow.

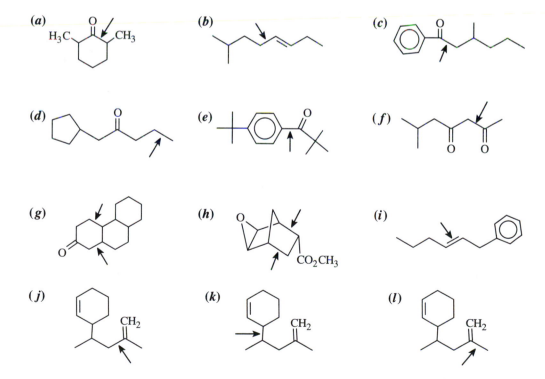

(*m*)

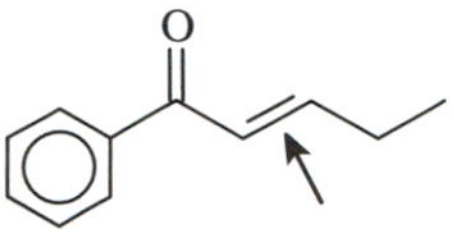

(*n*)

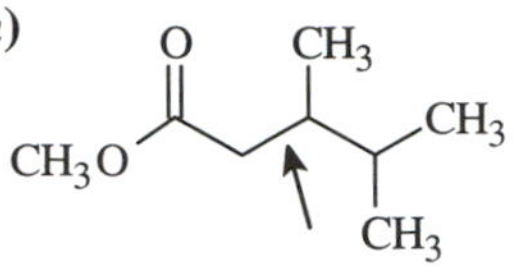

(*o*)

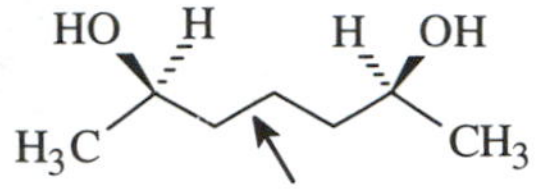

(*p*)

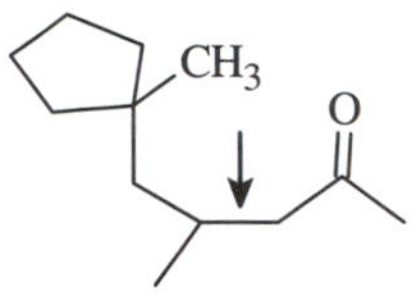

(*q*)

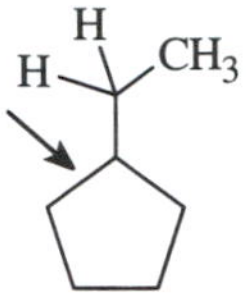

(*r*)

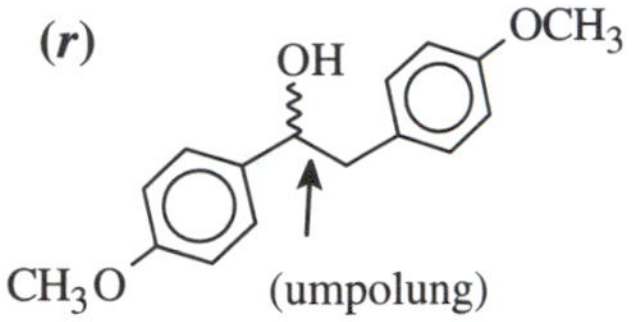

(umpolung)

(*s*)

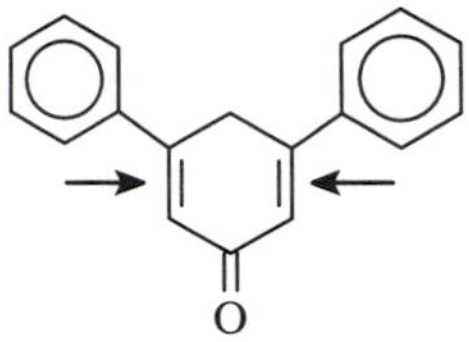

(*t*)

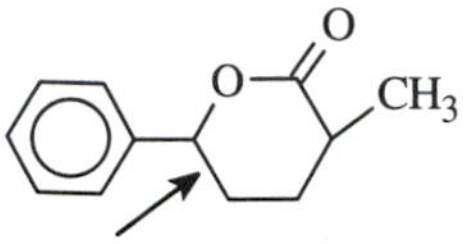

(*u*)

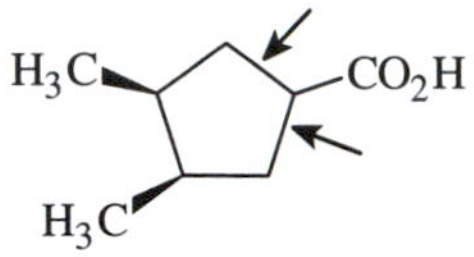

3. Provide retrosynthetic pathways for the following targets, from "simple, readily available" starting materials.

(*a*)

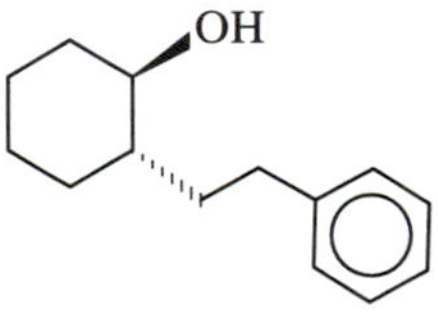

(*b*)

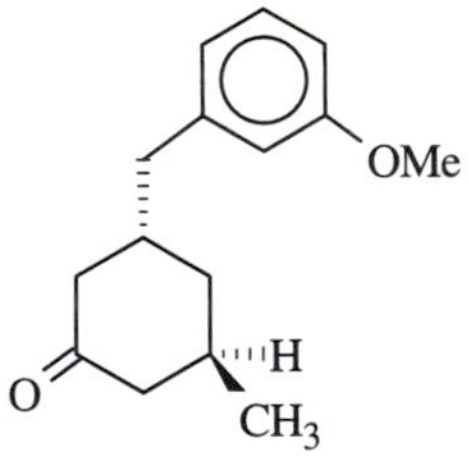

(*c*)

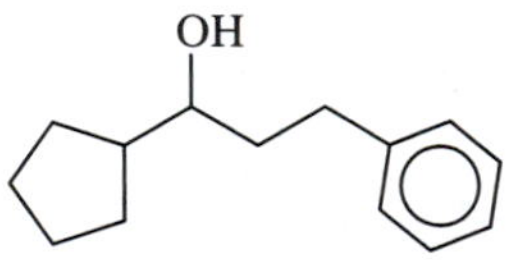

(*d*)

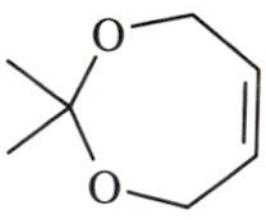

(*e*)

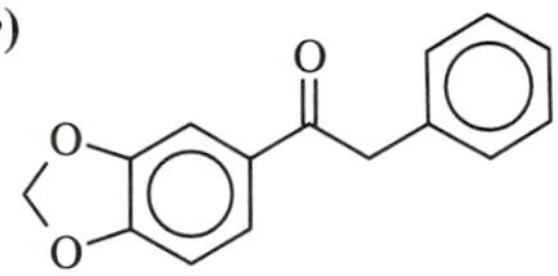

(*f*)

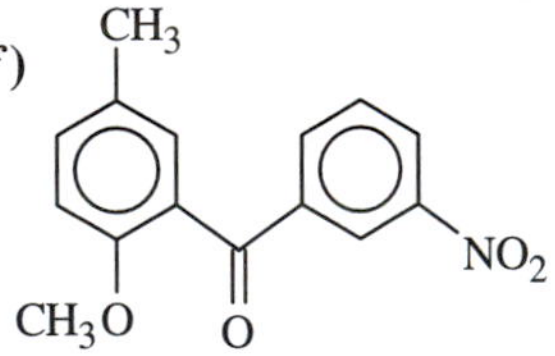

(*g*)

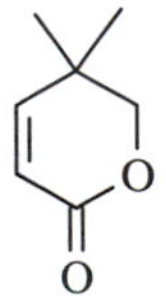

(*h*)

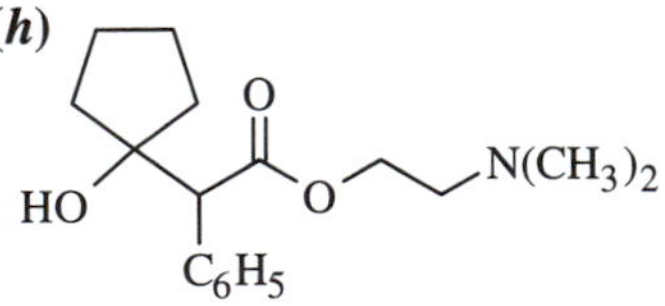

(*i*)

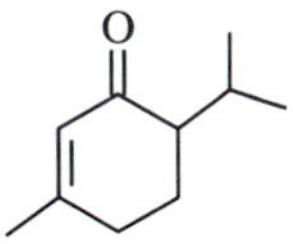

(*j*)

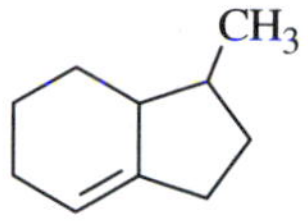
CH3

(free radical)

(*k*)

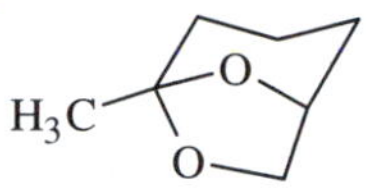
H3C
O
O

(*l*)

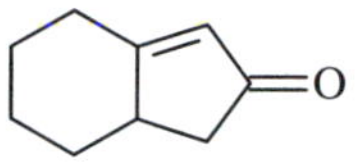
O

(*m*)

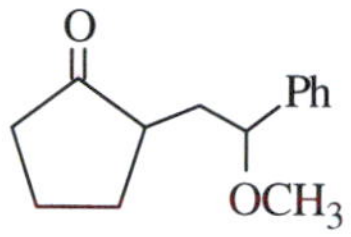
O
Ph
OCH3

(*n*)

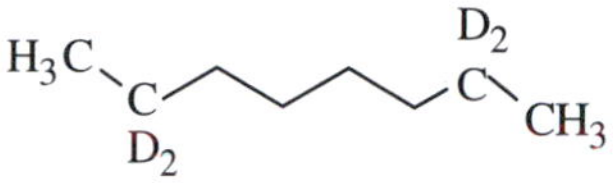
H3C
C
D2
D2
C
CH3

(*o*)

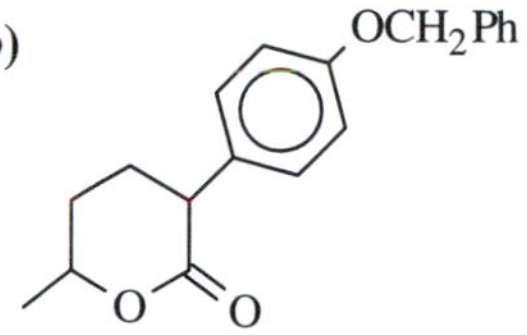
OCH2Ph
O
O

(*p*)

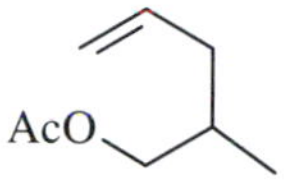
AcO

(*q*)

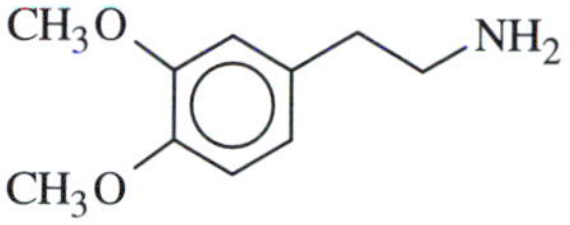
CH3O
NH2
CH3O

(*r*)

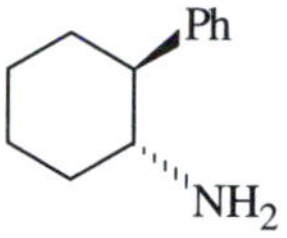
Ph
NH2

(*s*)

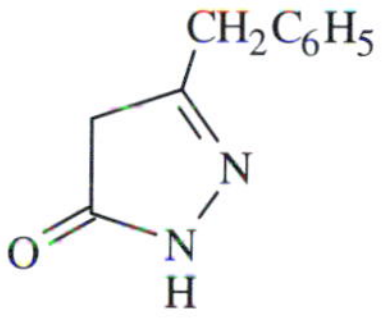
CH2C6H5
N
O
N
H

(*t*)

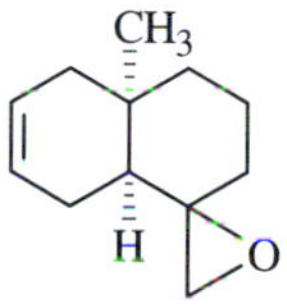
CH3
H
O

(*u*)

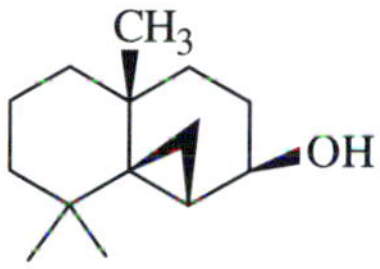
CH3
OH

(*v*)

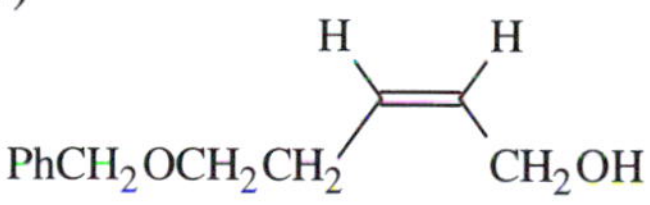
H
H
PhCH2OCH2CH2
CH2OH

(*w*)

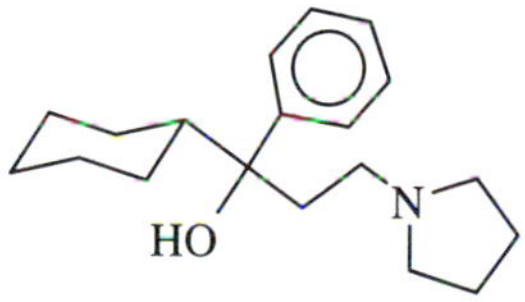
HO
N

(*x*)

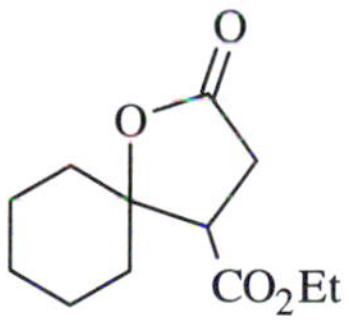
O
O
CO2Et

(*y*)

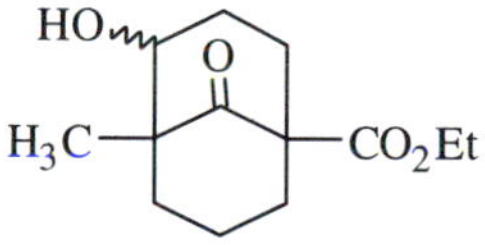
HO
O
H3C
CO2Et

10

MECHANISMS OF ORGANIC

REACTIONS

Activation Energy

This book has taken a mechanism-based approach to understanding the reactions of organic compounds. The movements of electrons, steric and stereoelectronic effects, anion and cation stabilities, and donor–acceptor interactions have all been used as a framework to understand the methods commonly used for manipulating organic molecules. The objective of this enlightenment is to make either reasonable predictions about the success of a projected reaction or an educated guess as to how the reaction might respond to proposed changes in the structure of the starting material or reagents. This jump in logic, from rationalizing what has happened during a reaction to predicting what will happen in a new reaction, requires a very clear understanding of mechanistic principles and the means by which they can be determined.

The first and most important fact about any chemical reaction or any elementary step in a chemical process is that there is an energy barrier separating the products from the reactants. In order for reactants to convert to products, the energy of the system must be raised until the total energy of the system is at least as high as the energy barrier. Then and only then can the reactants change into products by passing over the barrier and relaxing into the product state. The energy barrier is of prime importance for two reasons. First, the height of the barrier determines how fast reactants can be converted to products at a given temperature. Second, at the maximum height of the energy barrier (which is called the *transition state*), there is a

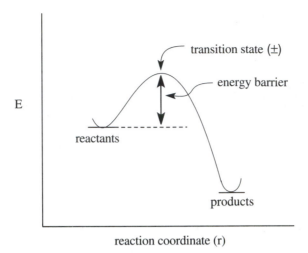

FIGURE 10.1

collection of atoms, called the *activated complex*, that is the structural bridge between reactants and products (see Figure 10.1).

Activated Complex

These considerations are related because the activated complex is an unstable and high-energy collection of atoms which has bonds and electron distributions distorted relative to those in both the reactants and products. The structure of the activated complex is determined by the bonding and structural changes that must occur in order for the reactants to be converted to products. Furthermore, the more the structure of the activated complex is distorted from normal geometries and bond distances, the higher is its energy. Consequently, more energy must be added to the system to distort the reactants to the structure of the activated complex.

As a structural bridge between reactants and products, the activated complex has structural features of both. For example, consider a very simple reaction: the ionization of a tertiary bromide to a carbocation and a bromide ion in methanol. The reactant is a fully covalent molecule with a complete, yet polarized, bond between the tertiary carbon and the bromine atom:

The products contain a trivalent, positively charged carbon atom surrounded by a solvent shell and a bromide ion. The bromide ion carries a negative charge and is surrounded and hydrogen-bonded to a shell of solvent molecules. The bond between carbon and bromine is broken as the bromide moves away from the carbon, and the pair of bonding electrons ends up in the valence shell of the bromide ion. One can depict this reaction using curved arrow notation which tracks electron movement:

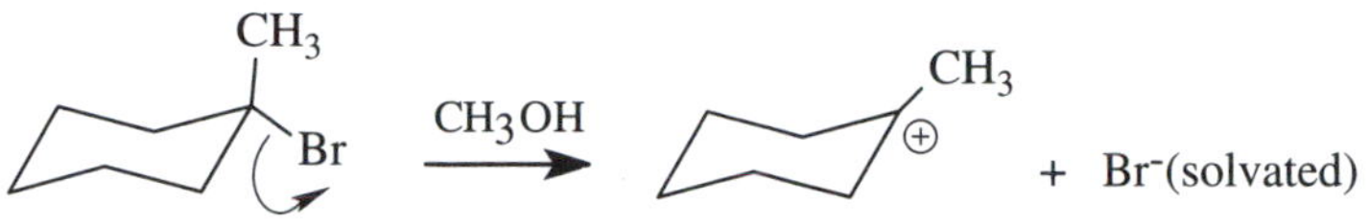

However, between reactants and products is an activated complex which must have structural features common to both. Qualitatively, the activated complex ($\ddagger$) can be pictured as a structure in which the bromine has started to move away from carbon, lengthening the C—Br bond.

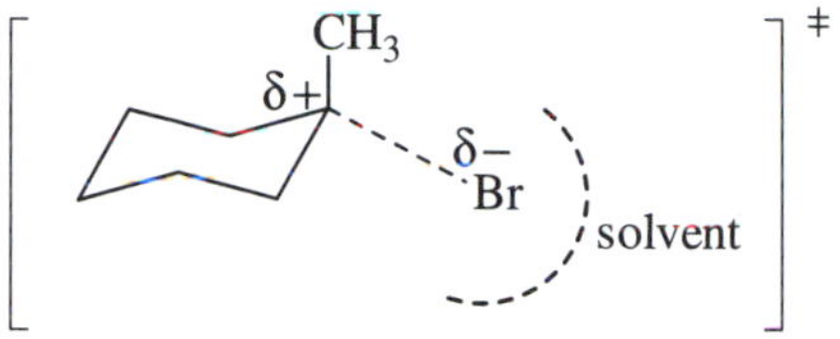

It has begun to develop a partial negative charge since the bonded pair of electrons is even more displaced onto the bromine. For the same reason, the tertiary carbon has begun to develop a partial positive charge. The bond connecting carbon and bromine is longer and weaker than in the reactant, but there is still some bonding energy between carbon and bromine that is completely absent in the products. In addition, due to the greater amount of negative charge on bromine in the activated complex (compared to the reactant), the solvent starts to organize around the bromine, but the interaction is not nearly so strong as in the bromide ion product. The solvation of the carbocation product is much weaker than the solvation of the bromide in a hydroxylic solvent and is often disregarded. However, to the extent that the carbocation product is solvated, the solvent will begin to solvate the partially charged carbon in the activated complex.

The foregoing has been a rather microscopic description of the activated complex (perhaps too detailed), but the idea is quite clear. At whatever level the activated complex is described, it must contain structural features of both the reactants and products it connects. That is, it can only be described in terms of the stable structures of reactants and products.

This is the dilemma. The activated complex occurs at the transition state and has a vanishingly short lifetime ($\sim$1 bond vibration or $\sim 10^{-13}$ second), yet its structure and energy must be described. Furthermore, changes in both its structure and energy must be evaluated if one is to compare different reactants in a predictive way.

Reaction Energetics

To attack this problem, we must begin with those species whose structures are known—the reactants and products. Each is a stable collection of atoms characterized by a free energy. The difference in free energy between the reactants and products is ΔG, the free energy of reaction. The ΔG indicates whether the reaction takes place with evolution of energy (exergic) or with the uptake of energy (endergic). Connecting reactants and products is an energy profile which contains the activation barrier ($\Delta G\ddagger$) that must be surmounted if the reaction is to proceed (see Figure 10.2).

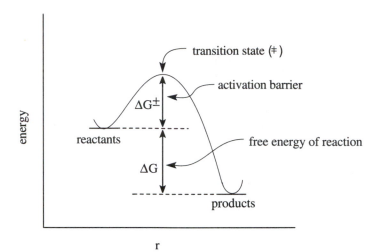

FIGURE 10.2

For a given chemical transformation, there may be more than one step. In such cases, each step of the the transformation will have an activation barrier. Furthermore, at each minimum on the energy curve between reactants and products are more or less stable collections of atoms with finite lifetimes. They usually are higher in energy than either the reactants or products and are called *reaction intermediates*. Many common reaction intermediates have been encountered in the reactions already presented. These include carbocations, carbanions, enolates, free radicals, and carbenes.

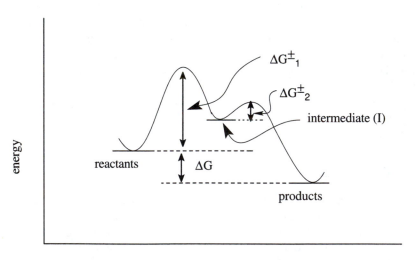

FIGURE 10.3

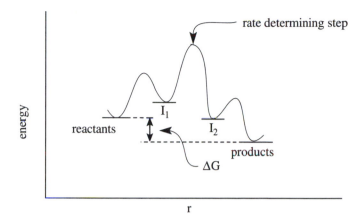

rate determining step

I_1

reactants

I_2

products

ΔG

energy

r

FIGURE 10.4

A multistep process involving intermediates still has an overall free energy of reaction (ΔG, reactants vs. products). However, each elementary step is a separate chemical reaction, and therefore each step has a free energy, an activation barrier, a transition state, and an activated complex corresponding to the chemical changes occurring in that step of the process. The step with the transition state of highest energy is called the rate-determining step because only by passing over the highest barrier can reactants proceed to products (see Figure 10.3).

The rate-determining step thus defines how fast reactants can be converted to products. Steps in the process before the rate-determining step contribute to the rate law for the reaction by contributing to the overall barrier height. Steps after the rate-determining step have no bearing on the rate of reaction (see Figure 10.4).

Structure of the Activated Complex

Even multistep processes can be simplified to a consideration of the energy and structure of the activated complex of the rate-determining step. One very common and successful approach for describing the structure of the activated complex is based on the notion that it must have structural features of both the reactant and product. Furthermore, it is generally assumed that the energy curve in the vicinity of the transition state is reasonably symmetrical. Thus the relative energies of the reactants and products give some indication of the structure of the activated complex. This is a very important concept since it relates energy terms to structural characteristics. Energy changes associated with a chemical reaction can be measured fairly easily, whereas structural information about the activated complex cannot be directly measured at all. Thus the connection between energy and structure is unique in its ability to provide insight into the structure of the activated complex.

In a typical reaction coordinate diagram, which describes energy changes as the reactants progress to products, the energy barrier between reactants and products is a symmetric curve. The abscissa (reaction coordinate) describes structural changes that occur on going from reactant to product. The reaction coordinate r is an arbitrary axis which cannot easily be defined

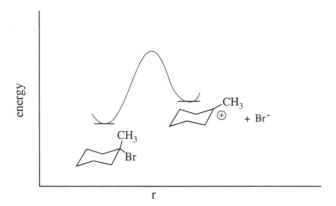

FIGURE 10.5

in simple units but corresponds to structural changes in many dimensions. These structural changes must, however, all be related to the *differences* in structure between reactants and products. Recalling the earlier example of the ionization of a tertiary bromide to a carbocation, it is seen that the reaction coordinate corresponds to several distinct types of changes. These include breaking of the C—Br bond, flattening of the carbocationic carbon, charge development on carbon and bromine, and change in the solvent shell around the bromide (see Figure 10.5). Thus the reaction coordinate cannot represent a single type of change, except by extreme oversimplification.

If a symmetric barrier is present and the reactants and products are of equal energy, it is easy to see that the transition state will lie halfway along the reaction coordinate (see Figure 10.6). Thus the structure of the activated complex is "halfway" between reactants and products. To illustrate, consider the substitution of one iodide for another in the reaction of methyl iodide with iodide ion (one can be sure this reaction occurs by using radioactive iodide):

$$^*I^- + CH_3-I \rightarrow {}^*I-CH_3 + I^-$$

The products are identical to the reactants and thus have the same energies (see Figure 10.7).

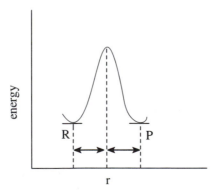

FIGURE 10.6

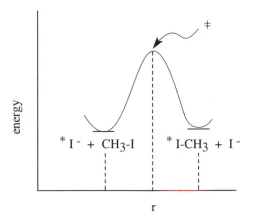

FIGURE 10.7

Consequently, this substitution reaction has a transition state which falls midway between the reactants and products, along the reaction coordinate. Thus the activated complex has a structure midway between that of reactants and products. This corresponds to a collection of atoms having a pentavalent carbon with trigonal bipyramidal geometry (the carbon is one-half inverted in geometry) with one-half bonds between each iodine and carbon and one-half of a full negative charge on each iodine. This activated complex, as shown below is a simplified but reasonable depiction of the structure of the activated complex.

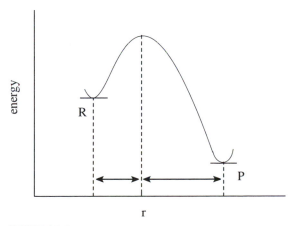

If the products in a reaction step are lower in energy than the reactants (exergic), the symmetry of the activation barrier causes the transition to lie less than halfway along the reaction coordinate.

FIGURE 10.8

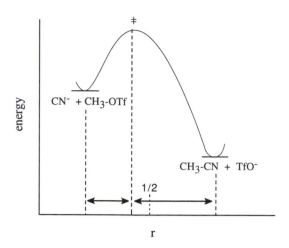

FIGURE 10.9

The structure of the activated complex is more closely related to the structure of the reactants than to the structure of the products (see Figure 10.8).

For example, the reaction of methyl triflate with cyanide ion is also a bimolecular substitution reaction, and it is a very exothermic process. Thus the transition state lies more toward the reactants, and the structure of the activated complex will more closely resemble the structure of the reactant than it will the structures of the products (see Figure 10.9).

Thus in the activated complex, the cyanide-to-methyl bond will be little-formed (< one-half) and the carbon-to-trifloxy bond will be largely intact (> one-half). Most of the negative charge will remain on the cyanide nucleophile and little will have developed on the triflate leaving group. The geometry at the carbon will still be tetrahedral-like, although some flattening will have occurred.

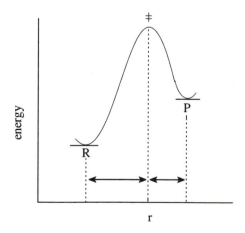

FIGURE 10.10

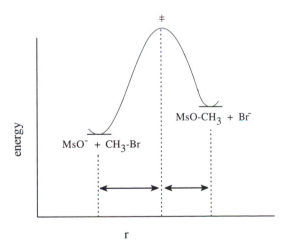

For a reaction step in which the products are less stable than the reactants (endergic), the transition state will lie farther along the reaction coordinate toward the products and the activated complex will have a structure more similar to the products than to the reactants (see Figure 10.10).

If we consider reaction of mesylate ion with methyl bromide, we find that this is an endergic reaction; thus the transition state lies closer to the products than to the reactants (see Figure 10.11). The activated complex will therefore have a structure more closely resembling the products. There will be significant bond formation between mesylate and carbon, and only a weak residual bond between carbon and bromine. The bromine will have acquired significant negative charge, and the carbon will be partially inverted in geometry.

It is possible, therefore, to gain significant insight into the structural characteristics of the activated complex from the structures of the reactants and products and their relative energies.

FIGURE 10.11

For an exothermic reaction step, the activated complex more closely resembles the reactants and is described as early. For an endothermic reaction step, the activated complex more closely resembles the products and is described as late. The more exothermic a process, the earlier is the transition state, and the more endothermic a process, the later is the transition state.

These connections between energy and structure provide a powerful method for characterizing the activated complex. Furthermore, changes in reactants and products produce energy changes which can be translated into changes in structure of the activated complex. Hence, it is possible to predict how changes in structure will influence changes in energy and rates of reaction. These considerations are central to understanding the mechanisms of chemical reactions and they permit us to make the best structural and reagent choices for a particular conversion.

The Hammond Postulate

With the development of a conceptual approach for identifying structural characteristics of the activated complex, the logical next step is to ask how structural features in reactants change the energy and thus the structure of the activated complex. The Hammond postulate provides one solution to this question. It states that, for endothermic reactions, features that stabilize and thus lower the energy of a product also lower the energy of the transition state leading to that product. This is shown schematically in Figure 10.12. If product 2 (P_2) is lower in energy than product 1 (P_1), transition state 2 ($\ddagger_2$) will be lower than transition state 1 ($\ddagger_1$). It will also be earlier. As a consequence, P_2 will have a lower activation barrier and will be formed faster than P_1. A simplified restatement of the Hammond postulate is that more stable products are formed faster. It must be remembered that this is applicable only to endothermic reactions that have reactants of the same or similar energies.

The ionization of alkyl tosylates to give carbocations is an endothermic reaction. Knowing that 3° carbocations are more stable than 2° carbocations, one would conclude that the activation barrier for ionization of the 3° tosylate is lower than that of the 2° tosylate and thus the 3° tosylate should ionize faster (see Figure 10.13).

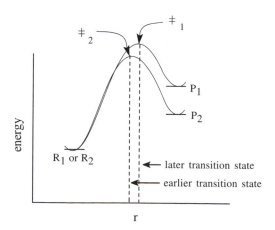

FIGURE 10.12

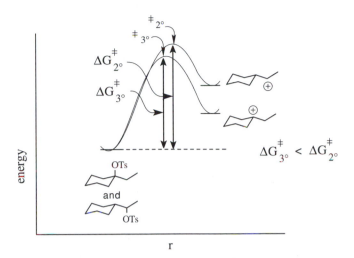

FIGURE 10.13

One would also predict that the transition state for ionization of the 3° tosylate would be earlier, so there should be less C—O bond breaking and less charge development than in the activated complex for ionization of the 2° tosylate.

The Hammond postulate provides a crucial relationship between the rate of reaction and the activated complex of that reaction. In practice, structural changes are made in the reactant(s), and the influence of those changes on the rate of reaction is measured. If the reaction is faster, the change in the reactant has led to a lower product energy, hence a lower activation energy and an earlier transition state (one which has more reactant character). If the reaction is slower, the change in the reactant has led to a higher product energy, hence a higher activation energy and a later transition state (one with more product character). The results of rate studies can thus be translated into structural changes (bonding, charge distribution, geometry) in the activated complex; this further translates into mechanistic information about the reaction (see Figure 10.14).

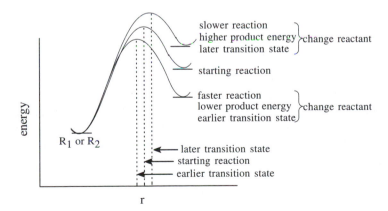

FIGURE 10.14

Reaction Kinetics

The first step in delineating the mechanism of a reaction is to determine what reactant species must come together to produce the activated complex of the rate-determining step. This can be done by determining the order of the reaction with respect to each of the reactants in the process. If the rate of a reaction is dependent on the concentration of a particular reactant, that reactant is involved in the transition state of the rate-determining step. This provides important structural information about the activated complex since it reveals which chemical species are present in the activated complex.

The order of a reaction is found by determining the relationship between rate and concentration for each reactant. Thus for the elementary process A $\rightarrow$ B, the rate of reaction v can be expressed as either the decrease in the concentration of reactant A with time or the increase in the concentration of product B with time.

$$v = \frac{-d[A]}{dt} = +\frac{d[B]}{dt}$$

Further, the rate of the reaction can be expressed as a function of the concentration of A, where k is the rate constant for the process. This differential rate expression

$$\frac{-dA}{dt} = +\frac{d[B]}{dt} = k[A]$$

shows that the reaction rate is directly dependent on the concentration of A; the greater is [A], the faster A is converted to B. The reaction is first-order with respect to A because the exponent of [A] is one.

The above expression is the first-order differential rate law for the conversion of A to B. The change in concentration of A over the complete course of the reaction is given by the integrated rate law, which is found by solving the differential rate law:

$$-\frac{dA}{dt} = k[A]$$

$$\frac{d[A]}{[A]} = k\,dt$$

$$-\ln\frac{[A]_0}{[A]_t} = kt \Rightarrow \ln[A]_t = -kt + \ln A_0$$

The integrated rate law shows that the natural log of the concentration of the starting material A decreases linearly with time. By determining the concentration of A at various times $[A]_t$ and plotting $\ln[A]_t$ versus t, a straight line with slope $-k$ will be obtained *if the reaction is first-order in A*. (Since the concentration of product at any time $[B]_t$ equals $[A]_0 - [A]_t$, a plot of the increase in the concentration of B with time, in the form of $\ln([A]_0 - [B]_t)$ versus time, would also give a straight line whose slope is $+k$.) If such linear dependence is observed, the reaction is first-order with respect to A and the rate constant for the reaction can be determined. If the rate plot is not linear, the reaction is not first-order with respect to A. That is, a first-order rate law does not correctly describe behavior.

Second-order reactions occur when two reagents must collide in solution to produce the activated complex. Thus, for the reactions

$$2A \rightarrow B \qquad A + B \rightarrow C$$

$$-\frac{dA}{dt} = k[A]^2 \qquad \frac{dC}{dt} = k[A][B]$$

each reaction is second-order because the sum of the exponents of species in the rate law is two. This means that, in the first instance, two molecules of A must collide to produce the activated complex. In the second case, one molecule each of A and B collide to produce the activated complex. In each case, the second-order dependence requires that *both* of the colliding molecules are part of the activated complex of the rate-determining step.

Integration of these differential rate laws gives

$$\frac{1}{[A]_t} - \frac{1}{[A]_0} = kt \quad \text{and} \quad \frac{1}{[A]_0 - [B]_0} \ln \left[\frac{[B]_0[A]_t}{[A]_0[B]_t} \right] = kt$$

Again, plotting concentrations versus time using these integrated second-order rate laws gives linear plots *only if the reaction is a second-order process*. The rate constants can be determined from the slopes. If the concentration versus time plots are not linear, the second-order rate equations do not correctly describe the kinetic behavior. There are integrated rate laws for many different reaction orders.

It is often possible to simplify the rate law of a second-order process by employing pseudo first-order conditions. For a second-order reaction where

$$A + B \rightarrow P$$

and

$$-\frac{dA}{dt} = \frac{dP}{dt} = k[A][B]$$

the integrated rate law for a second-order reaction can be used to plot the kinetic data. Alternatively, if the concentration of one of the reactants, for example A, is much larger than the other (10-fold excess minimum, 20-fold better), its concentration does not change significantly over the course of the reaction. Thus, $[A]_t \approx [A]_0$. The differential rate law can be approximated as

$$\frac{dP}{dt} = k[A][B] = k[A]_0[B] = k'[B]$$

where $k' = k[A]_0$. This new rate law is a first-order equation, which is much easier to deal with and plot. A plot of $\ln[B]$ versus t will give a straight line of slope k'. From k' and $[A]_0$ (which is known), the rate constant can easily be determined. This is a much simpler method than the normal second-order treatment for determining the rate constant. The order of the reaction, with respect to A, can be checked by using several different initial concentrations of $[A]_0$ and the relationship

$$k' = k[A]_0 \quad \text{therefore} \quad \log k' = \log[A]_0 + \log k$$

Plots of $\log k'$ versus $\log[A]_0$ will have unit slopes if the reaction is first-order in A. Alternatively, doubling $[A]_0$ should double the rate if the reaction is first-order in A.

In a multistep process involving a reactive intermediate, the rate law for the overall reaction cannot be written down a priori because the step in which the reactants disappear is different from the step in which the products are formed. In a large number of cases, the intermediate is of high energy and reacts very rapidly, either returning to reactant or going on to product (see Figure 10.15).

In such cases, the steady state approximation can be used to derive a rate expression that can be tested. Thus for a reaction process involving an intermediate [I]

$$A \underset{k_{-1}}{\overset{k_1}{\rightleftharpoons}} I \overset{k_2}{\rightarrow} P$$

and

$$v = \frac{dP}{dt} = k_2[I]$$

The concentration of the intermediate [I] which gives product is given by the difference between the rate of its formation from A and the rate of its conversion either back to reactant A or forward to product P:

$$\frac{dI}{dt} = k_1[A] - k_{-1}[I] - k_2[I] = 0$$

The steady state approximation assumes that since I is very reactive, its concentration will be very low at any time during the reaction and that it will not change appreciably. Therefore, $dI/dt = 0$. Solving the above expression for the concentration of I and substitution into the rate law for the formation of product gives

$$[I] = \frac{k_1[A]}{k_{-1} + k_2}$$

and

$$\frac{dP}{dt} = \frac{k_2 k_1}{k_{-1} + k_2}[A] = k_{obs}[A]$$

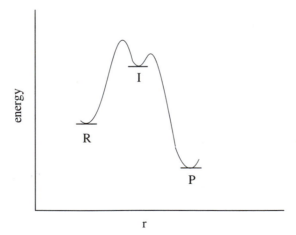

FIGURE 10.15

Here, the observed rate constant $k_{obs} = k_2 k_1 / k_{-1} + k_2$. This is a first-order rate expression in reactant A and can be integrated and plotted normally.

Several limiting cases can be envisioned for such a multistep process. If $k_2 \gg k_{-1}$, the intermediate goes on to product more rapidly than it returns to reactant:

$$\frac{dP}{dt} = \frac{k_2 k_1}{k_2 + k_{-1}} [A] \quad \text{simplifies to} \quad \frac{dP}{dt} = k_1[A] \text{ for } k_2 \gg k_{-1}$$

This represents the case where the first step is rate-determining. (If this situation is known beforehand, one need not work through the steady state approximation, but can merely write down $dP/dt = kA_1$, since the first step is rate limiting and irreversible.)

If $k_{-1} \gg k_2$, then

$$\frac{dP}{dt} = \frac{k_1}{k_{-1}} k_2[A] = K_{eq} k_2[A]$$

This represents the case where there is a fast preequilibrium preceding the rate-determining step. It is again not necessary to work through the steady state approximation. If the intermediate I is in equilibrium with the reactant A, and because

$$K_{eq} = \frac{[I]}{[A]} \quad \text{and} \quad [I] = K_{eq}[A]$$

the rate expression becomes

$$\frac{dP}{dt} = k_2[I] = K_{eq} k_2[A]$$

This is actually the correct way to think about this case since the steady state approximation requires that the concentration of I be low, which it may not be if there is a fast preequilibrium. If $k_2 \approx k_{-1}$, the full steady state rate expression is needed to describe the rate of reaction.

$$\frac{dP}{dt} = \frac{k_2 k_1}{k_2 + k_{-1}} [A]$$

Consider once again the solvolysis of a tertiary bromide in methanol:

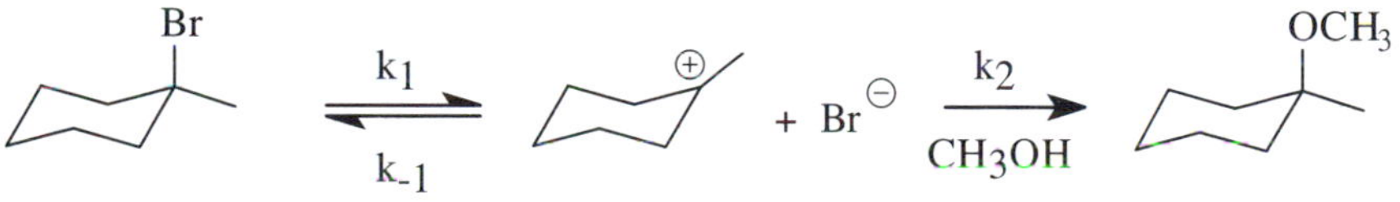

A   I   P

This reaction can be reduced to the kinetic scheme

$$A \underset{k_{-1}}{\overset{k_1}{\rightleftharpoons}} I + Br^{\ominus} \overset{k_2}{\rightarrow} P$$

The rate of product formation is given by the pseudo first-order expression in which $k_2 = k[CH_3OH]$. (Normally, k_2 is taken as the rate constant because the concentration of methanol is constant.) Therefore,

$$\frac{dP}{dt} = k_2[I]$$

and the steady state approximation is written as

$$\frac{dP}{dt} = O = k_1[A] - k_{-1}[I][Br^-] - k_2[I]$$

Since the return of intermediate I to reactants is a second-order reaction, solving for [I] and substitution into the rate law gives

$$\frac{dP}{dt} = \frac{k_2 k_1[A]}{k_{-1}[Br^-] + k_2}$$

If $k_2 \gg k_{-1}[Br^-]$, $v = k_1[A]$ and simple first-order behavior is found. If $k_{-1}[Br^-] \gg k_2$, there is a rapid ionization preequilibrium and

$$K_{eq} = \frac{[I][Br^-]}{[A]} \qquad \text{and} \qquad [I] = \frac{k_{eq}[A]}{[Br^-]}$$

then

$$\frac{dP}{dt} = k_2[I] = \frac{k_2 K_{eq}[A]}{[Br^-]}$$

The rate of product formation will be directly proportional to [A] but inversely proportional to [Br−]. By using an excess of Br^-, so that its concentration does not change appreciably over the course of the reaction, pseudo first-order behavior can be achieved with $k_{obs} = k_2 K_{eq}/[Br^-]_0$.

If $k_{-1}[Br^-] \approx k_2$, the full rate expression will be needed to describe the kinetic behavior. The rate will be first-order in [A], and the rate will slow down in the presence of added bromide, but not in a simple inverse relationship.

Determining Activation Energies

By using the above methods, the rate constants for most organic reactions can be obtained. Rate constants, by virtue of the fact that they relate directly to the passage of reactants over the barrier of the rate-determining step, can be used to probe the energy and structure of the activated complex. The energy of the activated complex corresponds to the height of the activation barrier for the rate-determining step. The barrier height can be calculated by the Arrhenius equation

$$\ln k_{rate} = -\frac{E_a}{RT}$$

where k_{rate} is the rate constant of a reaction, R is the ideal gas constant, T is the absolute temperature, and E_a is the Arrhenius activation energy. E_a is determined from plots of $\ln k$ versus $1/T$, determined at various temperatures, and largely corresponds to the enthalpy of activation since $E_a = \Delta H^{\pm} + RT$. The enthalpy and entropy of activation and hence the free energy of activation are determined by the Eyring equation

$$\ln \frac{k_{rate}}{T} = -\frac{\Delta H^{\pm}}{RT} + \frac{\Delta S^{\pm}}{R} - \ln \frac{k}{h}$$

where k is Boltzmann's constant and h = Planck's constant. Plots of $\ln(k_{rate}/T)$ versus $1/T$ give straight lines whose slopes are $-(\Delta H^{\pm}/R)$ and whose intercepts are $-(\Delta S^{\pm}/R) - \ln(k/h)$. Each can be numerically evaluated to give $\Delta H^{\pm}$, $\Delta S^{\pm}$, and finally $\Delta G^{\pm}$, by $\Delta G^{\pm} = \Delta H^{\pm} - T\Delta S^{\pm}$.

Isotope Effects

Besides the energy of the activated complex, structure and bonding in the activated complex can be probed in several ways using rate constant data. One very powerful way to investigate bonding in the activated complex is to use kinetic isotope effects. These are based on the fact that a heavier isotope of an element has a lower zero point energy, so that more energy is required to break a bond to a heavier isotope than a bond to a lighter isotope (that is, the activation energy is greater). At a given temperature, this means that the rate of reaction for a compond containing a heavy isotope is slower than the rate of reaction for the compound with a lighter isotope. This is *only* true if breaking that bond is involved at the transition state of the rate-determining step. If breaking that bond occurs prior to or after the rate-determining step, isotopic substitution does not give a large change in the rate. This effect is most pronounced for hydrogen/deuterium isotopes. These have the largest mass differences of any isotopes and thus the largest difference in zero point energies. If a bond to hydrogen (or deuterium) is being broken in the rate-determining step, k_H/k_D values of 2–8 are typical. These are termed *primary* kinetic deuterium isotope effects.

If the C—H(D) bond is not being broken in the rate-determining step, there are sometimes smaller effects on the rate resulting from isotopic substitution. These are termed *secondary* kinetic deuterium isotope effects. They result from differences in deformation energies, but they are small and typically $k_H/k_D = 1$–1.3 for these effects. If a kinetic deuterium isotope effect is found to be greater than about 1.5 , it is a primary kinetic deuterium isotope effect and C—H(D) bond breaking is occurring in the rate-determining step. If a kinetic deuterium isotope effect is found to be between 1 and 1.5, it is a secondary kinetic deuterium isotope effect and C—H(D) bond breaking is not occurring in the rate-determining step.

The largest values of primary kinetic deuterium isotope effects are found for reactions where the bond to hydrogen is about one-half broken ($k_H/k_D = 6$–8). Smaller values are found in reactions in which the bond to hydrogen is less than or more than one-half broken. Normally, k_H/k_D values less than maximum correspond to bond cleavage of less than one-half. Primary kinetic deuterium isotope effects thus provide insight into the extent of C—H bond cleavage in the activated complex.

For example, the free radical bromination of toluene by NBS proceeds with $k_H/k_D = 4.9$, while for the bromination of isopropyl benzene, $k_H/k_D = 1.8$. Both are primary kinetic deuterium isotope effects, indicating that hydrogen abstraction by a bromine atom is the rate-determining step. The much lower value of the isotope effect for isopropylbenzene suggests that the transition state is much earlier than for toluene. The lesser extent of hydrogen transfer is due to the more stable radical being produced, resulting in an earlier transition state (see Figure 10.16).

The electrophilic nitration of benzene using acetyl nitrate involves the replacement of a hydrogen on the benzene ring by a nitro group. The reaction is second-order overall, first-order in benzene and first-order in the nitrating agent, $v = k[C_6H_6][\text{acetyl nitrate}]$.

Use of fully deuterated benzene gives $k_H/k_D = 1$. These data suggest that the nitrating agent attacks the benzene ring in the rate-determining step, but C—H bond breaking is not involved in the rate-determining step. These observations are consistent with an electrophilic attack of

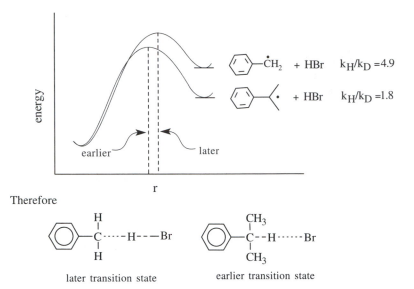

FIGURE 10.16

the nitrating agent of the π system. The proton is lost in a subsequent fast step, after the rate-determining step.

Thus even though loss of hydrogen is required for the product to be formed, its removal is not taking place in the rate-determining step of the reaction, so it must take place after the rate-determining step.

The base-promoted bromination of ketones is a second-order process, first-order in ketone and first-order in base; thus $\nu = k[\text{ketone}][\text{base}]$. The bromine concentration does not appear in the rate law; that is, the reaction is zero-order in $[\text{Br}_2]$.

Use of deuterated substrates gives $k_H/k_D = 6.5$. This is a primary kinetic deuterium isotope effect, indicating that proton removal is an essential component of the rate-determining step. The lack of rate dependence on bromine requires that bromine is added to the molecule after the rate-determining step. A mechanism consistent with these facts has proton removal and enolate formation rate-determining.

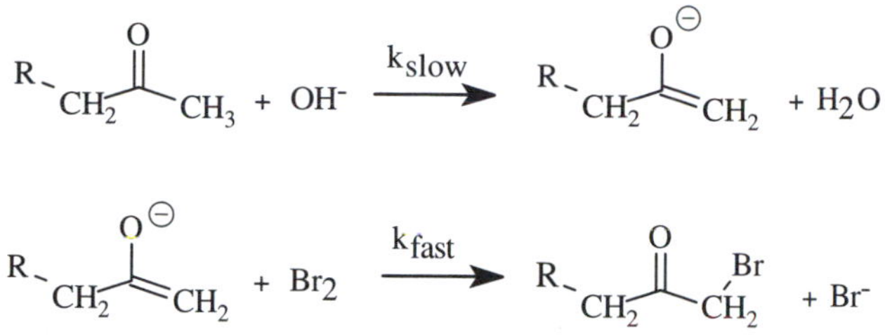

If we take this basic scenario and add the notions of electron movement to the picture, we can construct a detailed picture of electronic change that is consistent with the observed facts.

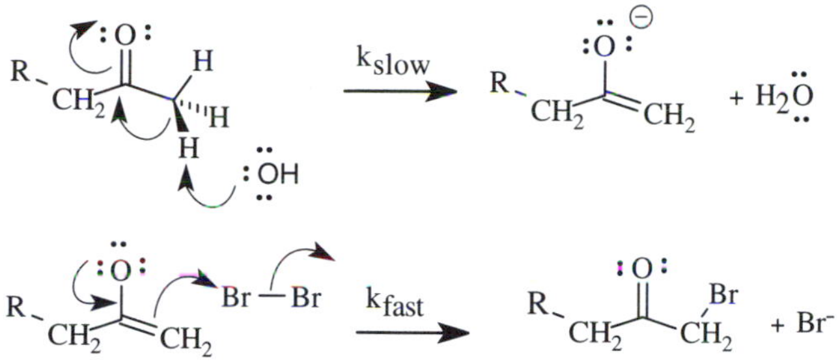

The activated complex for proton removal, the rate-determining step, can be envisioned as having a partial charge from proton removal delocalized into the carbonyl group (as it is in the product enolate). This also requires that the proton being removed have a dihedral angle of 90° with the plane of the carbonyl group, so the developing charge can overlap with the carbonyl π bond.

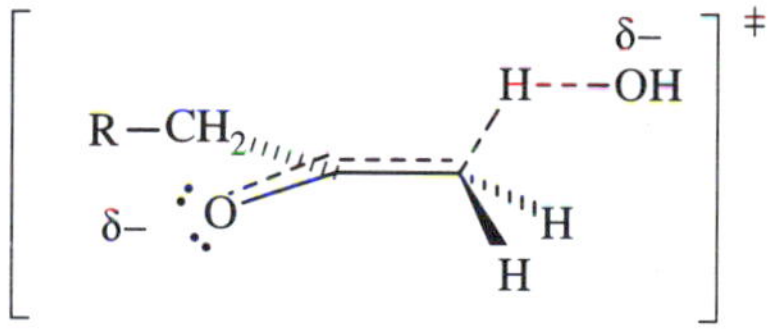

Base-promoted elimination in the two β-phenethyltrimethylammonium derivatives shown below is found to be second-order overall, first-order in substrate and first-order in base, that is, $\nu = k[C_6H_5CH_2CH_2N^+(CH_3)_3][CH_3CH_2O^-]$. This means that both the substrate and the ethoxide base are present in the transition state of the rate-determining step.

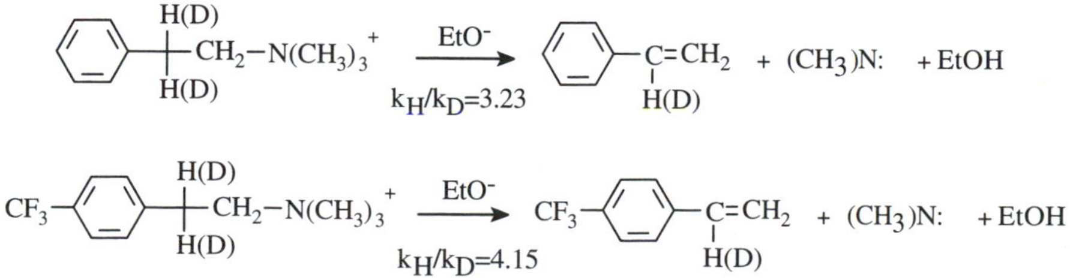

The rate constants for the deuterated and protio substrates were measured. The magnitudes

(3–4) of the kinetic deuterium isotope effect for both substrates are typical primary kinetic deuterium isotope effects. This means that C—H bond breaking is involved at the transition state of the rate-determining step. This suggests that proton removal by base in the activated complex is an essential element of the rate-determining step and is a key feature in the mechanism of the elimination reaction.

The difference between the k_H/k_D values, however, means that the extent of C—H bond breaking is different at the transition state in the second substrate compared to the first. (The transition state of the second substrate is actually earlier in terms of proton removal by base because, in both cases, proton transfer is greater than half completed.) Thus a change in structure of the substrate leads to a distinct change in the structure of the activated complex which can be detected and described by kinetic isotope effects.

From the above examples, it is clear that kinetic deuterium isotope effects are a powerful way to probe bonding changes in the activated complex. The magnitude of the isotope effect indicates whether bonds to hydrogen are being made or broken in the rate-determining step. Differences in kinetic isotope effects in closely related precursors can also be used to pinpoint whether one transition state is earlier than another. This is a direct measure of the effect of the substrate structure on the structure of the transition state.

Other elements can be used to measure isotope effects, however, the magnitudes of these isotope effects are much smaller than primary kinetic deuterium isotope effects. Substitution of ^{13}C for ^{12}C in a reaction could lead to a *maximum* kinetic isotope effect of $(k_{12c}/k_{13c}) = 1.05$, for a reaction in which a bond to carbon is broken in the rate-determining step. (Recall that maximum k_H/k_D's are 8 to 10.) Most standard kinetic methods are not capable of distinguishing such small rate differences reproducibly, so kinetic isotope effects for elements other than hydrogen (deuterium) are not very abundant in the literature. In some instances, isotopic abundances determined by mass spectrometry can be used to measure such differences accurately and isotope effects can be informative. The decarboxylation of malonic acid proceeds with $(k_{12c}/k_{13c}) = 1.045$. This large primary isotope effect (for carbon) indicates that C—C bond breaking is well developed in the transition state. This detailed information about the structure of the activated complex permits a shift in focus from a curved arrow–type of mechanism to a real structure of the activated complex (see Figure 10.17).

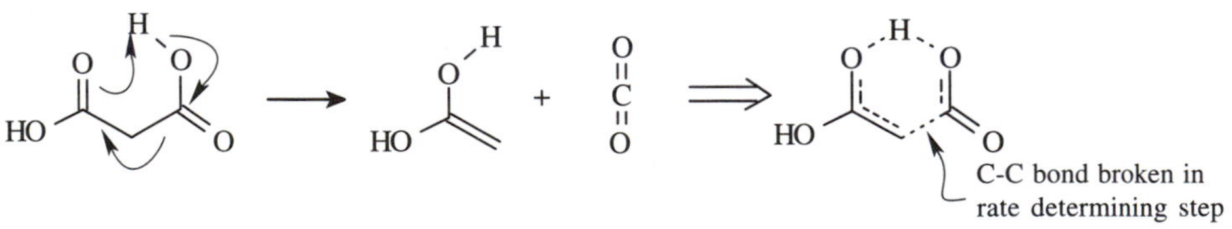

FIGURE 10.17

Electronic Effects

Besides bond breaking, another common feature of many reactions is the formation of charged species as intermediates. Carbocations, carbanions, and oxonium ions are commonly encountered intermediates that are formed in the rate-determining step of a multistep reaction. As a consequence, charge development in the activated complex is expected. In terms of the reaction mechanism, it is very important to know the charge type $(+, -,$ or none) and the extent of charge development in the activated complex.

The use of rate constants can also provide a clue to charge development. Changes in the rate constant of a reaction due to changes in structure can be indicative of the charge distributions present in the activated complex. For example, rate constants are much larger for the base-promoted deuterium exchange of phenylacetone than for acetone itself because the phenyl group stabilizes the negative charge on the enolate ion (and the transition state leading to it). Hence the α proton is removed more rapidly and deuterium exchange is speeded up correspondingly. This behavior is entirely consistent with an increase of electron density on the α carbon during the rate-determining step.

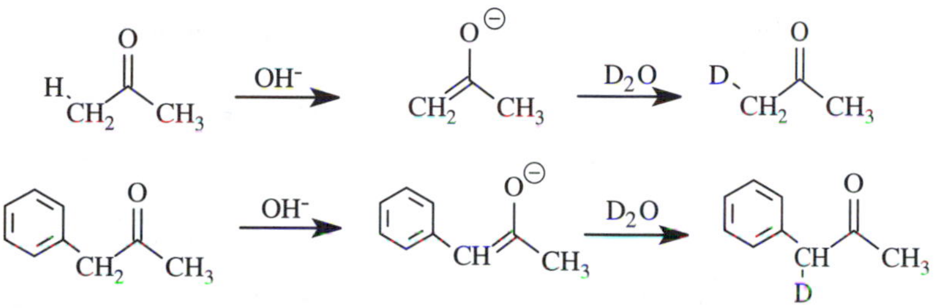

The hydration of styrene, α-methylstyrene, or α-trifluoromethylstyrene gives a benzylic alcohol product. However, α-methyl styrene reacts 10^5 more rapidly than styrene itself, whereas α-trifluoromethylstyrene reacts 10^7 less rapidly than styrene. This behavior is consistent with protonation of the double bond to give a carbocation as the rate-determining step. The developing positive charge in the transition state is stabilized by the inductive effect of the methyl group in α-methyl styrene; the transition state is of lower energy, and it thus reacts faster. The developing positive charge in the transition state is destabilized by the electron-withdrawing inductive effect of the trifluoromethyl group in α-trifluoromethylstyrene; the transition state is of higher energy, and thus it reacts more slowly (see Figure 10.18).

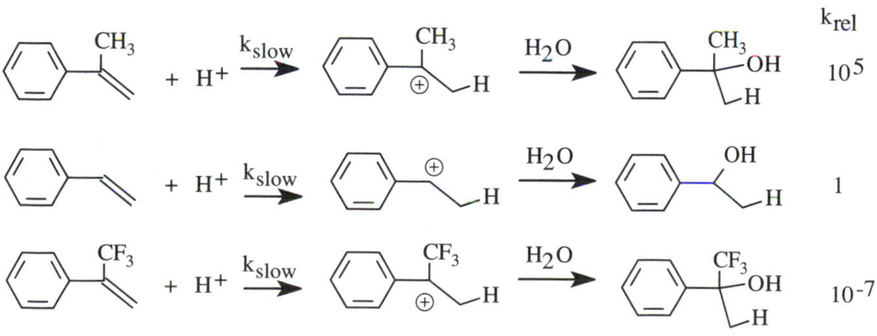

FIGURE 10.18

While changes in rate constants in response to changes in structure are extremely valuable for indicating the type of charge development which is occurring in the activated complex, the actual extent of charge development in the activated complex is another structural descriptor that would be very useful.

The Hammett Equation

The Hammett equation is a very useful tool for monitoring the extent of charge development in the activated complex. If a substituent is attached to the *meta* or *para* position of a benzene ring, it will change the ability of the aromatic ring to donate or withdraw electrons relative to benzene itself. Thus, a *p*-methylphenyl group should be electron-donating relative to phenyl, whereas a *p*-bromophenyl group should be electron-withdrawing relative to phenyl. Furthermore, placement of the bromo group in the *meta* position places it closer to the point of attachment of the phenyl ring, so the electron-withdrawing ability of the *m*-bromophenyl group is greater than that of the *p*-bromophenyl group. Substituents are normally attached only to the *meta* or *para* positions of the phenyl so that they do not sterically interfere with the site of attachment of the phenyl group.

By the use of a model reaction (ionization of benzoic acids), the ability of a substituent to modify the electron-donating or -withdrawing ability of the phenyl group and thus influence that reaction can be defined quantitatively by the Hammett equation:

$$\log \frac{K_Z}{K_H} \equiv \sigma$$

The result is a substituent constant (σ) which is a numerical description of the electronic effect of a substituent relative to a hydrogen atom. Stated a different way, a substituent constant σ is a quantitative way to describe the electron-donating or electron-withdrawing properties of a substituent when it is attached to a benzene ring. Different σ-values are obtained for a given substituent if it is in the *meta* or the *para* position, so σ values are positionally dependent. Table 10.1 lists some common σ values.

The model reaction used to evaluate σ constants is an ionization equilibrium in which a negative charge is produced upon going from reactants to products. Electron-withdrawing groups have positive σ values because they increase the ionization relative to hydrogen; K_Z/K_H is greater than one, and $\log K_Z/K_H$ is greater than zero. Electron-donating substituents have negative σ values because they decrease ionization relative to hydrogen; K_Z/K_H is less than one, and $\log K_Z/K_H$ is less than zero. Hydrogen itself is treated as a substituent with $\sigma = 0$ because $K_Z = K_H$ thus $K_Z/K_H = 1$ and $\log K_Z/K_H = 0$.

Furthermore, the absolute magnitude of the σ value provides a quantitative measure of the relative electron-donating or withdrawing ability. Thus a *m*-CF$_3$ group ($\sigma = 0.46$) is a stronger electron-withdrawing group than *m*-Cl ($\sigma = 0.37$), but is a weaker electron-withdrawing group than *m*-CN ($\sigma = 0.62$). On the other hand, a *p*-methyl group ($\sigma = -0.14$)

TABLE 10.1 σ-Values for Common Aromatic Substituents

Substituent	σ_m	σ_p	σ_m^+	σ_p^+	σ_p^-
NH$_2$	−0.09	−0.57	−0.16	−1.3	—
OCH$_3$	0.10	−0.28	0.05	−0.78	—
CH$_3$	−0.06	−0.14	−0.10	−0.31	—
H (reference substituent)	0	0	0	0	0
Cl	0.37	0.24	0.40	0.11	—
Br	0.37	0.26	0.41	0.15	—
CO$_2$R	0.35	0.44	0.37	0.48	0.68
CF$_3$	0.46	0.53	0.57	—	—
CN	0.62	0.70	0.56	0.66	1.00
NO$_2$	0.71	0.81	0.73	0.79	1.27

is a weaker electron donor than p-OCH$_3$ ($\sigma = -0.28$) but a better electron donor than p-Si (CH$_3$)$_3$ ($\sigma = -0.07$).

These substituent constants can be used with rate data to evaluate the type and extent of charge development in the activated complex of the rate-determining step for a wide variety of chemical reactions. The rates of reaction for a particular transformation are measured using a series of compounds which differ only by the phenyl substituents present.

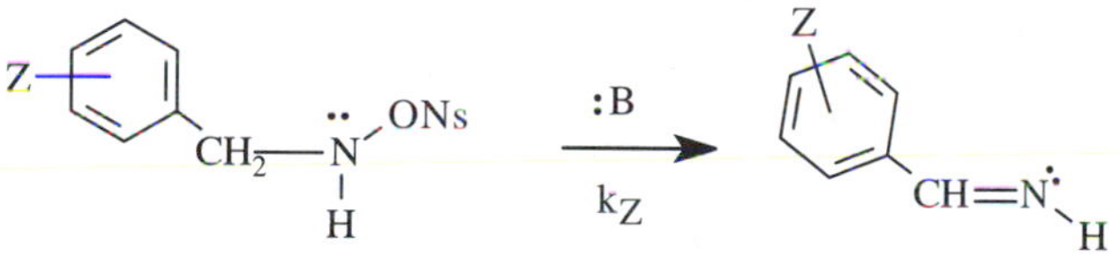

Here, the differences in rate caused by the electronic effect of the substituent are correlated by the Hammett equation, $\log k_Z/k_0 = \rho\sigma_Z$, where k_Z is the rate constant obtained for a compound with a particular *meta* or *para* substituent, k_0 is the rate constant for the unsubstituted phenyl group, and σ_Z is the substituent constant for each substituent used. The proportionality constant ρ relates the substituent constant (electron-donating or -withdrawing) and the substituent's effect on rate. It gives information about the type and extent of charge development in the activated complex. It is determined by plotting $\log k_Z/k_0$ versus σ_Z for a series of substituents. The slope of the linear plot is ρ and is termed the reaction constant. For example, the reaction shown above is an elimination reaction in which a proton and the nosylate group are eliminated and a C—N π bond is formed in their place. The reaction is second-order overall, first-order in substrate and first-order in base. The rate constants were measured for several substituted compounds (see Table 10.2).

These data were plotted according to the Hammett equation to give the plot shown in Figure 10.19. The first thing to note is that the Hammett plot is linear. The linearity of the plot implies that the substituent constants are correctly modeling electronic changes taking place in the reaction under consideration. That is, the influence of a substituent on the model reaction is of the same type as for the reaction under investigation.

Next, the sign and absolute magnitude of the ρ value determined from a Hammett plot gives information about charge development at the transition state. The sign of ρ tells whether

TABLE 10.2 Rate Constants for Substituted Compounds

Substituent	k_Z	σ_Z
H	8.69×10^{-3}	0
p-CH$_3$	7.39×10^{-3}	-0.14
m-CH$_3$	7.94×10^{-3}	-0.06
p-Cl	1.17×10^{-2}	0.24
m-Cl	1.50×10^{-2}	0.37
m-CF$_3$	1.54×10^{-2}	0.46

a positive or negative charge is being developed in the activated complex relative to the re-actants. A positive ρ value means that electron density is increased (negative charge is being produced) in the activated complex. A negative ρ value means that electron deficiency is being produced (often a positive charge) in the activated complex. Generally, ρ values have absolute magnitudes between 0 and 3, but values as high as 10 or 12 are known. A value of $\rho = 0$ means that substituents have no electronic effect on the reaction rate and thus no charge is being de-veloped at the transition state. Large absolute values of ρ mean that substituents influence the rate greatly and thus the amount of charge developed in the activated complex is large and is influenced significantly by the electronic properties of the substituents. For example,

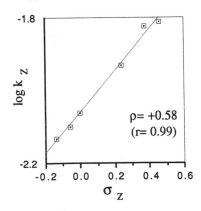

This $\rho = 1.4$ means that negative charge is being developed on the arenesulfonate group at the transition state, consistent with this group departing as an anionic-leaving group as part of the rate-determining step. The magnitude of 1.4 means that a significant amount of charge is

FIGURE 10.19

developed on the leaving group. This interpretation is possible only by comparing $\rho = +1.4$ with ρ values for sulfonate-leaving groups in other reactions in which a range of normal values is known. The normal range is about $+0.8$ to $+1.65$. Therefore, the value of 1.4 indicates significant charge development at the transition state. In addition to the positive ρ value, it is also known that the reaction is second-order overall, first-order in substrate and first-order in triethylamine. Moreover, there is a primary kinetic deuterium isotope effect for the benzylic position. These observations allow a detailed structure to be drawn of the activated complex in which concerted 1,3 elimination results in formation of the three-membered ring product:

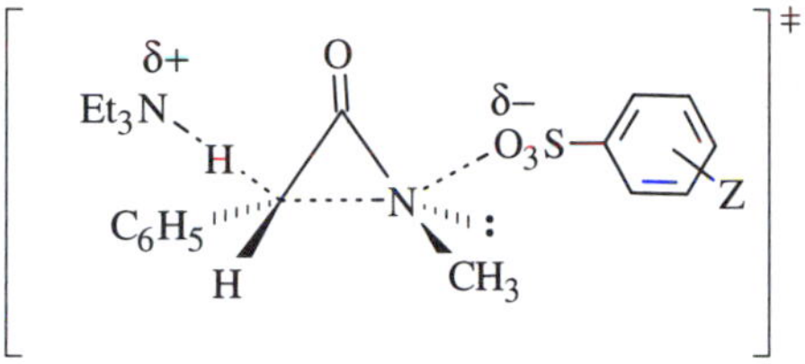

The reaction of *N, N*-dimethylbenzylamines with methyl iodide is found to have $\rho = -1.0$. The negative sign indicates that partial positive charge is being produced in the transition state. The modest size of ρ is consistent with the charge being developed on the nitrogen, which is insulated from the aromatic ring by the saturated CH_2 group. The ability of substituents to influence rate is reduced by the insulating methylene group and therefore ρ is smaller. In comparison, the methylation of *N, N*-dimethylaniline has $\rho = -3.3$ because the nitrogen atom, on which positive charge is developed in the transition state, is directly attached to the phenyl ring and the substituents have a greater influence on the stability of the charge being developed. Hence, the magnitude of ρ is much larger.

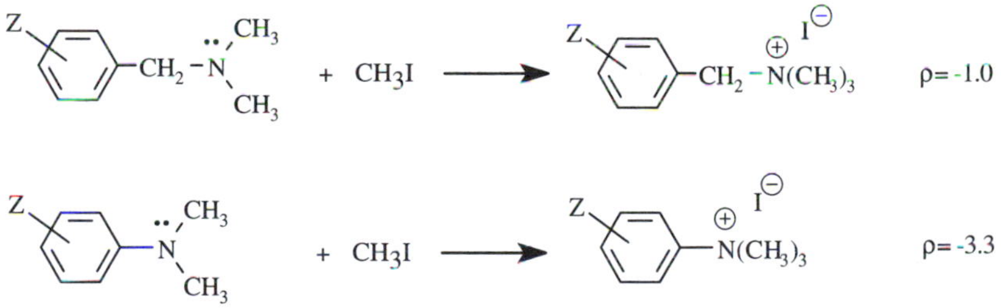

For similar reactions, comparison of the ρ values can be used to determine which reaction has a greater charge development. Comparison of the olefin-forming eliminations below reveals which reaction has greater charge development at the benzylic position and, thus, which has a greater degree of proton removal in the activated complex.

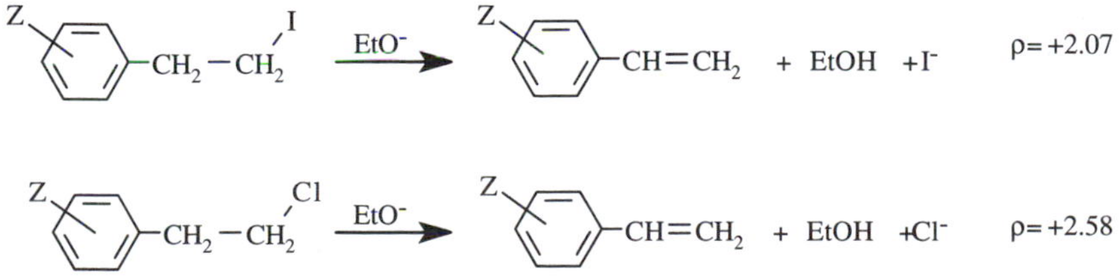

The larger value of $\rho = 2.58$ for the chloride-leaving group, compared to the iodide-leaving group ($\rho = 2.07$) suggests that greater negative charge is developed on the benzylic position when chloride is the leaving group than when iodide is the leaving group. Base removes the proton to a greater extent for the poorer chloride-leaving group than for the better iodo-leaving group. Greater electron density (on the benzylic position) is required to expel the chloride-leaving group than to expel the iodide-leaving group (i.e., chloride needs a greater "push" than iodide). The transition state for proton removal is later with the chloride-leaving group and earlier with the iodide-leaving group. These differences in ρ values are thus very useful for discussing changes in the structures of activated complexes and reaction mechanisms caused by structural changes in the reactants.

Sometimes, Hammett plots of rates versus σ_Z values are nonlinear. When this occurs, it usually indicates that the model reaction from which σ_Z values were determined (the ionization of benzoic acids) does not accurately model the electronic changes occurring in the reaction being studied. In most cases, this happens when a positive or negative charge is being developed at a position where direct resonance interactions with the substituent magnify the electronic effect of substituents. Usually, this happens when charge is developed at a benzylic position or directly on the aromatic ring (see Figure 10.20).

For these cases, new model reactions were developed and the electronic effects of substituents were obtained as σ^+ and σ^- substituent constants. The use of σ^+ constants is applicable for reactions in which a positive charge is generated on, or in direct conjugation with, the aromatic ring. However, σ^- constants are used for reactions in which a negative charge is generated on, or in direct conjugation with, the phenyl ring. A listing of these constants is found in Table 10.1. These constants are used in the same way as σ_Z constants. Rates are measured and correlated with the appropriate set of σ values (σ, σ^+, or σ^-) in typical Hammett fashion:

$$\log \frac{k_Z}{k_H} = \rho^+ \sigma_Z^+ \quad \text{and} \quad \log \frac{k_Z}{k_H} = \rho^- \sigma_Z^-$$

As before, the sigma constants are positionally dependent; that is, σ^+ *meta* is different from σ^+ *para*.

A better linear correlation of the rates with a particular set of σ values means that the set of sigma values used is a better model for the electronic character of the reaction being studied. Knowing that a better correlation is found for σ^+ than σ means that the reaction probably involves positive charge development which can be delocalized directly onto the aromatic ring. For example, the hydration of substituted styrenes gives a much better linear correlation of $\log(k_Z/k_H)$ versus σ^+ than $\log(k_Z/k_H)$ versus σ. This is consistent with the rate-determining step being protonation of the double bond giving a benzylic cation. At the transition state, resonance delocalization of the developing positive charge into the ring accounts for the better correlation with σ_Z^+. The value $\rho^+ = -3.57$ means that a significant amount of positive charge is developed at the transition state.

$\rho^+ = -3.57$

The rates of rearrangement of aromatic amine oxides are found to be correlated much better with σ^- than with σ, and have $\rho^- = 3.6$. A mechanism consistent with this finding has the

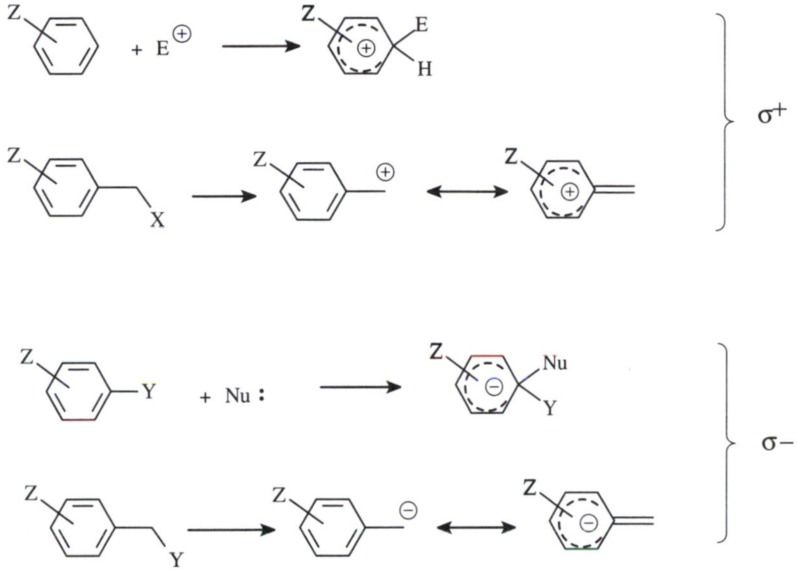

FIGURE 10.20

oxygen attacking the substituted aromatic ring, thus increasing the electron density on the ring. This charge is delocalized over the aromatic ring by resonance and thus it is in direct resonance interaction with substituents in the *para* position. The process is much better modeled by σ^- than by σ constants.

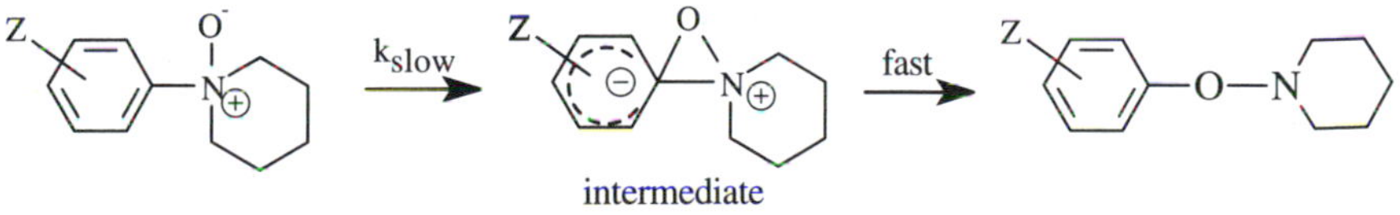

Because of the way in which they were developed, only σ_p^- substituent constants are available. If a substituent is *meta*, normal σ_m values are used because there is no direct resonance interaction with a substituent in the *meta* position.

Stereochemistry

In addition to bond breaking and charge buildup in the transition state, stereochemical changes during a reaction can also provide insight into the structural requirements of the activated complex of the rate-determining step. If a reaction is stereospecific (that is, if only one stereoisomer is formed in a reaction), there is likely to be a particular spatial relationship between groups that is required for efficient product formation. (The key here is the term *efficient* because reactions can sometimes proceed even if the correct spatial relationship is not obtainable, but they will go much more slowly.) If a reaction is stereoselective (that is, if one stereoisomer is the major but not exclusive product), one particular spatial relationship is favored over another in the product-forming step. If the stereoselectivity of a reaction can be understood, key structural elements in the activated complex can often be identified:

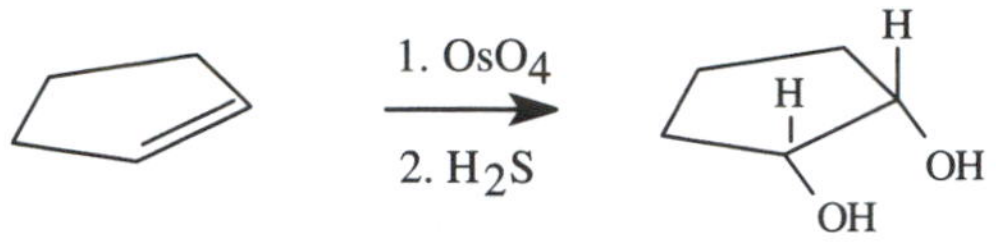

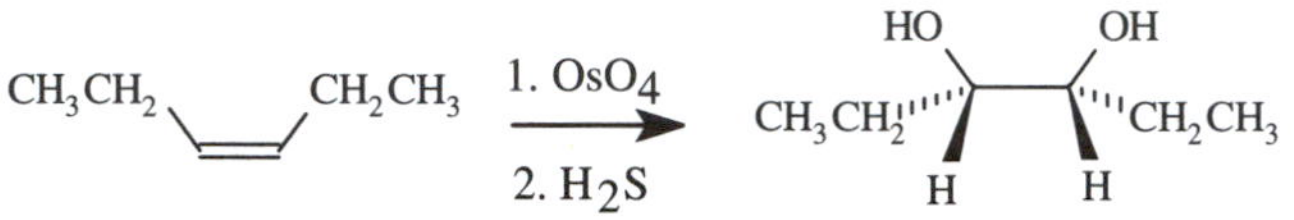

Several common examples show the power of this reasoning. The reaction of osmium tetroxide with olefins, followed by reduction, gives diols resulting exclusively from *syn* addition to the double bond. The reaction is hence stereospecific. The addition stereochemistry is clearly seen in cyclic olefins, but it is also seen in acyclic olefins where single diastereomers are produced.

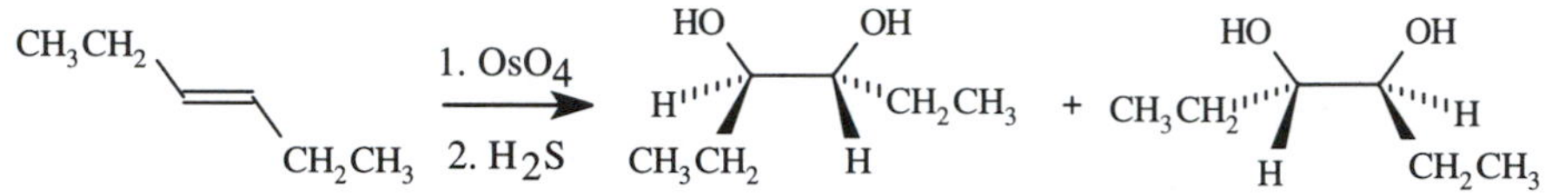

The results from the cyclic series show that both oxygens come from the same side of the double bond, probably from a single OsO_4 molecule. The results in the acyclic series demonstrate that both oxygens add to the ends of the double bond at the same time. If one oxygen added first, an intermediate with a single carbon–carbon bond would be formed which could isomerize by rotation around that bond. The observation of complete diastereoselectivity requires that both C—O bonds be formed simultaneously. Thus a concerted addition across the double bond is the most reasonable pathway consistent with these results. The stereochemical analysis for the *cis* starting material is shown in Figure 10.21. An analogous analysis for the *trans* olefin would predict only the *d,l* diastereomer for concerted addition of OsO_4 to the π bond, but the same *meso-d,l* mixture would be obtained if the addition were stepwise.

Contrast the *syn* addition of osmium tetroxide with the well-known anti stereochemistry found in the addition of bromine to alkenes. Cyclic systems give only *trans* addition, in most cases, and acyclic olefins give single diastereomers that depend on the geometry of the starting olefin. These results are consistent with one bromine adding to one face of the olefin to give a bridged ion which maintains the stereochemistry of the original olefin. Bromide ion adds from the opposite face to give a single diastereomeric dibromide product.

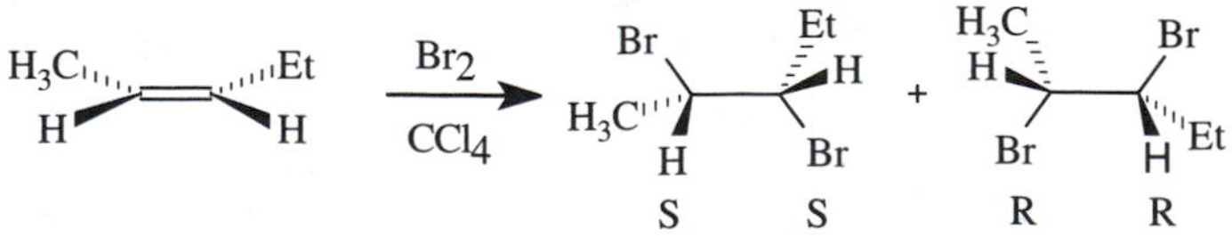

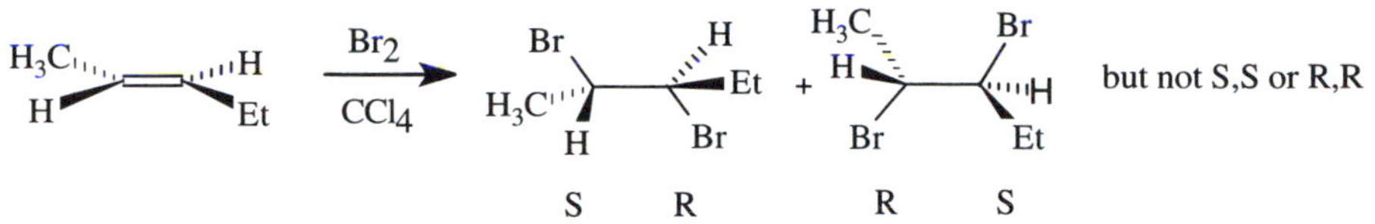

but not S,S or R,R

Walden inversion is the term given to the change in stereochemistry observed in bimolecular nucleophilic substitutions. For example, reaction of (2S)-2-triflyloxyesters with sodium azide gives (2R)-2-azidoesters.

Inversion of configuration requires that the nucleophile adds electrons to the σ* orbital of the carbon–triflate bond from the side opposite that bond. As required by the stereochemistry,

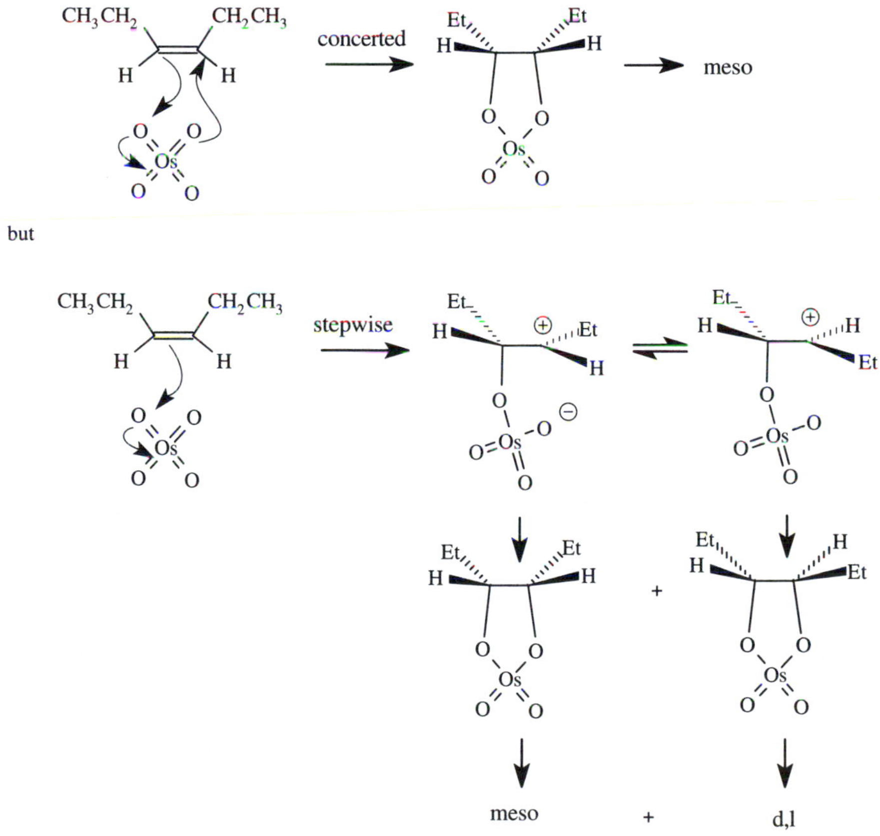

but

FIGURE 10.21

formation of the bond from azide to carbon is concurrent with cleavage of the carbon-leaving group bond.

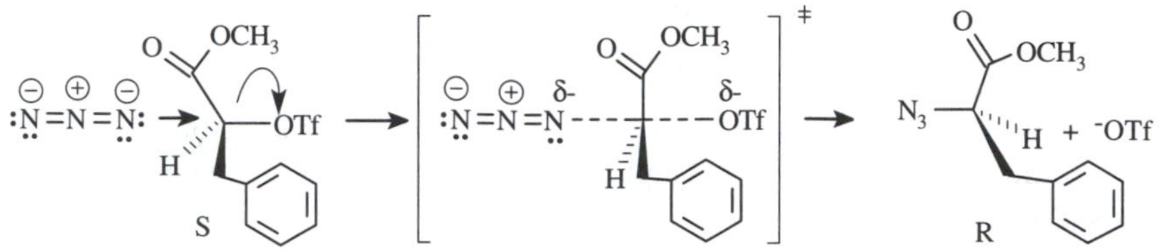

If racemization is observed, it could only be due to a cleavage of one of the bonds to the chiral center either prior to carbon–nitrogen bond formation or subsequent to it. This could occur by (a) enolization of the starting triflate, (b) an ionization of triflate to a carbocation (and then nucleophilic attack by the azide), or (c) enolization in the azido product (see Figure 10.22). The fact that clean inversion occurs means that not only does the substitution by azide occur with inversion, but also that none of these other processes are significant under the reaction conditions since they would lead to racemized product.

The stereoelectronic requirements of groups undergoing base-promoted elimination are also easily seen by considering the reaction products. Treatment of *trans*-2-methylcyclo-hexyl tosylate gives 3-methylcyclohexene as the major product, whereas treatment of *cis*-2-methylcyclohexyl tosylate gives the more stable 1-methylcyclohexene as the only product.

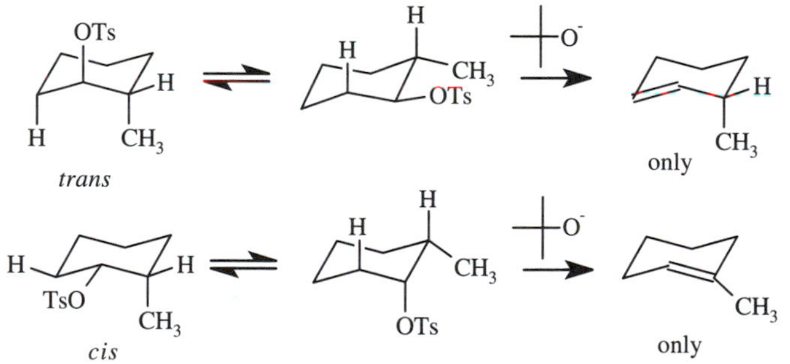

These data are consistent with the favored transition state having an antiperiplanar relationship between the proton being removed and the leaving group. In a six-membered ring, this can only occur when they are both diaxial. In the *trans* isomer, the conformation in which the tosylate is axial only has a proton at C_6 axial and is antiperiplanar. Thus elimination occurs across C_6 and C_1 to give only 3-methyl cyclohexene. In the *cis* isomer, the conformation in which the tosylate group is axial has antiperiplanar Hs at *both* C_2 and C_6. Elimination could proceed in either direction. However, removal of the proton at C-2 is favored, because the more stable olefin product is produced.

These examples show the power of stereochemical information in pinpointing structural elements of activated complexes. Combined with other types of mechanistic information, even the most intimate mechanistic details can be clarified in many cases. For example, consider the solvolysis in ethanol of 3-phenyl-2-tosyloxy butane, in which the replacement of the tosylate group by a solvent nucleophile is noted.

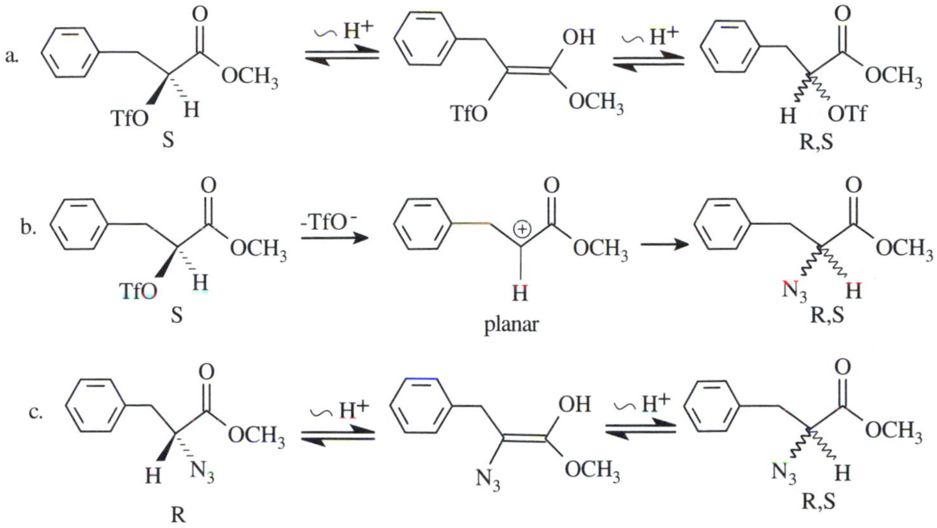

FIGURE 10.22

While this appears to be a simple substitution reaction, the details can be further explored. It was found that this reaction proceeded by a first-order rate law, which suggests an ionization pathway (Sn1) for the substitution. However, when a group of substituted aromatic compounds were investigated, plots of the rate constants (log k_Z/k_H) gave a much better correlation with σ_Z^+ than with σ_Z and $\rho^+ = -1.3$. This Hammett study reveals that a positive charge is developed on the aromatic ring in the transition state of the rate-determining step. One of the only ways for this to happen is for the phenyl ring to interact with the positive charge produced by the ionization of the leaving group.

The use of the 2R, 3R-isomer of the starting material led to formation of only 2R, 3R-2-ethoxy-3-phenylbutane. Thus the configuration at each chiral center was retained in the product. This stereochemical data rules out simple ionization and solvent capture as a reaction mechanism since this would lead to a mixture of 2R and 2S configurations. From these observations, it has been postulated that the phenyl group assists ionization of the leaving group by electron donation to produce a bridged ion.

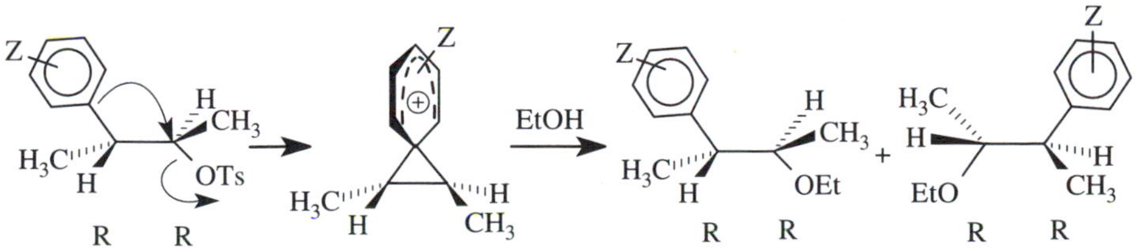

The bridged ion has a positive charge delocalized over the aromatic ring, as required by the σ_Z^+ correlation. Furthermore, the solvent nucleophile can only add from the side opposite the bridging phenyl group, leading to retention of configuration as the stereochemical results demand.

Stereochemical studies can be an indispensable adjunct to other types of mechanistic investigations for unraveling the details of reaction processes. They allow the positions of atoms or groups in a molecule to be tracked through a reaction, thereby revealing the spatial requirements of the reaction.

Bibliography

For a complete, yet succinct, discussion with references see J. A. March, *Advanced Organic Chemistry, Reactions, Mechanism, and Structure*, 4th ed., Wiley, New York, 1992, Chapters 6 and 8.

An excellent summary is found in F. A. Carey and R. J. Sundberg, *Advanced Organic Chemistry, Part A. Structure and Mechanisms*, 3d ed., Plenum, New York, 1990, Chapter 4.

Another excellent summary is found in T. H. Lowry and K. S. Richardson, *Mechanism and Theory in Organic Chemistry*, 3d ed., Harper & Row, New York, 1987, Chapter 2.

Problems

1. The base-promoted elimination of quaternary ammonium ions (Hofmann elimination) has been proposed to proceed by an E2-like mechanism. Tell how each of the following observations supports this mechanistic classification. Be specific about exactly what each piece of information reveals.

(a)

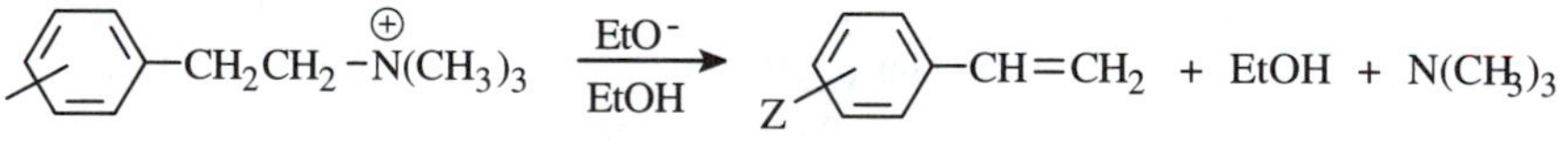

$$Z-C_6H_4-CH_2CH_2-\overset{\oplus}{N}(CH_3)_3 \xrightarrow[\text{EtOH}]{\text{EtO}^-} Z-C_6H_4-CH=CH_2 + EtOH + N(CH_3)_3$$

$\rho = +3.58$ for a series of Z-substituents when plotted against σ_Z.

(b)

$$C_6H_5-CH_2CH_2-\overset{\oplus}{N}(CH_3)_3 \quad \text{vs} \quad C_6H_5-CD_2CH_2-\overset{\oplus}{N}(CH_3)_3$$

$$\frac{k_H}{k_D} = 3.23$$

(c)

$$C_6H_5-\overset{12}{C}H_2CH_2-\overset{\oplus}{N}(CH_3)_3 \quad \text{vs} \quad C_6H_5-\overset{14}{C}H_2CH_2-\overset{\oplus}{N}(CH_3)_3$$

$$\frac{k_C^{12}}{k_C^{14}} = 1.044$$

(d)

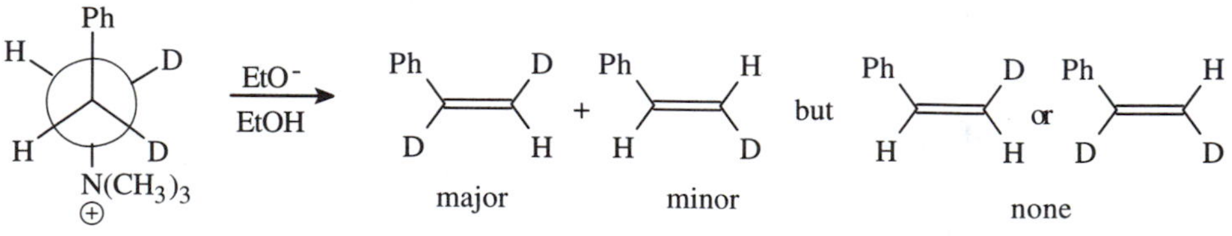

$$\xrightarrow[\text{EtOH}]{\text{EtO}^-}$$

major   minor   none

2. Explain the mechanistic significance of the ρ values for the elimination of 1 under the two different sets of conditions.

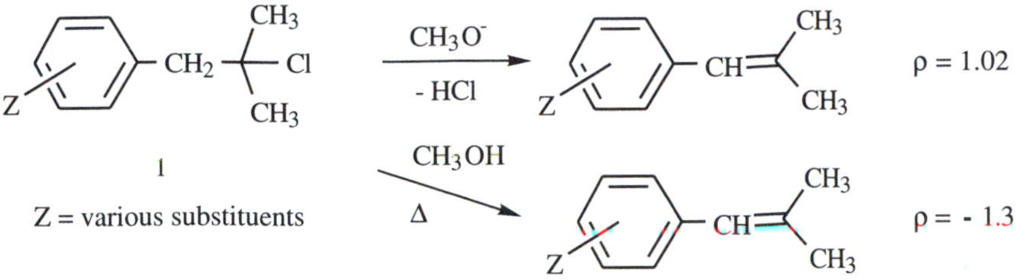

1

Z = various substituents

3. Propose a mechanism for the following reaction given that $k_H/k_D = 6.1$ and $\rho^- = 2.02$ (there is a better correlation with σ^- substituent constants than with σ values).

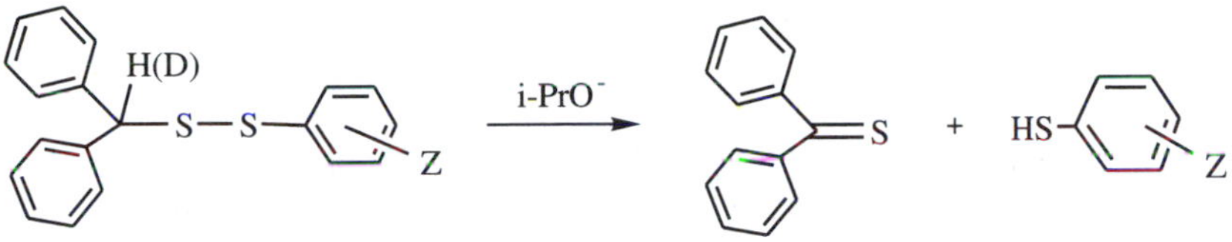

4. Tell how isotopes can be used to distinguish between the following two mechanisms:

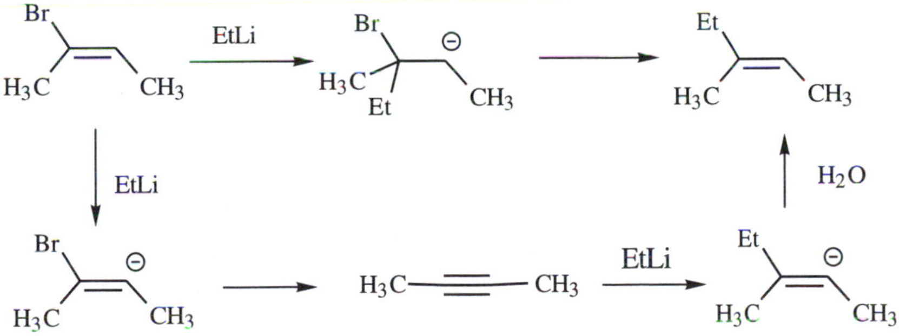

5. The following transformations occur by different mechanisms. Show them and predict what a Hammett study would show for each of them.

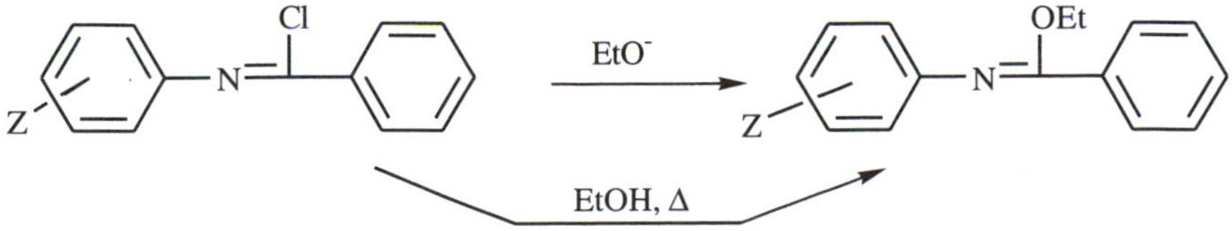

6. The base hydrolysis of a series of substituted phenoxy esters gave a much better correlation when the rate constants were plotted against σ_Z^- than against σ_Z. Give the mechanistic significance of this behavior.

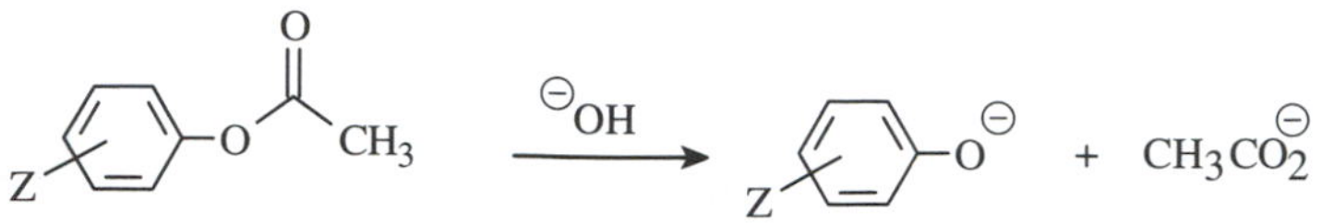

7. Based on the better correlation of rates versus σ_Z^+ and $\rho^+ = -1.03$ give a likely mechanism for the following substitution:

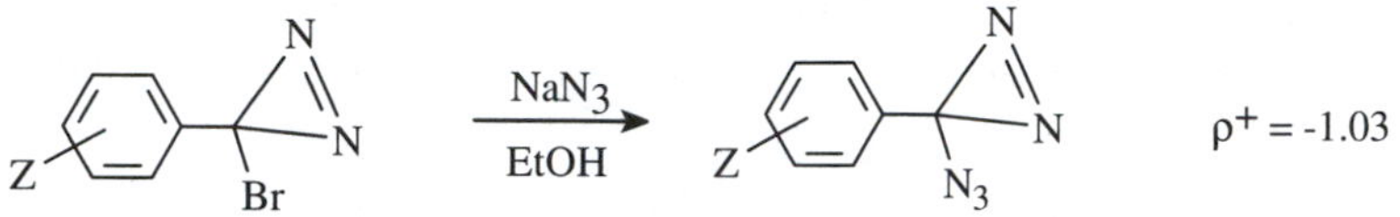

$\rho^+ = -1.03$

8. When 17α-methyl-5α-androstan-3β, 17β-diol and its deuterated analog were tested, it was found that the protio compound is three times more active than the deutero analog. It is also known that the diol itself is not biologically active. What is the likely metabolically active material? Explain.

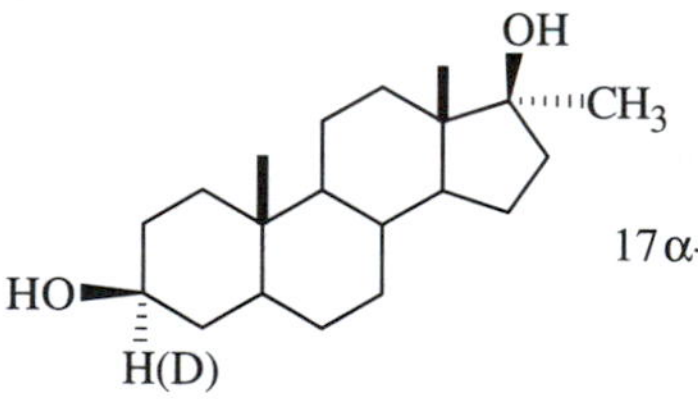

17α-methyl-5 α-androstan-3 β, 17β-diol

9. The oxidation of substituted benzaldehydes to substituted benzoic acids by pyridinium fluorochromate has been studied and it has been found that the reaction is first-order with respect to pyridinium fluorochromate but is of complicated order with respect to the aldehyde. The following scenario was proposed to account for this behavior:

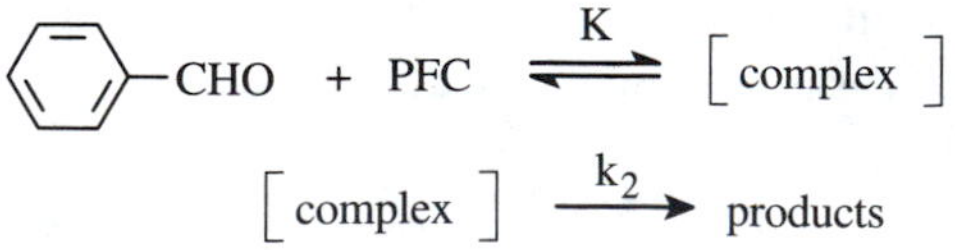

Using the steady state approximation, derive the rate law for this mechanistic scenario.

It was further found, using α-deutero benzaldehyde, that $k_H/k_D = 5.33$ and, using substituted benzaldehydes, that $\rho^+ = -2.2$.

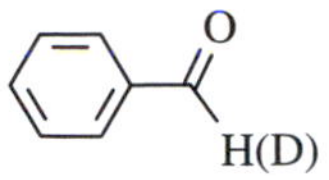

$$\frac{k_H}{k_D} = 5.33 \qquad\qquad \rho^+ = -2.2$$

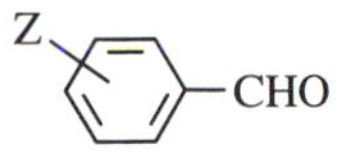

Use these data to delineate the nature of the rate-determining step and propose plausible mechanism for the reaction. Explain how the data was used to arrive at the mechanism.

10. The rate of reaction of substituted aromatic chlorides with methoxide to produce substituted anisoles is found to give a linear correlation with σ^- but not with σ_Z. Explain in terms of a reaction mechanism.

$$\rho^- = +8.47$$

11. The addition of phenylsulfenyl chloride to cyclohexene in the presence of silver fluoride gives a β-fluoro thioether in which the coupling constants of the α and β protons of the product requires them to be trans diaxial. What is the stereochemistry of the addition and how is it achieved?

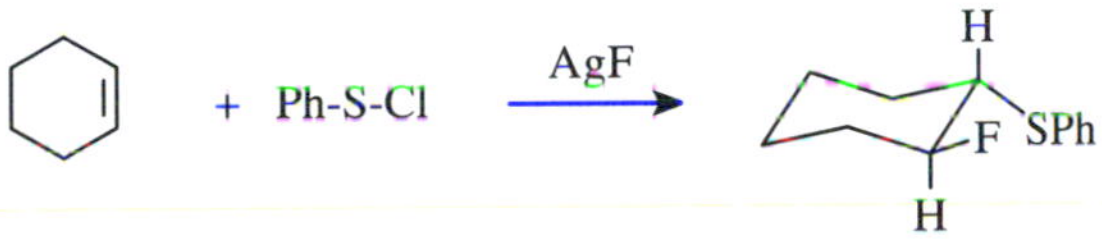

12. The Beckman rearrangement could occur by either a stepwise or a concerted mechanism.

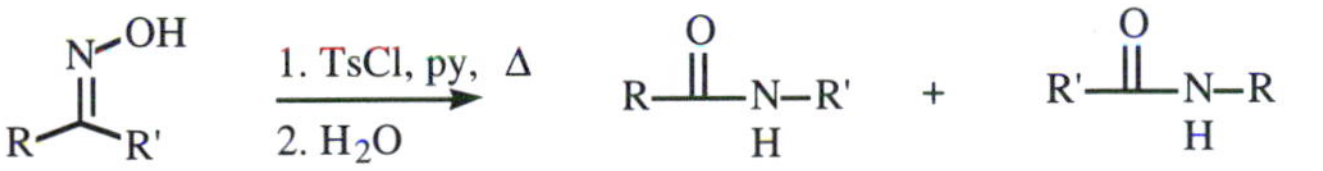

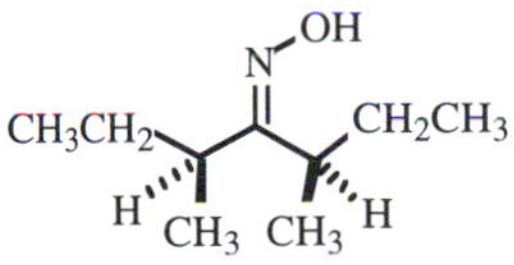

4

(a) Show both mechanisms using curved arrow notation.
(b) Suppose one had made optically active oxime **4**.
    (1) Would it rotate plane polarized light?
    (2) Label the configurations of the chiral centers in **4**.
    (3) Show how **4** could be used to help distinguish the mechanisms given.

STRUCTURE DETERMINATION OF
ORGANIC COMPOUNDS

Structure Determination

Because of the wide variety of reactions available to synthetic chemists, it is possible to devise synthetic strategies for just about any target. Nevertheless, the planning and execution of any synthesis must be verified by showing that the product of each step is in fact the predicted compound, and that the target compound was actually obtained. Thus a critical part of any synthesis involves determining and proving the structures of synthetic intermediates and final products. Although careful planning will generally result in the formation of the expected product, there are always enough exceptions to make structure proof an imperative step.

In earlier times, proof of structure was based largely on wet methods. The first step was to rigorously purify the compound by crystallization, distillation, sublimation, and the like. The functional groups present in the material were established by classification tests. The elemental analysis gave the molecular formula, and from a knowledge of the starting materials, a tentative structure could be written. Confirmation of the structure was obtained either by degradation to known compounds or by an alternative synthesis of the compound from known starting materials. Thus whereas the presence of functional groups could be determined rather straightforwardly, the connectivity of atoms and groups in a molecule was much more difficult to establish.

Today, structure proof involves the same components: purification, functional group identification, and establishment of atom and group connectivity. However, the ways in which these are accomplished are more efficient, sensitive, and reliable. They are also much faster. The ability to run reactions, purify products, and determine structures on milligram scales in a matter of hours has greatly increased the rate at which structural information can be obtained. This has resulted in an exponential growth of chemical knowledge and is directly responsible for the explosion of information being continually published in the chemical literature.

Chromatographic Purification

The first step in the identification of any compound is to obtain that material in pure form. The most common way to achieve this goal today is to use chromatography. Although a discussion of the many separation and purification techniques which utilize some form of chromatography is outside the scope of this book, all these techniques rely in one way or another on the interaction of molecules with a surface. Such interactions depend much more on the chemical properties of a molecule (functional groups, polarity, unsaturation, etc.) than on physical properties of the bulk substance (boiling point, vapor pressure, etc.). Furthermore, the interactions of a compound with a surface allow it to be resolved (separated) from other molecules by placing it in a flowing system (mobile phase). When the molecule is not adsorbed to the surface, it moves over the surface at the same velocity as the mobile phase (see Figure 11.1). When it is adsorbed to the surface, it does not advance with the mobile phase. Since adsorption is an equilibrium process, those compounds which are only weakly adsorbed (M in Figure 11.1) to the surface spend a greater portion of time in the mobile phase. These compounds move over the surface faster than compounds which are more strongly adsorbed (M*) and thus spend more time immobilized on the surface. Because different compounds are adsorbed differently on the surface, each can travel at a different rate over the surface. By collecting the effluent from the surface as a series of fractions, individual compounds can be separated cleanly from other components in the original mixture because each component exits the surface at a different time.

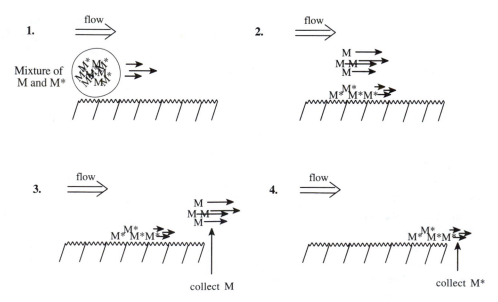

FIGURE 11.1

Many different mobile phases have been utilized to provide the forward velocity for non-adsorbed molecules. If the mobile phase is a gas, the technique is called gas chromatography (GC). In gas chromatography, the surface to which the molecules adsorb can be a wide variety of materials which are often prepared by coating an inert surface with a polymer whose properties are related to its structure. In this way, the surface properties and hence adsorption of the solid surface can be varied to give the best chromatographic resolution.

If a liquid is used as the mobile phase, the technique is termed liquid chromatography (LC). The solid adsorbent is constrained in a tube or column, through which the liquid mobile phase flows. Any number of solvents, buffer solutions, or supercritical fluids can be used as liquid mobile phases. High-pressure liquid chromatography is used if pressure is needed to force the liquid phase through the tube. If the liquid phase moves over a thin adsorbent surface propelled by capillary action, the technique is termed thin layer chromatography (tlc). Two types of surfaces are generally used as the solid phase.

In "normal phase" LC systems, the solid phase is a polar solid such as silica gel (most common) or alumina, and the liquid is generally an organic solvent of low polarity. In such a case, polar compounds bind more strongly to the polar silica gel surface and thus travel more slowly along the surface. Nonpolar components have a lower affinity for the polar surface and a greater affinity for the nonpolar eluting solvent and consequently elute from the column more rapidly. In "reversed phase" systems, the surface of silica gel is modified to produce a nonpolar hydrocarbon-derivatized surface, and the mobile phase is often a polar, aqueous solvent mixture. In this case, polar compounds have a low affinity for the nonpolar surface; they remain dissolved in the polar mobile phase and elute more rapidly. Nonpolar components have a higher affinity for the nonpolar surface than for the polar mobile phase, and they elute more slowly. Using various chromatographic techniques, it is possible to separate most mixtures into the individual components efficiently, and very rapidly.

In addition to chromatographic techniques, traditional purification methods such as recrystallization, distillation, or sublimation are also employed. Such methods often require much more material than do chromatographic techniques.

Instrumental Methods

When the reaction product is obtained in pure form, modern instrumental methods of structure determination, rather than traditional wet methods, provide the fastest way to determine the functionality and connectivity present. Today's chemist has a large number of tools available with which to probe the structure of molecules. For determining the structures of organic molecules, the "big three" are nuclear magnetic resonance (nmr) spectroscopy, infrared (IR) spectroscopy, and mass spectrometry (MS). The frequency of use of these techniques generally falls in the same order (nmr > IR > MS).

The first two of these methods (nmr and IR) are spectroscopic techniques in which the molecule is interrogated with electromagnetic radiation in a particular range of frequencies (nmr uses radio frequencies and IR uses infrared radiation). Only certain frequencies will be absorbed by a particular compound, and those frequencies that are absorbed can be used to infer structural details about the compound.

Mass spectrometry is not a true spectroscopic technique because absorption of electromagnetic energy is not involved in any way. In mass spectrometry, the molecule is fragmented into ions and these charged pieces are separated on the basis of their mass-to-charge ratio. Since the usual charge is $+1$, the masses of the pieces are determined. Knowledge of the masses of the pieces allows the structure of the compound to be reconstructed.

Another distinction between these techniques is the structural information they are capable of revealing. Both nmr spectroscopy and mass spectrometry establish connectivity between atoms and groups in a molecule (albeit slightly differently). In addition, the functional groups present are suggested. Infrared does not generally establish connectivity, but it is unmatched for identifying functional groups present in the molecule. By combination of these three major methods, the functional groups, the molecular weight, and the connectivity of the atoms and groups can be established rapidly and efficiently. It is normally a trivial step to then write the structure of the molecule.

Nuclear Magnetic Resonance Spectroscopy

The speed with which nuclear magnetic resonance spectroscopy has been incorporated into scientific inquiry is amazing. The first commercial spectrometers became available in the 1950s. By the middle 1980s, whole bodies could be placed in the probes of nmr spectrometers (magnetic resonance imaging) and the structures of body parts could be determined in exquisite detail. Today, structures of proteins and other macromolecules in solution or in the solid state are determined routinely. What was unthinkable in the 1960s is routinely practiced in the 1990s, even by undergraduates. The power of the method and the structural detail it provides have no doubt fueled its rapid development.

Nuclear magnetic resonance spectroscopy is possible due to the absorption of energy at particular frequencies by atomic nuclei that are placed in a magnetic field. Most atomic nuclei are characterized by a property termed spin, and this gives rise to a magnetic moment associated with that nucleus. The magnitude of the magnetic moment of the nucleus, which is also quantized by the spin quantum number, is characteristic of that nucleus. Nuclei such as ^{12}C, ^{16}O, and ^{32}S have nuclear magnetic moments of zero. Nuclei such as ^{1}H, ^{11}B, ^{13}C, ^{15}N, ^{17}O, ^{19}F, and ^{31}P have finite magnetic moments, spin quantum numbers of $I = 1/2$, and are most useful in nmr measurements. Still other nuclei, such as ^{2}D and ^{14}N, have finite magnetic moments but spin numbers of $I > 1/2$, and are much more difficult to deal with, although today's nmr instruments handle these elements routinely as well.

Fortunately for organic chemists, hydrogen and carbon are the most common nuclei found in organic compounds, and the ability to probe these nuclei by nmr is invaluable for organic structure determination. Since proton magnetic resonance (pmr) is the most common type, the behavior of ^{1}H nuclei in magnetic fields will serve as a model for other nuclei which have spin quantum numbers $I = 1/2$ and thus behave similarly (^{13}C, ^{19}F, etc.).

The proton has a nuclear magnetic moment (denoted as a vectoral quantity) which, under normal circumstances, can adopt any spatial orientation. Since this magnetic moment is a nuclear property, each hydrogen in a molecule has an identical nuclear magnetic moment. When placed in a strong magnetic field, the magnetic moment of the nucleus interacts with the magnetic field. The strength of the interaction depends on the strength of the applied field (H_0) and the nuclear magnetic moment, characterized by the magnetogyric ratio γ (same for all hydrogens, but different for other nuclei).

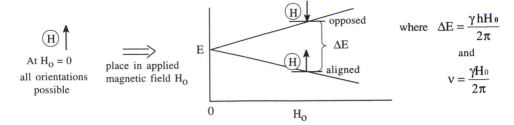

FIGURE 11.2

In a strong magnetic field, the nuclear magnetic moment is no longer free to adopt just any orientation. Instead, the spin quantum number ($I = 1/2$ for the hydrogen nucleus) results in only two allowed orientations ($2I + 1$) of the nuclear moment relative to the direction of the applied field: either aligned with H_0 (lower energy) or opposed to it (higher energy). The difference in energy (ΔE) between the two states is given by $\Delta E = \gamma h H_0/2\pi$ and is dependent on the cross product of the strength of the applied field H_0 and the magnetic moment of the hydrogen γ. Since γ is the same for all hydrogen nuclei, the energy difference between the two allowed orientations is proportional only to the strength of the applied field (see Figure 11.2).

If the magnetic field H_0 is fixed and held very constant, the energy gap between the two spin states of the hydrogen nuclei will remain constant. Irradiation of the system at the appropriate frequency ($\Delta E = h\nu$) will cause the energy to be adsorbed and the spin of the nucleus will flip from the low energy state (aligned) to the higher energy state (opposed). It is this absorption of energy which is used to probe the structural features of the molecule (see Figure 11.3).

Since the magnetic moment of a nucleus (γ) is an atomic property, for a given magnetic field H_0 all hydrogens should absorb energy at the same frequency. However, examination of a molecule such as 1,2,2-trichloropropane reveals that the two different types of hydrogens (H_1 and H_3) absorb at two different frequencies (ν_1 and ν_3), as seen in Figure 11.4.

Since the applied field H_0 is constant, and all hydrogen nuclei have the same magnetic moment γ, and since H_1 and H_3 absorb at two different frequencies, the magnetic field that is actually experienced by each set of nuclei (H_{eff}) must be different. Stated differently, even though a constant magnetic field H_0 is applied to the sample, each type of hydrogen, H_1 and H_3, experiences a unique magnetic field H_{eff} (where $H_{eff} \neq H_0$) and consequently absorbs energy at a unique frequency, ν_1 and ν_3. Thus, different types of protons are distinguished by the different frequencies at which they absorb energy. Furthermore, integration of the absorption

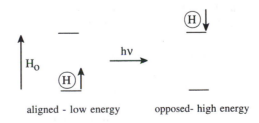

aligned - low energy   opposed- high energy

FIGURE 11.3

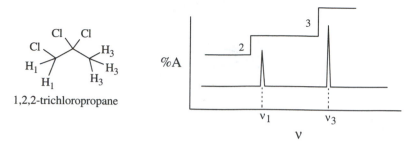

1,2-trichloropropane

FIGURE I 1.4

intensities of the two signals gives a 3 : 2 ratio, which corresponds to the smallest whole number ratio of each type of proton present. (The integrated area of the peak is given by the step height on the integration curve.) Thus ^{1}H nmr is able to distinguish different types of protons in a molecule and tell how many there are (at least by ratio).

Chemical Shift

The range of frequencies over which protons absorb in most organic molecules depends on the applied field. For example, for an applied field of 14,000 gauss, most protons will absorb over a range of 600 Hz, beginning at the value of 60×10^6 Hz (60 MHz), or from 59,999,400 Hz to 60,000,000 Hz. At 23,486 gauss, this range is 1000 Hz from the value of 100 MHz, or from 99,999,000 Hz to 100,000,000 Hz. Thus the actual range of frequency of absorption depends on the magnetic field of the instrument. (This is exactly as expected, since the energy gap between the spin states and hence the frequency of absorption are both dependent on the applied field.) To compare absorption values from different instruments, a dimensionless scale must be devised that is independent of the magnetic field of the instrument. This is accomplished by using the absorption of tetramethylsilane (TMS) as a spectral anchor. The frequency of absorption of a given set of protons is measured relative to the frequency of absorption of TMS. This absorption frequency difference ($\Delta \nu$) in hertz (cps) is expressed as δ, the chemical shift of the protons in parts per million, where

$$\delta = \frac{\Delta \nu (\text{Hz})}{\text{operating frequency of the spectrometer (Hz)}} \times 10^6 \text{ ppm}$$

The chemical shift δ is dimensionless and independent of the spectrometer. Since normal absorption ranges $\Delta \nu$ are about 0–600 Hz for an operating frequency of 60×10^6 Hz, or 0–1000 Hz at 100×10^6 Hz, chemical shifts range from 0 to 10 ppm for most protons.

In practice, a small amount of TMS (< 1 percent) is added to the nmr sample, the TMS signal is set at 0 ppm, and the protons of the sample are then measured in parts per million relative to TMS. The choice of TMS as a standard is useful because nearly all other protons absorb at frequencies lower than TMS. It is routine to present nmr spectra with low frequency on the left and high frequency on the right, as in Figure 11.5. Thus the TMS signal defines δ = 0 ppm on the right side of the spectrum and other proton signals are found to the left, or

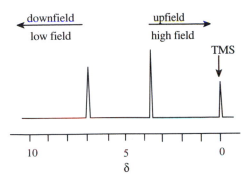

FIGURE 11.5

downfield, from TMS, from 0 to about 10 ppm. It is also normal to describe signals having larger chemical shifts as being downfield from protons with smaller chemical shifts. The left side of the spectrum is termed low field, and the right side is described as high field.

With a method available to measure differences in chemical shifts between protons, it is appropriate to ask why different protons experience different H_{eff}s even though a single H_0 is applied to the sample. The explanation lies in the fact that nuclei are surrounded by electron clouds (see Figure 11.6). In the applied field H_0, electron pairs in bonds surrounding the hydrogens act to counter the applied field by induced fields H_{ind}. The result is that the nucleus is shielded from the applied field by its electron cloud. (Nuclei which are more shielded come at higher field and have lower chemical shifts.)

Thus it is the electron density around the nucleus which shields the nucleus from the applied field. It follows that the greater the electron density around a proton, the larger will be the induced field H_{ind}, and the more shielded the proton will be. It will appear more upfield and will have a smaller chemical shift (δ value) Conversely, the lower the electron density around a proton, the less shielded it will be, the more downfield it will be, and the larger will be its δ value (see Figure 11.7).

Structural features that withdraw electrons from protons cause downfield shifts and larger δ values, whereas structural features which increase electron density around protons

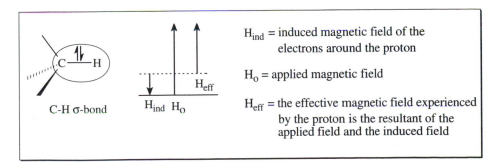

FIGURE 11.6

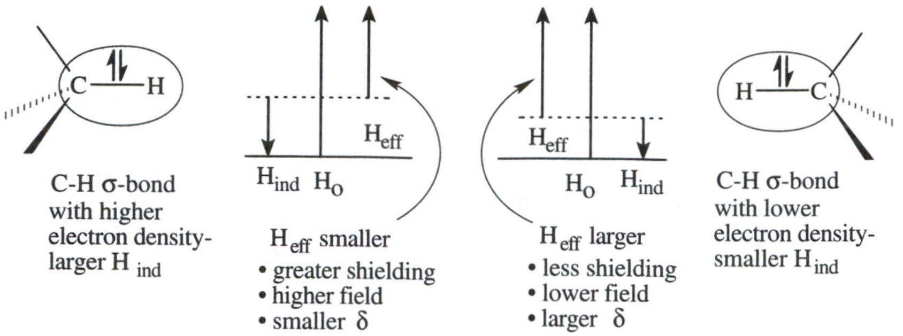

C-H σ-bond
with higher
electron density-
larger H $_{ind}$

H$_{ind}$ H$_O$

H$_{eff}$ smaller
• greater shielding
• higher field
• smaller δ

H$_O$ H$_{ind}$

H$_{eff}$ larger
• less shielding
• lower field
• larger δ

C-H σ-bond
with lower
electron density-
smaller H$_{ind}$

FIGURE 11.7

cause upfield shifts and lower δ values. For example, chemical shifts for methyl chloride, dichloromethane, and chloroform are δ = 3.0, δ = 5.5, and δ = 7.1, respectively. The inductive effects of increasing numbers of chlorine atoms decrease the electron density about the hydrogens and result in increasing chemical shifts.

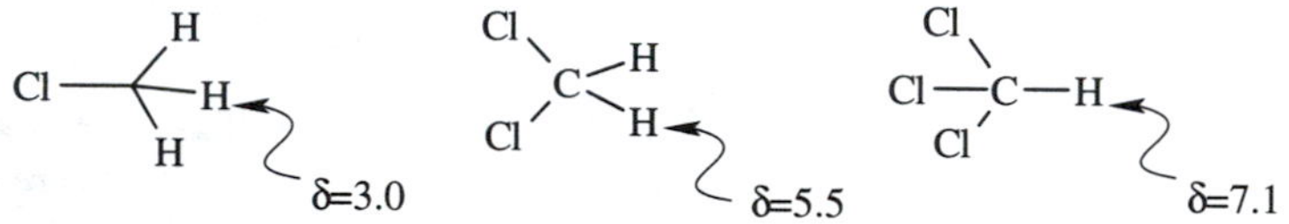

δ=3.0   δ=5.5   δ=7.1

Likewise, 1,2,2-trichloropropane (discussed previously) has the two-proton signal downfield from the three-proton signal. This happens because the methylene protons are influenced by the inductive effects of three chlorine atoms, two vicinal and one geminal, while the methyl group is influenced by only two vicinal chlorine atoms. The electron density is higher at the methyl hydrogens, they are more shielded, and they occur at higher field than the two protons of the methylene group.

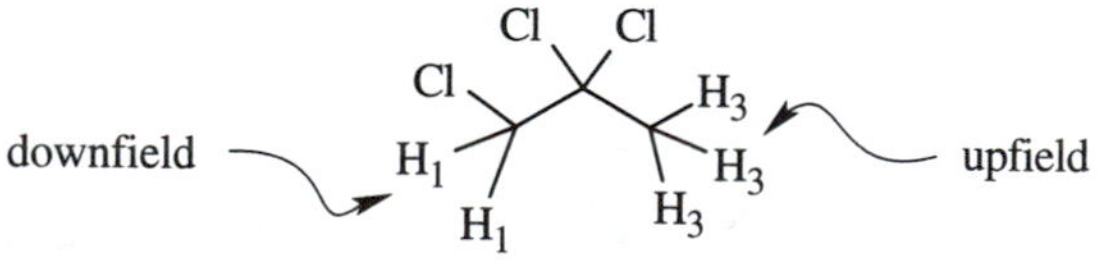

downfield   upfield

1,2,2-trichloropropane

Consideration of a series of compounds containing methyl groups illustrates clearly the influence of the electron density on chemical shift (see Figure 11.8). As the electron-withdrawing abilities of groups attached to the methyl group increase, progressive downfield shifts are evident and δ values increase. Conversely, TMS comes very far upfield because silicon–carbon bonds are polarized toward carbon and result in very high electron density about the methyl hydrogens of TMS.

Although the influence of electron density on chemical shift is clear, it is not the only factor which determines the chemical shift, as seen from the following series of compounds:

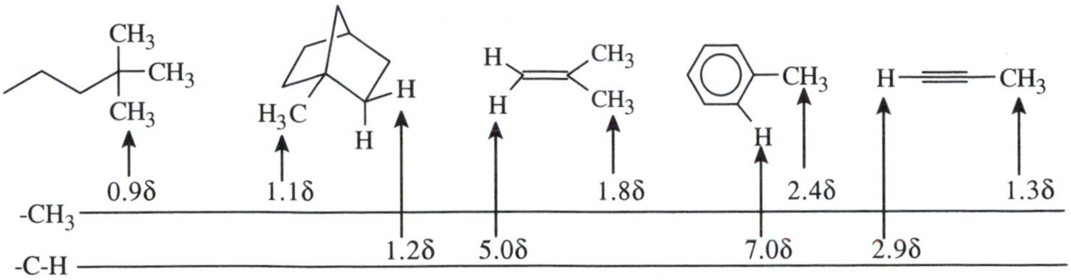

CH_3
$—CH_3$
CH_3
0.9δ

H_3C H H
1.1δ

H CH_3
H CH_3
1.8δ

$—CH_3$
H
2.4δ

H$≡$$—CH_3$
1.3δ

-CH$_3$

-C-H   1.2δ   5.0δ   7.0δ   2.9δ

Comparing the methyl groups, we find that typical saturated aliphatic methyl groups come at 0.9–1.1 ppm. However, attaching a methyl group to a double bond gives a change to 1.8δ. Attaching the methyl group to an aromatic ring moves it further downfield, to 2.4δ. Attachment to a triple bond moves it back upfield, to 1.3δ. Analogous but even larger changes in chemical shift are seen for protons directly attached to double bonds, aromatic rings, and triple bonds. Simple electron density shielding arguments cannot satisfactorily account for these large changes in chemical shift.

For example, the greater s character of sp^2 orbitals and hence greater effective electronegativity of sp^2-hybridized carbon might account for the downfield shift of the protons of a methyl group when it is attached to an olefinic carbon rather than to a saturated sp^3 carbon. However, the sp^2 carbons of aromatic rings should induce the same downfield shift. In fact, aromatic methyl groups are shifted significantly further downfield. By the same argument, attachment of a methyl group to the sp-hybridized carbon of an acetylene, which has even greater s character, should cause the chemical shift to move even further downfield. In fact, propargylic methyl groups are found at higher field than allylic methyl groups. Clearly, there are other factors at work which influence the chemical shifts of different types of protons.

Simple shielding of the hydrogen nucleus by its surrounding cloud of electrons is *isotropic,* in that the induced magnetic field is the same for any orientation of the hydrogen relative to the magnetic field. This is due to the fact that the electron cloud around the hydrogen nucleus behaves as though it is spherical (or nearly so). Other types of electron clouds (double bonds, aromatic clouds, triple bonds) are not spherically symmetric. The induced fields for these types of bonds are consequently not the same at different orientations of the functional group in the magnetic field. This *anisotropic shielding* or *anisotropy* leads to regions of shielding and deshielding around the functional group that are averages of the orientations possible.

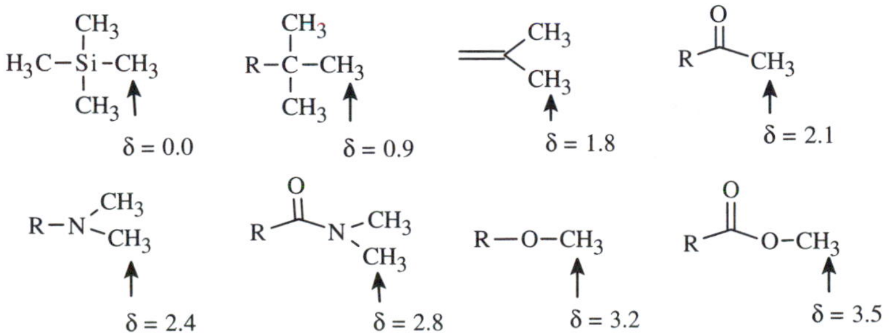

CH_3
$H_3C—Si—CH_3$
CH_3
δ = 0.0

CH_3
$R—C—CH_3$
CH_3
δ = 0.9

CH_3
CH_3
δ = 1.8

O
R CH_3
δ = 2.1

CH_3
$R—N$ CH_3
δ = 2.4

O
R N CH_3
CH_3
δ = 2.8

$R—O—CH_3$
δ = 3.2

O
R $O—CH_3$
δ = 3.5

FIGURE 11.8

Aromatic rings have among the strongest anisotropy of any group. Above and below the ring, there is a strong shielding region (H_{ind} is in opposition to the applied field) while in the plane of the ring there is a strong deshielding region (H_{ind} is in the same direction as the applied field). This phenomenon is termed ring current and has been used as a criterion to establish whether a compound is aromatic (see Figure 11.9). Consequently, protons and groups attached to the ring are in the plane of the ring and are thus strongly deshielded. These come at low field relative to a comparable proton in a nonaromatic compound. Aromatic protons normally come at $\delta > 7$ ppm, and benzylic methyl groups come at $\delta \approx 2.4$. Both are significantly shifted downfield due to the anisotropy of the aromatic ring. (The shift of benzylic protons is less than the shift of aromatic protons because benzylic protons are further from the aromatic ring than are the protons directly attached to the ring.)

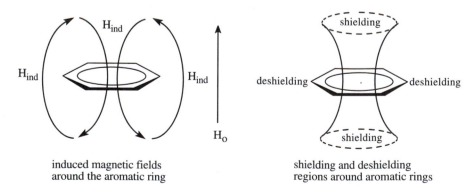

18-annulene   p-cyclophane

If protons can be positioned either in the center of or above the aromatic ring, they will fall in the shielding region and should come at high field. For example, 18-annulene is an aromatic compound ($4n + 2$, $n = 4$). The protons on the outside of the ring lie in the deshielding region ($\delta = 9.3$ ppm), while those on the inside of the ring fall in the shielding region and have $\delta = -3.0$. They come at higher field than TMS due to the anisotropic shielding from the ring current. For the same reason, the central protons in p-cyclophanes come at higher fields because they are placed over the aromatic ring in the shielding region.

Double bonds contain one σ bond and one π bond, which results in anisotropic shielding, as shown in Figure 11.10. There is a conical shielding region normal to the molecular plane and a deshielding region in the molecular plane. This is true for all double-bonded functional groups such as olefins, carbonyl groups, and imines, and it explains why olefinic protons ($\delta \approx 5$) and aldehyde protons ($\delta = 9$–10) absorb at such low fields.

induced magnetic fields
around the aromatic ring

shielding and deshielding
regions around aromatic rings

FIGURE 11.9

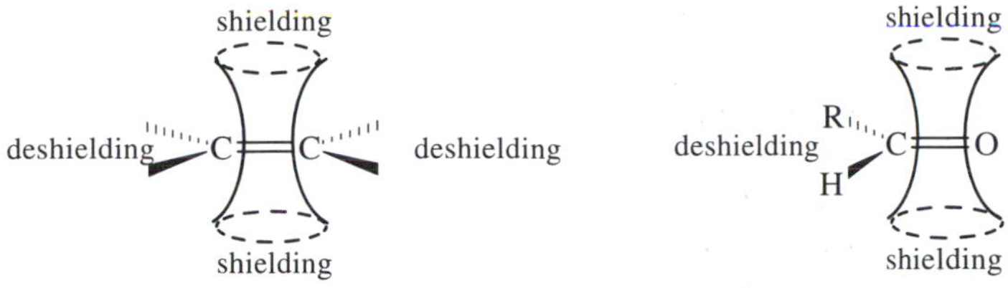

FIGURE 11.10

Both acetylene and nitriles, because of their cylindrical symmetry, have shielding regions along the triple bond axis. Thus groups attached to the triple bond are constrained to the shielding region and are shifted upfield relative to similar vinyl protons. Acetylenic protons come at $\delta = 2$–3 and propargylic methyl groups are upfield from allylic methyl groups.

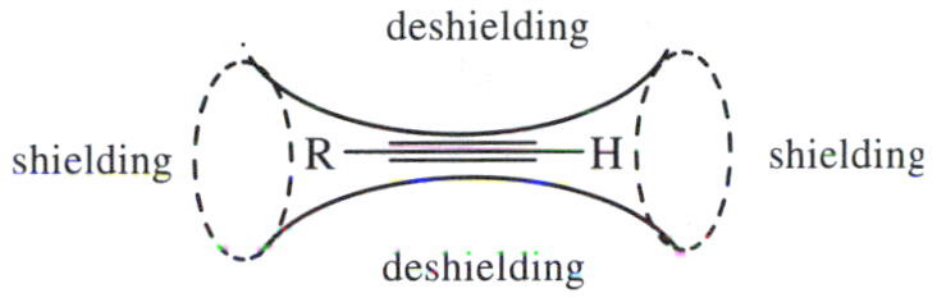

The chemical shift of a given proton is thus determined by a combination of isotropic shielding by the electron cloud surrounding the proton, and by anisotropic shielding from nearby functional groups which are strongly anisotropic. These factors are usually sufficient to give unique chemical shifts for most protons in a molecule and they can normally be distinguished using modern high field nmr spectrometers (200–300 MHz). Furthermore, the integration of these signals gives the number of different types of protons.

Spin–Spin Coupling

Determining the number of different kinds of protons in a molecule is a very important use of nmr spectroscopy. However, it does not establish the connectivity of the carbons bearing those protons, and the connectivity is crucial in correct structure determination. However, nmr spectroscopy can also give insight into the connections between functional groups by the presence of spin–spin coupling in the nmr spectrum.

Consider the nmr spectrum of 1,1-dichloro-2,2-dibromo ethane (Figure 11.11). Based on the differing electronegativities of chlorine and bromine, the two protons in the molecule are nonequivalent and should thus give signals at different chemical shifts with the same integrated areas. The dichloromethyl proton should appear downfield relative to the dibromomethyl proton. The actual nmr spectrum indeed shows two different signals, one for H_a and one for H_b, but each absorption consists of two lines and is termed a doublet. The signal for each proton is thus "split" into two resonances. This splitting is due to the fact that each proton can sense the spin state of the neighboring proton and is called spin–spin splitting.

In the magnetic field of the nmr spectrometer, both H_a and H_b have distinct absorption frequencies based on the H_{eff} each proton experiences. This gives rise to individual signals for H_a and H_b. Focus now on one hydrogen, H_a. The hydrogen which is next to H_a (namely,

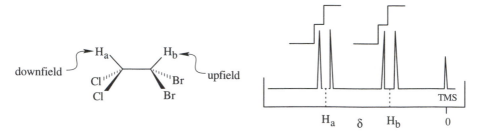

FIGURE 11.11

H_b) has two spin states (aligned or opposed to H_0) that are nearly equally populated. (Actually, there are slightly more in the lower energy spin state than in the upper, but the difference is very small.) Thus the magnetic moment of the neighboring proton H_b (μ_{H_b}) either adds or subtracts an incremental amount (μ_{H_b}) to H_{eff}, the field experienced by H_a. As a consequence, H_a will experience two distinct magnetic fields: $H_{eff} - \mu_{H_b}$ and $H_{eff} + \mu_{H_b}$. Consequently, H_a will absorb energy at two distinct frequencies, and the signal for H_a will be split into two lines—a doublet—due to the presence of the adjacent proton, H_b. Focusing on H_b, the same analysis leads to the prediction that H_b will also experience two distinct magnetic fields, $H_{eff} - \mu_{H_a}$ and $H_{eff} + \mu_{H_a}$, absorb energy at two different frequencies, and thus be split into a doublet by the presence of H_a (see Figure 11.12).

The middle of the doublet corresponds to the actual chemical shift of the proton due to H_{eff}. The total integrated area under both lines of the doublet corresponds to the signal intensity of one proton, and the width between the two lines in hertz (cps) is called J, the coupling constant. The coupling constant is a measure of the strength of the interaction between the coupled nuclei that leads to spin–spin splitting. J values for proton–proton coupling can range from 0 to 20 Hz, but most coupling constants fall in the range of 0–10 Hz.

A value of $J = 0$ means there is no significant interaction with neighboring protons; thus the absorption is not affected by the spin states of neighboring protons. This normally occurs when there are *more than three bonds* separating different types of protons.

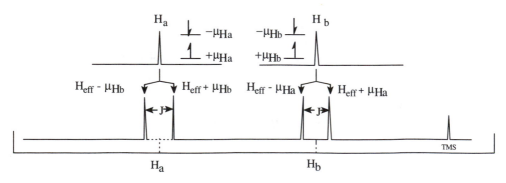

FIGURE 11.12

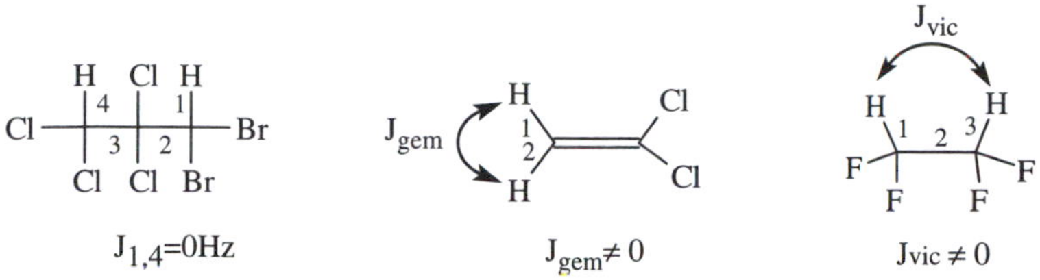

Geminal coupling (two bonds, J_{gem}) and vicinal coupling (three bonds, J_{vic}) are the types of spin–spin splitting normally encountered. In addition, the interaction between protons is reciprocal: if two protons are coupled, they are coupled equally and $J_{1,2} = J_{2,1}$. That is, if H_a is split by H_b by some amount, say $J = 6$ Hz, H_b *must* be split by H_a by $J = 6$ Hz. Finally, equivalent protons do not split each other. Thus the *t*-butyl hydrogens of *t*-butanol are a singlet:

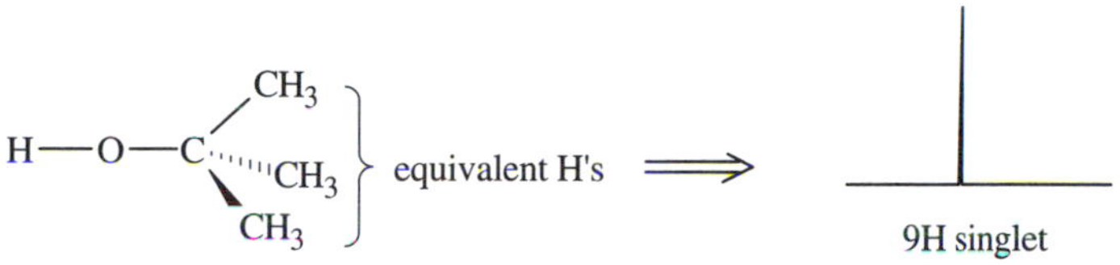

With more than one neighboring hydrogen atom, different splitting patterns are observed. For example, 1,1,2 trichloroethane has two signals, a two-proton doublet ($J = 6$ Hz) upfield (H_a) and a one-proton triplet ($J = 6$ Hz) downfield (H_b), as seen in Figure 11.13. The equivalent protons H_a of the CH_2—Cl group give the upfield absorption, which is split into a doublet by the single vicinal proton, (H_b). The single methine proton, (H_b) gives the downfield absorption, which is split into a triplet by the two adjacent, equivalent methylene protons, H_a. The triplet splitting is due to the three-spin distributions possible for the two equivalent CH_2 protons: both aligned, one aligned (two possibilities), both opposed. The H_b triplet has three lines in a 1 : 2 : 1 ratio, which reflects the number of spin states of the neighboring CH_2 group. Because the two sets of protons are coupled, the spacing between each line of the triplet ($J = 6$ Hz) must be the same as the doublet splitting ($J = 6$ Hz). The middle line of the triplet corresponds to the chemical shift of the CH_2 group, whereas the middle of the doublet corresponds to the chemical shift of the methine proton H_b (see Figure 11.14).

Diethyl ether (see Figure 11.15) has a three-proton triplet at 1.2δ ($J = 7$ Hz) for the methyl protons, which are split by the two protons of the CH_2 group. The methylene protons absorb at 3.3δ and are split into four lines (quartet) in a 1 : 3 : 3 : 1 ratio (see Figure 11.16).

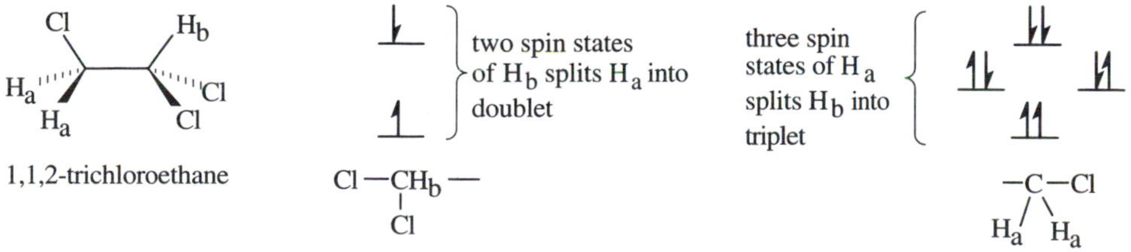

FIGURE 11.13

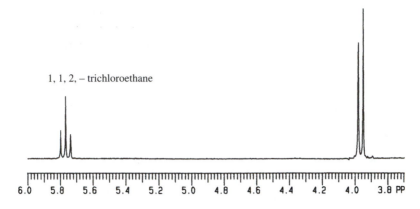

1, 1, 2, – trichloroethane
6.0 5.8 5.6 5.4 5.2 5.0 4.8 4.6 4.4 4.2 4.0 3.8 PP

FIGURE 11.14

diethyl ether
0 3.5 3.0 2.5 2.0 1.5 1.0 PPM 0.5

FIGURE 11.15

$H_3C—CH_2$
gives rise to
-CH_2-
J J J
1 3 3 1
δ

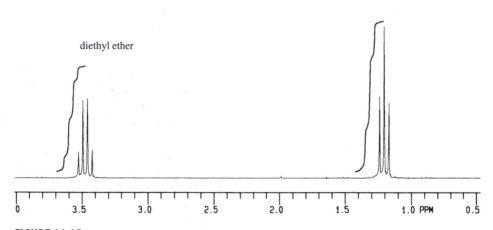

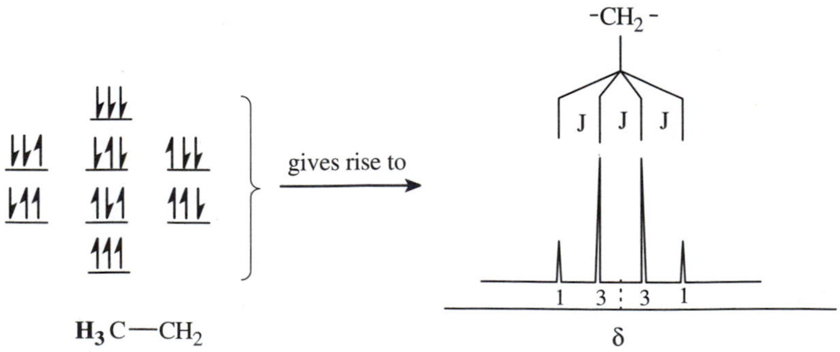

FIGURE 11.16

TABLE 11.1 Pascal's Triangle Applied to Multiple Splitting

Nearest Neighbors	Lines/Intensity	Splitting
0	1	singlet
1	1 1	doublet
2	1 2 1	triplet
3	1 3 3 1	quartet
4	1 4 6 4 1	pentet
5	1 5 10 10 5 1	sextet
6	1 6 15 20 15 6 1	septet

This splitting occurs because the three equivalent protons of the methyl group can have four possible spin distributions, which are nearly equally populated. They are three aligned; two aligned, one opposed (three possibilities); one aligned, two opposed (three possibilities); and all opposed. The center of the quartet is the chemical shift of the CH_2 group and the coupling constant of the quartet ($J = 7$ Hz) must be the same as the coupling constant of the methyl triplet ($J = 7$ Hz) since the two sets of protons are coupled. (When protons are coupled, the signal for each set is split by the same coupling constant.)

If these considerations are generalized, it is seen that the signals for protons coupled equally to n equivalent vicinal protons will be split into multiplets having $n + 1$ lines. The intensities of the individual lines in the multiplets follow Pascal's triangle (Table 11.1). The middle of the multiplet is the chemical shift of the protons responsible for that absorption, and the total integrated area under the multiplet corresponds to the total number of protons of the signal. However, the integrated area of the individual lines of the multiplet are in the ratio of Pascal's triangle. Several examples of simple splitting patterns are shown in Figure 11.17.

The $n + 1$ rule for predicting the multiplicity of a given proton signal holds when the coupling constants with all of the nearest neighbors are the same. For example, the multiplicity of the central methylene group of 1-bromo-3-chloropropane (see Figure 11.18) is a pentet which requires that $J_{12} = J_{23}$. That is, the central methylene group has the same coupling constant to the protons of the bromomethyl group (J_{12}) as to the protons of the chloromethyl group (J_{23}).

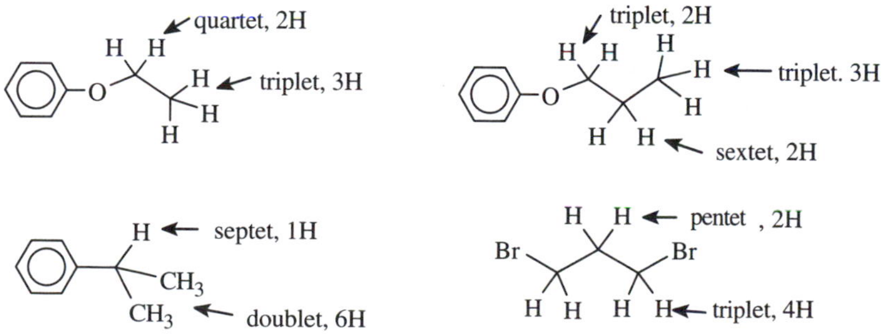

FIGURE 11.17

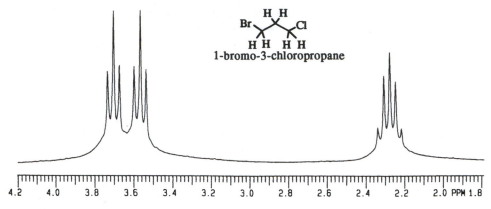

FIGURE 11.18

Those groups are not equivalent and have different chemical shifts, but each signal is split into a triplet by the C—2 methylene group by the same J value.

If a proton, or set of protons, is not coupled equally to neighboring protons, the $n + 1$ rule is not adequate to describe the multiplicity of the absorption. Instead, one observes multiplets of multiplets as the splitting pattern (e.g., doublet of doublets or triplet of doublets). The multiplicity can be understood by carrying out sequential splitting diagrams. For example, consider a proton H_b split by two neighboring vicinal protons H_a and H_c by $J_{ab} = 2$ Hz and $J_{bc} = 7$ Hz, respectively. This is shown schematically in Figure 11.19, where the H_b signal is split into a doublet by H_c ($J_{bc} = 7$ Hz) and each line of that doublet is split into a doublet by H_a ($J_{ab} = 2$ Hz). The result is a doublet of doublets. The spacing between the small doublet splitting is $J = 2$ Hz and the splitting between the centers of the two doublets is $J = 7$ Hz. The same diagram is produced by first splitting the H_b signal by $J_{ab} = 2$ Hz and then splitting each line into a doublet by $J_{bc} = 7$ Hz.

Because of the requirement that $J_{ab} = J_{ba}$, H_a will be split into a doublet ($J = 2$ Hz) by H_b, and H_c will also be split into a doublet ($J = 7$ Hz) by H_b. By taking into account these different splitting patterns, the connectivity relationships between H_a, H_b, and H_c are

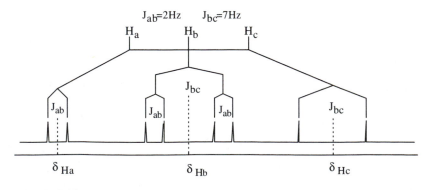

FIGURE 11.19

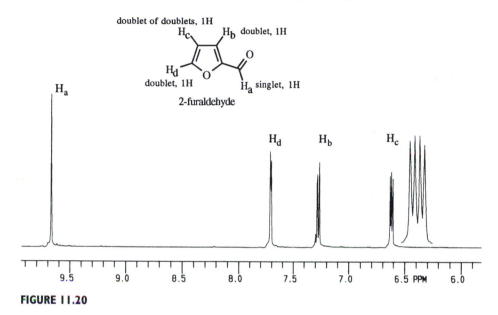

FIGURE 11.20

clear. Because H_a and H_c are both doublets, but they are split by different coupling constants, they cannot be coupled to each other. The signal for H_b, however, is seen to be a doublet of doublets, with $J = 2$ Hz and $J = 7$ Hz. Since these values are the same as the couplings of H_a and H_c, H_b is coupled to both H_a and H_c. The connectivity is thus H_a, H_b, H_c. Splitting patterns are powerful ways to establish connectivity in molecules. The patterns seen in Figures 11.20 and 11.21 are typical of the types of connections encountered in various organic compounds.

Descriptions of Spin Systems

It is often helpful to categorize spin systems in terms of the chemical and magnetic equivalence of coupled protons. Protons are chemically equivalent if they have the same chemical environment and thus the same chemical shift. Chemical equivalence can result from either identical environments or rapid rotations which yield an "average" environment for a group of protons. Considering toluene, it is seen that the two *meta*-protons are found in the plane of the ring between the *ortho* and *para* protons:

They have the same chemical environment and thus absorb at the same frequency. The methyl group is a singlet, indicating that the three methyl protons absorb at the same frequency and are chemically equivalent. Yet in the conformation shown it is clear that the environment of

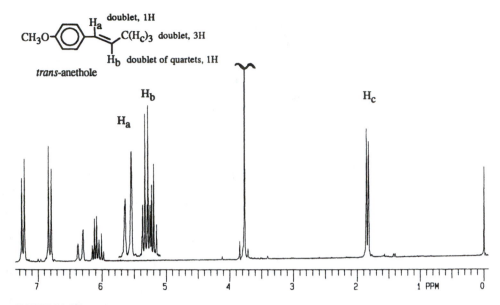

FIGURE 11.21

each proton is not the same. One is found in the plane of the ring, while a second is above, and the third below, the plane of the ring. However, due to rapid rotation of the methyl group, these hydrogens rapidly exchange positions and thus all absorb at an "average" frequency, and all are chemically equivalent by rotation.

Protons are *magnetically* equivalent if they have the same chemical shift and are coupled equally to other equivalent nuclei in the molecule. This is similar to chemical equivalence but is a more rigorous definition of equivalence. For example, the methyl protons of isobutane are chemically and magnetically equivalent, since they absorb at the same frequency and are all coupled equally to the methine proton (which should be split into a 10-line multiplet). Likewise, the two methyl groups of *p*-xylene are chemically and magnetically equivalent because they are coupled equally ($J = 0$) to the aromatic protons both *ortho* and *meta* to them.

isobutane

p-xylene

On the other hand, the *ortho* protons of *o*-dibromobenzene are chemically equivalent because they have the same environment. However, they are magnetically nonequivalent, since a given *ortho* proton is not coupled equally to the two *meta* protons (there is a 1,2 interaction with one and a 1,3 interaction with the other, and $J_{1,2} \neq J_{1,3}$). By the same arguments, the methylene groups of 1,4-dibromo-*cis*-2-butane are chemically equivalent but magnetically nonequivalent, because the methylene groups are not coupled equally to the chemically equivalent vinyl hydrogens (again $J_{1,2} \neq J_{1,3}$). Of course, this also means the vinyl hydrogens are magnetically nonequivalent, since a given vinyl H is not equally coupled to the two methylene groups.

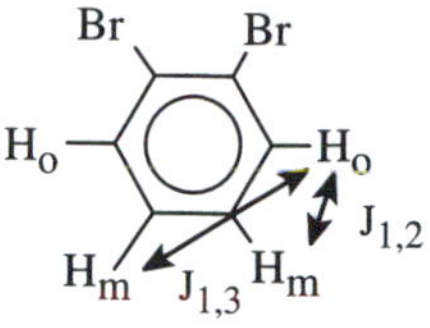

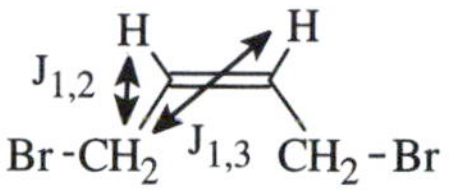

We would expect that the spectrum of the latter compound would consist of two signals: a two-proton triplet in the vinyl region and a four-proton doublet in the allylic region. We expect this because the coupling constant $J_{1,3}$ is zero. It if were not zero, a more complicated spectrum would result. Thus, magnetic nonequivalence can lead to much more complicated spectra.

Using chemical and magnetic equivalence, it is possible to designate the number and type of different protons in a spin system. This is done by choosing letters of the alphabet to indicate protons of similar chemical shift: ABC, MNO, or XYZ. For only two types of protons, letters from the first part and last past of the alphabet are chosen (A, X). If three types are present, then letters from the middle group are also used (e.g., A, M, X). A subscript is used to indicate how many of each type of proton are present in the spin system. Several spin systems are shown below with their designations:

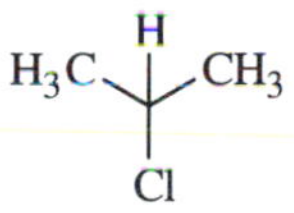

| AX_6 | A_3 | A_2X_3 | A_3 | A_2X_3 | AM_2X_3 | AB |

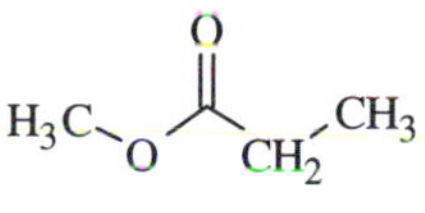

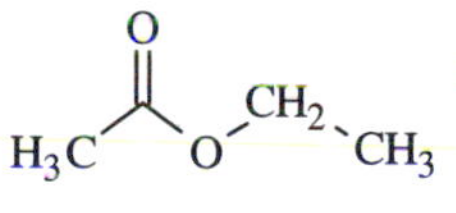

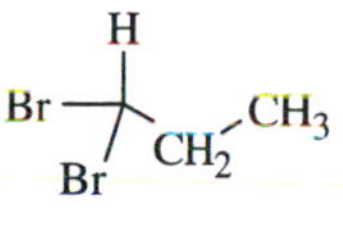

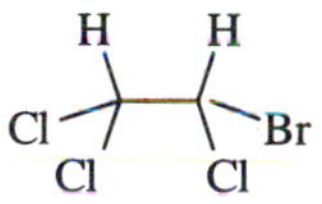

All are examples of groups of chemically and magnetically equivalent protons. A molecule can contain more than one spin system if the systems are isolated from each other. Protons that are chemically equivalent but magnetically nonequivalent are indicated by AA′, and so on. This system for designating spin systems is merely a labeling device. The appearance of actual spectra will depend on the magnitude of the various J values. Nevertheless, this is a convenient and common way of categorizing coupled proton systems.

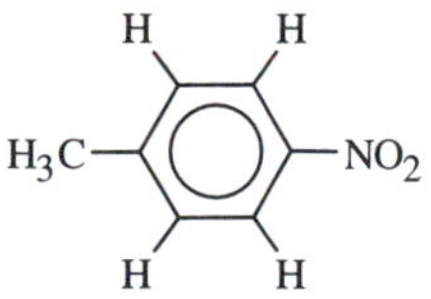

| AA′XX′ | AA′BB′ | AA′XX′ | ABX |

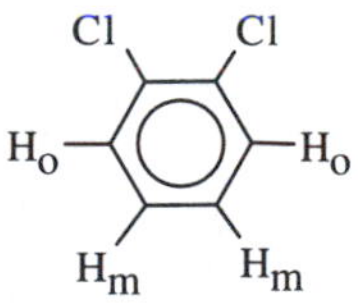

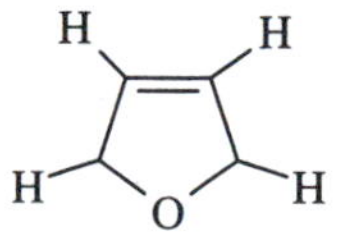

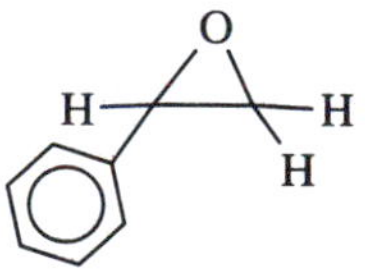

Another structural factor which can lead to nonequivalence of aliphatic protons is the symmetry of protons:

1. Aliphatic protons which are interconvertible by a rotational axis are termed homotopic and are chemically and magnetically equivalent. For example, the methylene protons of

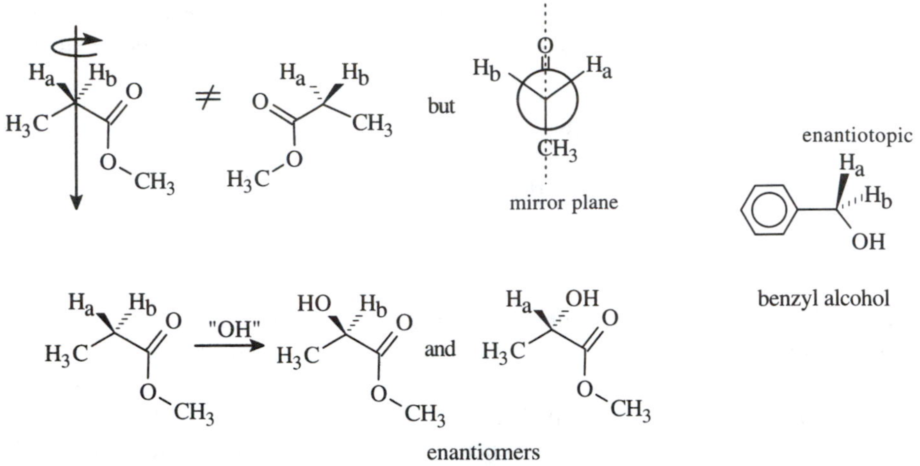

but

mirror plane

enantiotopic

benzyl alcohol

"OH"

and

enantiomers

FIGURE 11.22

diphenylmethane are homotopic, as are the methylene protons and the methyl protons of propane.

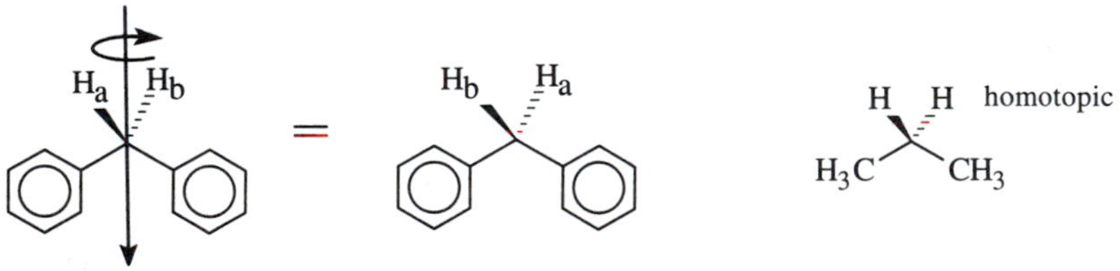

H H homotopic

2. Methylene protons which are not interconvertible by rotation, but which are interconvertible by reflection through a plane of symmetry, are *enantiotopic* and are chemically and magnetically equivalent in an achiral environment. Alternatively, protons are enantiotopic if sequential replacement by a different group gives a pair of enantiomers. The methylene protons of methyl propionate are enantiotopic because they are interchangeable by reflection but not rotation. (The protons of both methyl groups are interchangeable by rotation and are thus homotopic.) Replacement of H_a and H_b by another group, such as hydroxy, gives R- and S-methyl lactate respectively. The benzylic protons of benzyl alcohol are enantiotopic by the same criteria (see Figure 11.22).

Methylene protons which are not interconvertible by either reflection or rotation are *diastereotopic* and are chemically and magnetically nonequivalent. The presence of one or more chiral centers in a molecule leads to diastereotopic methylene groups, since the replacement of each proton by another group gives a pair of diastereomers. Since diastereotopic protons are not related by symmetry, they have unique environments and thus unique chemical shifts and coupling constants.

The C-2 methylene protons, H_a and H_b, in ethyl 3-azido-4-oxopentanoate are diastereotopic because of the chiral center at C-3 (see Figure 11.23). H_a and H_b have slightly different chemical shifts and split each other, and they are not coupled equally to the C—3 methine proton, H_c. Thus, H_a and H_b split each other into an AB quartet,

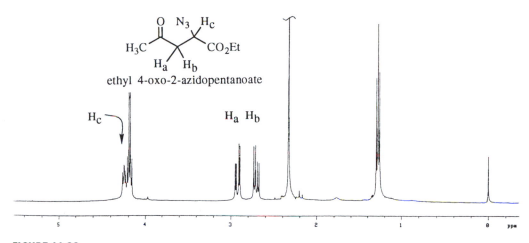

FIGURE 11.23

which is further split into doublets by H_c. Note that the splitting for each proton of the AB quartet has a different coupling constant with H_c. Although H_c is slightly obscured by the CH_2 protons of the ethyl group, the signal for this proton is not a triplet but looks like a doublet of doublets, as expected from the fact that $J_{ac} \neq J_{bc}$.

Second-Order Splitting

Our discussions about spin–spin splitting and multiplicity have been based on first-order, or weakly coupled, spectra which are spin systems where $\Delta\nu/J \geq 10$. That is, the difference in chemical shift in hertz of coupled protons, divided by the coupling constant, is 10 or more. In such cases, clean 1 : 1 doublets, 1 : 2 : 1 triplets, and the like, are observed, and coupling constants and chemical shifts can be read directly from line positions in the spectrum.

As $\Delta\nu/J$ decreases, the simple multiplets observed in weakly coupled spectra become increasingly distorted; new lines can appear and others merge or disappear. Such spectra are termed second-order, or strongly coupled, spectra. In these cases, the chemical shift does not lie in the center of the multiplet, and coupling constants are not always obvious. A simple example of such a change is seen when the chemical shifts of a first-order AX system become much closer and the spectrum becomes a second-order AB system (Figure 11.24). This is not a 1 : 3 : 3 : 1 quartet, but an AB quartet in which the intensities of the inner and outer lines depend on the difference in chemical shifts.

The treatment of such systems is outside the scope of this book, but it is possible to calculate the chemical shifts and coupling constants from line positions and intensities. There are also experimental methods by which chemical shifts and coupling constants can be determined in complex spectra. These include isotope exchange, decoupling techniques, lanthanide shift reagents, and the use of higher field nmr spectrometers. Since $\Delta\nu$ increases with the strength of the magnetic field, while J values do not change with magnetic field strength, the ratio $\Delta\nu/J$ increases as the field strength increases. Thus the higher the field strength, the larger is the ratio $\Delta\nu/J$, and the greater the chance to observe first-order coupling. Currently, spectrometers of 300 MHz to 500 MHz are routinely accessible, so the problems of second-order spectra

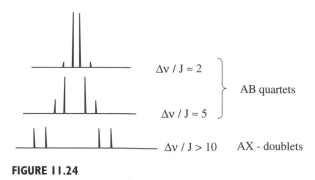

$\Delta v / J \approx 2$

$\Delta v / J \approx 5$ } AB quartets

$\Delta v / J > 10$ AX - doublets

FIGURE 11.24

are becoming much less common. With the advent of even higher field instruments, first-order spectra will be available for most compounds. (The first commercial 750 MHz spectrometer was delivered in 1994.)

Structure Identification by H^1 nmr

The nmr spectrum of a compound is generally used in conjunction with other available in-formation for identification purposes. The reactants, the reagents, and the reaction conditions can serve as a guide to the types of products that might be expected. Structure identification often merely confirms the structures of products that were predicted from the chemistry em-ployed in the synthesis. In other cases, products are obtained whose spectra do not match the predicted products. In such cases, more information is usually required to solve the structure. Thus although nmr is an extraordinarily powerful tool, it is not sufficient to solve all structural problems. This latter fact must be kept in mind.

The reaction between 1-phenyl-1-propanol and chromic acid gives a liquid product $\mathbf{P_1}$. The spectrum of the reactant 1-phenyl-1-propanol (Figure 11.25) contains a five-proton broad

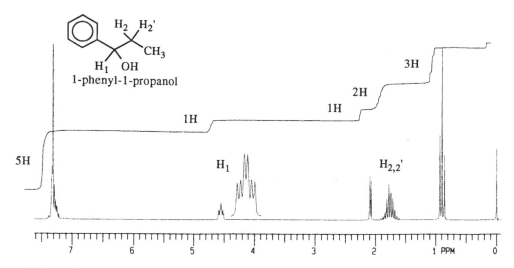

FIGURE 11.25

singlet for the aromatic protons. The methine proton H_1 is split into a triplet by $H_{2,2'}$ and is further split into doublets by the OH proton, which must be exchanging slowly in this sample to give splitting. The OH proton is split by H_1 into the doublet at 2.1δ. Protons H_2 and $H_{2'}$ are a multiplet rather than a pentet, as expected by the $n + 1$ rule. This is due to the fact that they are diastereotopic. Thus they have different chemical shifts, and split each other, and each is split into pentets by the $n + 1$ rule. The multiplet at 1.75δ is the result.

The nmr spectrum of the product P_1 (Figure 11.26) shows the five-proton aromatic signal at 7.3δ in the reactant has been converted to a three-proton multiplet over the range 7.35–7.6δ and two-proton doublet at 7.94δ (5H total). The product also has an A_2X_3 system (an ethyl group). The chemical shift of the methylene group at 2.99δ in the product is reasonable for a methylene group next to an aromatic carbonyl group. Furthermore, the one-proton multiplet at 4.55δ for the methine proton and the one-proton doublet for the OH proton of the starting material are not present in the product. The nmr is consistent with an oxidation of the alcohol to propiophenone, as predicted by the chemistry.

When 1-methylcyclohexanol is heated with anhydrous copper sulfate, two products P_2 and P_3 are isolated in an $85 : 15$ ratio. The nmr spectra of these products are shown in Figures 11.27 and 11.28. The appearance of signals in the olefinic region of both products indicates that both are elimination products. Furthermore, the major product has an intact methyl group (s, 1.63δ, 3H) whose chemical shift indicates it is likely allylic. The vinyl signal integrates for a single proton. The remaining protons are multiplets that have an integrated area corresponding to 8H. From these data, it is clear the major product P_2 is 1-methyl cyclohexene.

The minor product (P_3) does not have a methyl signal, and the vinyl signal is integrated for 2H. The remaining protons have a total integrated area of 10H and there is a 4H multiplet downfield from the remaining 6H multiplet. These data are consistent with *exo*-methylenecyclohexane as the minor product. The 4H signal is due to the allylic protons of the ring.

Treatment of 3-pentanone with isopropenyl acetate is reported to give 3-acetoxy-2-pentene.

The product isolated from the reaction has the ^1H nmr spectrum shown in Figure 11.29. Preliminary examination shows the isolated product to be a mixture; however, the products appear

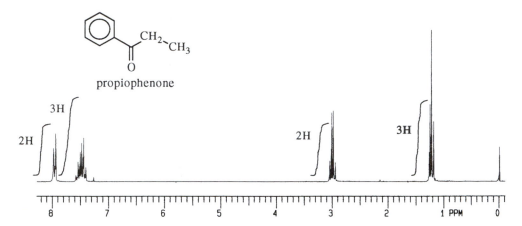

propiophenone

FIGURE 11.26

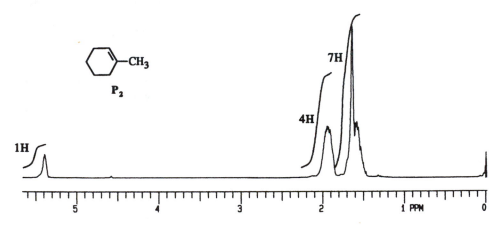

FIGURE 11.27 1-methylcyclohexene.

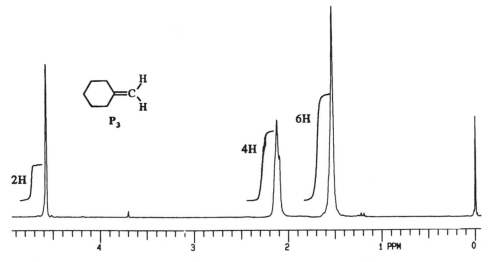

FIGURE 11.28 1-methylenecyclohexane.

278

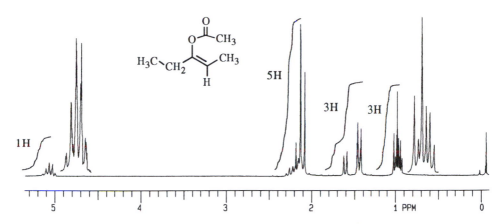

FIGURE 11.29 Vinyl acetate.

to be similar. The spectrum includes two singlet methyl groups (3H) and two A_2X_3, shown by overlapping triplets at 1.05δ. (Can you pick them out?) Also, two AX_3 groups are shown by the allylic methyl doublets at 1.5δ and 1.65δ. Particularly revealing is the vinyl signal at 5.1δ. Its relatively high field results from the fact that enol derivatives are electron-rich by resonance interaction of the oxygen lone pairs with the olefinic π system. This causes the vinyl proton β to the oxygen group to be shielded. Furthermore, it is not a simple quartet but is actually overlapping quartets, due to splitting by a methyl group. This is indicative of a partial structure.

This partial structure, along with the ethyl groups and acetate methyl singlets, confirms the structure assignment. The allylic methylene group is at 2.1–2.3 and is overlapped by the two acetate singlets. The product is a mixture of the Z and E isomers in the ratio of 2 : 5, as determined by the integrated areas of the methyl doublets. It is not possible to unambiguously assign the isomers from the nmr. However, it is likely the minor isomer is the Z isomer, since the allylic methyl group would be sterically deshielded by the acetate group and absorb downfield from the E isomer. The smaller allylic methyl doublet is found downfield from the major isomer.

The carbodiimide coupling of *N*-methylphenylglycine with benzylamine gives a product whose ^1H nmr is shown in Figure 11.30. The expected product is the amino amide.

The nmr spectrum shows that both reactants are incorporated in the product. The methyl singlet is indicative of the $-NHCH_3$ group, and the aromatic signal has increased to 10H indicating

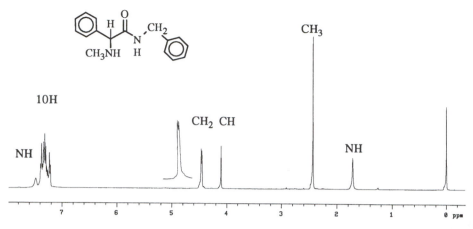

FIGURE 11.30 *N*-methylamino benyzlamide.

that two phenyl rings are present in the product. The signal at 4.4δ is proper for the benzyl group. The splitting pattern is problematic until it is recognized that, because there is a chiral center at C-2, the benzyl protons are diastereotopic and thus nonequivalent. They are part of an ABX spin system and thus give the complex splitting pattern seen. Actually, this is a two-proton multiplet that looks like a doublet or a very close AB quartet.

The two N—H protons in this compound illustrate different exchange behavior. The N—H proton at C-2 comes upfield at 1.74δ as a broadened singlet, due to fairly rapid exchange, and does not split either the C-2 proton (at 4.07δ) or the N-methyl group. Conversely, the amide N—H proton is a much broader singlet (at 7.55δ) and splits the benzylic protons by a small amount because the exchange is slower. When protons exchange rapidly, as they do on the NH of the amino group, the spin state of the proton is blurred and coupling information is lost. The neighboring proton cannot actually feel one spin state or the other because the protons with different spin states are exchanging rapidly.

When the protons do not exchange rapidly, as on the N—H of the amide group, normal coupling is observed. Since the rates of proton exchange are often critically dependent on the solution conditions, coupling to acidic protons is variable and thus may or may not be observed.

These examples illustrate how nmr spectra are routinely used to answer questions about reactions and products. Spectra are usually examined in conjunction with other information that permits a broad-based structure identification to be carried out. Outside of structure questions in texts and on exams, one is almost never handed an nmr spectrum and asked to identify the compound in the absence of other supporting information.

Carbon-13 nmr

Although proton magnetic resonance (pmr) is the most common type of nmr, it is also possible to observe other nuclei which have spin quantum numbers unequal to 0. Of greatest interest to organic chemists is ^{13}C nmr spectroscopy. Carbon-13 has a spin quantum number of $I = 1/2$, the same as a proton. When placed in a magnetic field, two possible orientations with respect to the field are possible: one of lower energy and one of higher energy. Transitions between these two spin states occur at discrete frequencies in the radio frequency region. Absorption of energy at the resonant frequency causes nuclei in the lower energy level (aligned) to undergo a transition to the higher energy level (opposed). This process is the same as discussed previously for protons, and the equations which govern the absorption are the same.

There are significant differences between a ^{13}C nucleus and a proton which must be addressed:

1. Low ($\sim$ 1 percent) natural abundance of ^{13}C

2. Lower magnetogyric ratio of ^{13}C, making the signal for ^{13}C much lower than that of a proton

3. Strong coupling to protons which, although first-order, gives complex multiplets that often overlap, making peak assignments difficult

These limitations made the development of ^{13}C nmr spectroscopy lag significantly behind the development of ^{1}H nmr. In the earliest work, the relatively weak sensitivity of ^{13}C was a major stumbling block and compounds specifically labeled with ^{13}C had to be prepared in order to obtain usable spectra. Today, it is possible to obtain excellent ^{13}C spectra on natural abundance samples of < 25 mg in less than 30 minutes. The hardware and software advances which have enabled such progress to be made lie in three areas:

1. Improved signal detection

2. Fourier transform techniques

3. Digital signal averaging

Suffice it to say, modern nmr spectrometers are capable of obtaining ^{13}C spectra quickly and easily and ^{13}C nmr is now a routine tool for structure identification.

These instrumental improvements do not solve the problem of coupling between protons and carbon that complicate ^{13}C spectra, but other techniques have. The proton-coupled ^{13}C spectrum of 2-octanone demonstrates that proton–carbon coupling significantly complicates the spectrum due to the large number of lines produced. From this spectrum, it is impossible to tell (because of overlapping multiplets in the spectrum) how many carbons are present or what their chemical shifts are. To solve this problem, broadband proton decoupling is used to remove all proton–carbon couplings. This leaves proton-decoupled or fully decoupled spectra which have only singlet absorptions for each carbon present. For example, 2-octanone has eight lines in the fully decoupled ^{13}C spectrum, as predicted by the fact that each of the carbons is in a unique chemical environment and has a unique chemical shift. A distinct advantage of ^{13}C nmr is that ^{13}C absorbs over a range of $\sim$ 250 ppm (compared to 10 ppm for ^{1}H). This means that each carbon can be distinguished by a unique chemical shift. It is thus possible to tell exactly how many nonequivalent carbons are present in a molecule merely by counting the lines in the fully decoupled spectrum (see Figure 11.31). (The small three-line signal at 77.3δ is from the solvent CDCl$_3$. This appears in all ^{13}C spectra run in CDCl$_3$ and is normally ignored.)

In addition to the number of nonequivalent carbons present, the chemical shifts of the carbons can reveal a great deal about the types of bonding patterns and substituents which are present. Because of the great range of chemical shifts observed for carbon-13 ($\approx$ 250 ppm), even small changes in the environment around carbon can result in a significant change in chemical shift. Figure 11.32 is a brief compilation of ^{13}C chemical shifts for representative

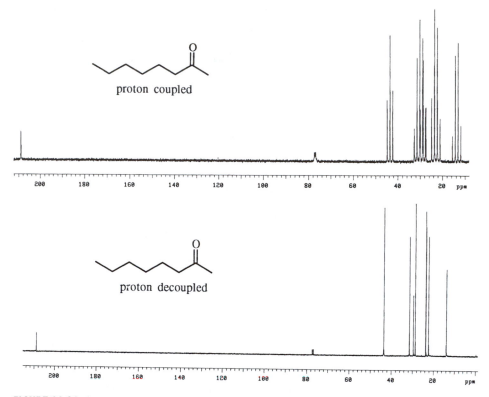

FIGURE 11.31 2-octanone.

classes of organic compounds. By assigning the chemical shifts in many series of compounds, it has been possible to develop correlation values. Thus it is now routine to test possible structures by calculating ^{13}C chemical shifts and comparing them with the observed spectra. For example, in Figure 11.33 the calculated and observed ^{13}C chemical shift values for cocaine are seen to be in remarkable agreement for most of the carbons in this reasonably complex compound.

    The value of fully decoupled ^{13}C nmr spectra is primarily tied to determining both the number of nonequivalent carbons present and their chemical shifts. Unfortunately, the integrated areas of ^{13}C signals are not directly proportional to the number of carbons responsible for those signals, under most circumstances. Thus both *n*-heptane and 4-(1-propyl) heptane have four signals in their ^{13}C nmr spectra, but it is not possible to determine if the ratio of different carbon types is 1 : 2 : 2 : 2, as expected for *n*-heptane, or 1 : 3 : 3 : 3 expected for the branched compound.

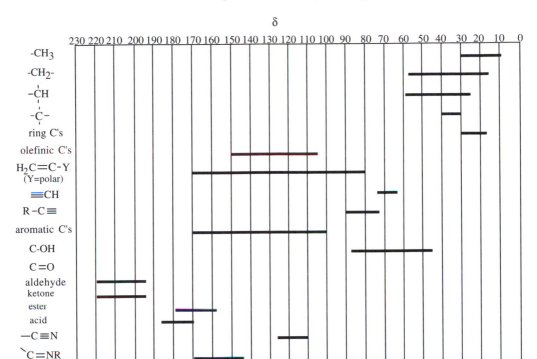

FIGURE 11.32 Representative Carbon-13 Chemical Shifts for Various Classes of Organic Compounds

It *is* possible to distinguish them based on the chemical shift of C4, which is 36.5 ppm in *n*-heptane but 53.1 ppm in the branched compound.

A second difficulty of fully decoupled ^{13}C nmr spectra is that the connectivity in the molecule is difficult to establish (except by chemical shift correlation) because coupling patterns are absent. This dilemma is partially resolved by the use of a technique called off-resonance decoupling. In off-resonance decoupled ^{13}C spectra, the carbons are coupled only to those protons directly attached to them and the coupling is first-order. Thus, quaternary carbons are singlets, methine carbons are doublets, methylene carbons are triplets, and methyl

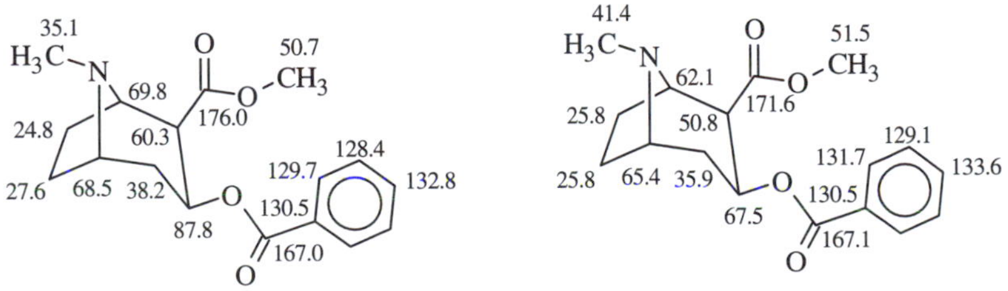

FIGURE 11.33. ^{13}C chemical shifts for cocaine (*a*) calculated and (*b*) observed.

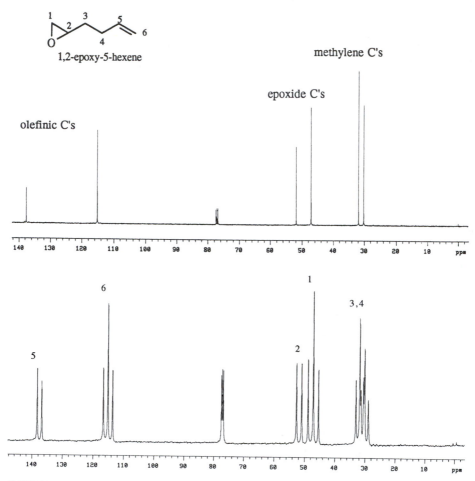

FIGURE 11.34. Fully decoupled and off-resonance decoupled spectra of 1,2-epoxy-5-hexene.

carbons are quartets. It is possible to use this information to establish proton–carbon connectivity, which can be used to add protons to partial structures determined by ^{13}C chemical shift data.

The carbons of 1,2-epoxy-5-hexene can be assigned from the off-resonance decoupled spectrum (see Figure 11.34). In the fully decoupled spectrum it is clear that the olefinic carbons (≈ 115 and 138δ) are distinct from the epoxide carbons (≈ 47 and 52δ) and from the methylene carbons (≈ 30 and 32δ), but it is not possible to assign which is which. In the off-resonance decoupled spectrum, both the olefinic and epoxide carbons are distinguished by their splitting patterns from the numbers of directly attached protons. The methylene carbons, however, are both triplets and cannot be distinguished.

A final feature of importance in ^{13}C nmr spectra is the notion of equivalency. Because some type of decoupling is normally done, either broadband or off-resonance, magnetic equivalency is not an issue in ^{13}C nmr, but chemical equivalence remains an issue. If two carbon atoms share the same chemical environment, they will have the same chemical shift. Thus it is important to recognize local or molecular symmetry elements. In a previous example, *n*-heptane is seen to

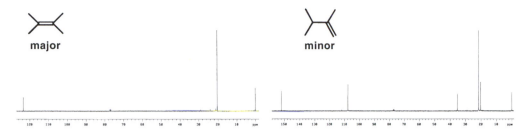

FIGURE 11.35. Major and minor butene products.

have four signals. The internal plane of symmetry results in three equivalent pairs of carbons, in addition to the unique central carbon. Toluene (or any monosubstituted benzene) has four signals for the aromatic protons, in addition to the methyl carbon signal. The xylenes offer another example of equivalency.

5 lines   4 lines   5 lines   3 lines

Ortho xylene has four signals, *m*-xylene has five signals, and *p*-xylene has only three. In general, the more symmetric a molecule, the fewer ^{13}C signals it will have. For example, adamantane has only two absorptions, as does buckminsterfullerene (C_{60}).

adamantane

C_{60}

Thus when the number of ^{13}C signals is less than the number of carbon atoms present in the molecule, there must be symmetry elements present that make some carbon atoms equivalent. The pyrolysis of 2-acetoxy-2,3-dimethylbutane in a hot tube at 200° gives two products which are both found to have the formula C_6H_{12}. The major product has only two ^{13}C absorptions while the minor product has five ^{13}C signals (see Figure 11.35). Thus the major product is likely to be the symmetric olefin while the minor product is the less symmetric olefin. Note that even the minor product has a pair of equivalent carbons giving rise to five, rather than six, lines, but the symmetry is still significantly less than that of the major olefin.

2 lines - major   5 lines - minor

The foregoing has been a brief introductory discussion of nuclear magnetic resonance that has concentrated on some basic principles that are very useful in understanding the technique. The actual practice of nmr today is much more advanced. The incorporation of Fourier transform techniques has revolutionized nmr spectroscopy. All types of pulse sequences and two dimensional (2-D) techniques have been developed to provide even greater structural detail than has been previously discussed. A discussion of such techniques belongs in a more specialized text, but it must be remembered that, although these techniques are faster, more sensitive, and much more sophisticated, they are still largely based on the principles presented here, as is the interpretation of the results.

Infrared Spectroscopy

Infrared (IR) spectroscopy is a very useful spectroscopic tool for determining the presence of functional groups and bonding sequences in a compound, by the absorption of light in the infrared region of the electromagnetic spectrum. The infrared region comprises light with wavelengths from about 1×10^{-4} to 8×10^{-7} meters (100–0.8 μm) and lies between the microwave region and the visible region of the spectrum. The wavelengths of greatest interest to organic chemists range from 2.5–15 μm, the so-called mid-IR region, because the greatest amount of structural information can be obtained by spectroscopy in this spectral region.

Infrared radiation, like any electromagnetic radiation, is characterized by properties of frequency ν and wavelength λ that are related by the speed of light c:

$$c = \lambda\nu \quad \text{or} \quad \nu = \frac{c}{\lambda}$$

To scale frequency to a more convenient range, IR spectroscopists have defined a frequency unit called wave number $\bar{\nu}$ given by $\bar{\nu} = 1/\lambda$, where λ is the wavelength in centimeters. The units of $\bar{\nu}$ are cm^{-1}, or reciprocal centimeters. The wave number is the number of vibrations which occur over a 1-cm distance. The higher the wave number, the more vibrations occur in a one-centimeter distance and the higher the frequency. Normally, IR spectra are recorded between 4000 to 650 cm^{-1} (2.5–15 μm).

The energy of infrared radiation is given by $E = h\nu$ and thus

$$E = hc\bar{\nu}$$

The energy of the infrared light absorbed by molecules during infrared spectroscopy is typically in the range of 9–2.5 kcal/mol (for 4000–650 cm^{-1} light). This amount of energy is not enough to break bonds in molecules, but it is enough to cause transitions in vibrational modes in the molecule. Thus infrared spectroscopy is best described as nondestructive, vibrational spectroscopy.

As was discussed in Chapter 1 on chemical bonding, a molecule can simplistically be thought of as a collection of bonds which hold the nuclei together in spatial relationships that achieve the lowest possible energy for the molecule. Deformations from these optimal angles and distances correspond to bond stretchings and bendings, and these are examples of vibrational motion in the molecule. Since vibrational motions, which include both bond-stretching and bond-bending modes of vibration, are quantized, each vibrational mode in the molecule absorbs energy at a particular frequency (which happens to fall in the IR region). This provides the basis for IR spectroscopy. By determining which frequencies of IR radiation are absorbed by a molecule, it is possible to conclude what types of vibrational modes are absorbing energy in the molecule. Consequently, one determines the atoms and bonds (functional groups) in the molecule which give rise to these vibrational modes.

A molecule which contains n atoms will have $3n - 6$ fundamental modes of molecular vibration. These $3n - 6$ fundamental vibrational motions can be divided into two types: stretching modes, of which there are $n - 1$, and bending modes, of which there are $2n - 5$. Stretching vibrations are those in which the internuclear distances between bonded elements change. Bending vibrations are those in which bond angles change. In general, it takes more energy to stretch a bond than to deform bond angles. Therefore, absorption frequencies which correspond to bond-stretching modes are often higher (higher energy) than absorption frequencies which correspond to bending modes (lower energy). Stretching frequencies are normally found in the higher frequency portion of the spectrum ($4000–1200$ cm^{-1}) whereas bending frequencies are found in the lower frequency region ($\sim1200–600$ cm^{-1}). Furthermore, the stretching frequencies give the most clearcut structural information about a compound.

IR Stretching Frequencies

As mentioned, stretching modes are those in which internuclear distances change. For such a distance change to occur, the bond between the nuclei must be either stretched or compressed. One way to think about this is to consider a chemical bond as a spring connecting two masses (see Figure 11.36). Each spring has a particular force constant which corresponds to the force required to compress or stretch that spring. Furthermore, the frequency of vibration is dependent on the masses that the spring connects. For a given spring, heavy masses lead to lower frequency vibrations while light masses lead to higher frequency vibration. This analogy is actually very close to correct in describing stretching vibrations of bonds, because the potential curves for springs and bonds have the same general shape and behavior at lower energies. The major difference is that the vibrational energies of a bond are quantized, whereas those of a spring are not. As a result, there are discrete vibrational energy levels that are allowed for a bond in a molecule. At room temperature, the vast majority of molecules are in the lowest vibrational energy levels.

To go from the lowest vibrational level, ν_1, to the next higher vibrational level, ν_2 (which is the only type of vibrational transition that is allowed), energy must be absorbed by the molecule. The energy absorbed must correspond exactly to ΔE, the energy gap between the two vibrational levels (see Figure 11.37). If the energy required to cause a vibrational transition is ΔE, only light of the particular frequency (and hence energy) which corresponds to the energy of the vibrational transition will be absorbed. Moreover, the frequency of light which causes the

k = spring force constant

k - bond force constant

FIGURE 11.36

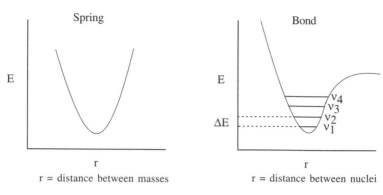

FIGURE 11.37

transition is given by

$$\nu = \frac{\Delta E}{h}$$

For most bonds, these frequencies occur in the infrared region. Each stretching mode in a molecule has its own potential curve and associated energy levels. Thus each stretching mode in a molecule will absorb infrared energy at the particular frequency required to cause the transition from the lowest energy level to the next higher energy level. Since each of these energy levels is dependent upon the force constants of the bonds and the masses of the atoms they connect, each type of bond has a characteristic IR absorption corresponding to the stretching frequency of that bond.

Because functional groups are recurring groups of atoms connected by similar bonding patterns, a given functional group tends to give characteristic IR absorptions, due to the vibrational frequencies of bonds present in that functional group. IR spectroscopy thus provides a fast and effective way to identify functional groups present in a molecule by noting the presence of absorptions corresponding to the bond types present in those functional groups. Table 11.2 is a compilation of IR absorptions for commonly encountered bonds and functional groups. Normal ranges are given, since the absorption frequency of a given bond type can vary somewhat depending on the structure. The frequencies shown are all stretching frequencies; bending frequencies are much more numerous, and usually harder to interpret. They are not included in this work.

It is also possible to identify structural features in molecules which strengthen or weaken bonds and thus lead to shifts in IR frequencies. Several general effects on IR frequencies are summarized as follows:

1. Multiple bonds are stronger than single bonds, and thus have larger force constants and absorb at higher frequencies than single bonds.

$$\begin{array}{ccccc}
\overset{|}{-}\!\overset{|}{\underset{|}{C}}\!-\!O\diagdown & \diagdown\!C\!=\!O\diagup & C-C & C=C & C\equiv C \\
1000\text{-}1200 \text{ cm}^{-1} & 1650\text{-}1750 \text{ cm}^{-1} & 1000 \text{cm}^{-1} & 1640\text{-}1675 \text{cm}^{-1} & 2150 \text{cm}^{-1}
\end{array}$$

TABLE 11.2 Functional Group Stretching Frequencies

Functional Group	Frequency (cm^{-1})
Alcohol O—H (free)	3640–3610
Alcohol O—H (H-bonded)	3500–3200 (variable)
Amine N—H	3500–3300 (1° doublet, 2° singlet)
Terminal Alkyne C—H	3315–3270
Olefinic and Aromatic C—H	3080–3020
Aliphatic C—H	2990–2850
Aldehyde C—H	2900–2700
Nitrile —C≡N	2300–2200
Terminal —C≡C	2260–2210
Internal —C≡C	2140–2100 (weak)
Ester C=O	1750–1740
Aldehyde C=O	1740–1720
Ketone C=O	1700–1720
Amide C=O	1715–1650
Unsaturated Ketone C=O	1680–1660
Alkene C=C	1675–1640
Aliphatic C—O	1280–1000 (strong)

2. More polar bonds are generally stronger than less polar bonds and consequently absorb at higher frequencies.

C—H	N—H	O—H	F—H
2862	3300	3650	4138 cm^{-1}

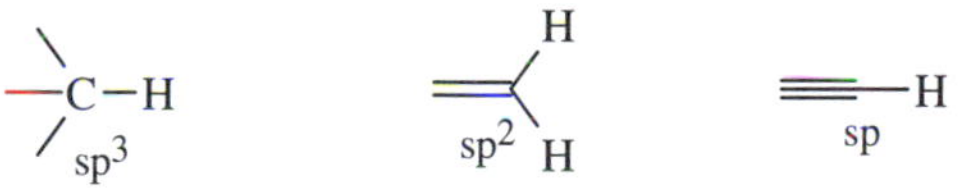

2890	3040	3300 cm^{-1}

3. Conjugation lowers the absorption frequency of each conjugated group (due to a lowering of the bond order) because of the contributions of resonance forms with lower bond orders.

1715 cm^{-1}	1647 cm^{-1}	C=O 1685 cm^{-1} C=C 1623 cm^{-1}	1710 cm^{-1}	1695 cm^{-1}

4. Hydrogen bonding causes the absorption frequency of acidic protons to vary widely, depending on the solution environment. In general, the greater the H bonding, the lower is the absorption frequency. For example, normal alcohol OH groups in dilute, nonbasic

solvents come at 3610–3650 cm^{-1} (termed the free OH stretch). As the concentration is increased, and H bonding increases, the OH absorption becomes broad and moves to lower frequencies.

| free OH - 3640-3610 cm $^{-1}$ | H-bonded - 3200-3500 cm $^{-1}$ | very strong H-bonds 2500 cm $^{-1}$ |

The very strongly H-bound carboxylic acid dimer has a very broad absorption at ~ 2500 cm^{-1} for the OH bond. Because H bonding is dependent on concentration and on the polarity and H bonding properties of the solvent, frequency shifts due to H bonding are quite variable.

Although other structural effects on absorption frequencies are known, the preceding factors are most commonly encountered in routine organic structure determination.

Use of IR for Structure Determination

IR spectroscopy is most commonly used to identify functional groups and bonding patterns in molecules from the higher energy portion of the spectrum (1200–4000 cm^{-1}), where absorptions are primarily due to bond stretching vibrations. Some information on atom connectivity in the molecule can also be deduced from the frequency shifts caused by structural factors. In general, however, it is not possible to completely deduce the structure of a molecule by examination of its IR spectrum. IR spectroscopy is, however, a powerful complement to NMR spectroscopy for structure determination.

For example, the reaction of 3-chlorocyclohexanone with DBU in toluene gives a product which is seen to have two vinyl protons by nmr and thus is an elimination product, probably either **A** or **B**.

Although one might rationalize, by both chemical intuition and by the splitting pattern, that conjugated isomer **A** is the product, examination of the IR spectrum shows a carbonyl group at (1680 cm^{-1}) and an olefin band at (1630 cm^{-1}). A typical cyclohexanone comes at 1710 cm^{-1} and cyclohexene comes at (1643 cm^{-1}). Clearly, the observed frequencies of the product are at lower frequencies than a simple ketone or olefin and are indicative of a conjugative interaction between these two functions. Thus **A** and not **B** is the product.

Treatment of propiophenone with *m*-CPBA in dichloromethane gives a single product. The carbonyl absorption of propiophenone is at 1695 cm^{-1}, whereas the product has a carbonyl absorption at 1745 cm^{-1}. This information reveals that the carbonyl group is intact but is no longer a ketone. The shift to higher frequency is consistent with the conversion to an ester, so the product could be either **C** or **D**.

Since ethyl benzoate **C** has a carbonyl stretch at 1725 cm^{-1}, the likely product is **D**, phenyl propionate. This is confirmed by the nmr spectrum that has the methylene group as a quartet at 2.42δ. This chemical shift is typical for a methylene group next to an ester carbonyl, but is a much too high field for the —CH2— group in ethoxy ester **C**, which comes at about 3.6δ.

Reaction of cinnamic acid, which has the ir and ^{1}H nmr shown in Figures 11.38–11.40, with BH$_3$·THF gives rapid consumption of the starting material. A single product **P$_4$** is formed. Comparison of the IR spectra shows that the double bond in the reactant (1630 cm^{-1}) is intact in the product (1654 cm^{-1}), but moved to higher frequency. The product **P$_4$** also has both vinylic (3026 cm^{-1}) and saturated (2861 cm^{-1}) C—H bonds. Both reactant and product have an O —H adsorption, but the strong H bonding in the acid (broad absorption at 3200–2600 cm^{-1}) is replaced by a shift to higher frequency in the product (3349 cm^{-1}), indicative of weaker H bonding found in an alcohol. Furthermore, the carbonyl group in the reactant acid (1681 cm^{-1}) is missing in the product. The IR data suggest that the carboxylic acid has been reduced by BH$_3$ · THF, in preference to hydroboration of the double bond.

The ^{1}H nmr corroborates this conclusion since two vinyl protons are observed both in the reactant and product. However, a new two-proton doublet appears at 4.15δ for the newly produced allylic methylene group. The acid O—H proton is moved far upfield as well. The coupling constants of the vinyl protons (J = 16 Hz) show the starting compound to be *trans* and the large splitting for the downfield vinyl doublet of the product (J = 16 Hz) shows the *trans* stereochemistry to be maintained in the unsaturated alcohol product. Moreover, the splitting between the methylene group and the upfield vinyl proton clearly supports its allylic position.

The ^{13}C spectrum (see Figure 11.41) is consistent with these structural assignments. The carbonyl carbon in the reactant (172.5δ) is gone and the product contains a new signal at 63.3δ, typical for a change to sp^3 hybridization and an allylic group.

In the preceding example, several types of spectroscopy are brought to bear. Although the product structure could probably be deduced from IR or nmr (either ^{1}H or ^{13}C), the use of all three methods confirms the assignment. It is often prudent to use more than a single technique for structure determination so that the results reinforce each other. If a structure assignment is not consistent with *all* the data, the structure is probably incorrect.

This lesson is brought home in the following example taken from a recent experiment. Ample chemical precedent suggested that the treatment of **E** with methyl amine should give **F**:

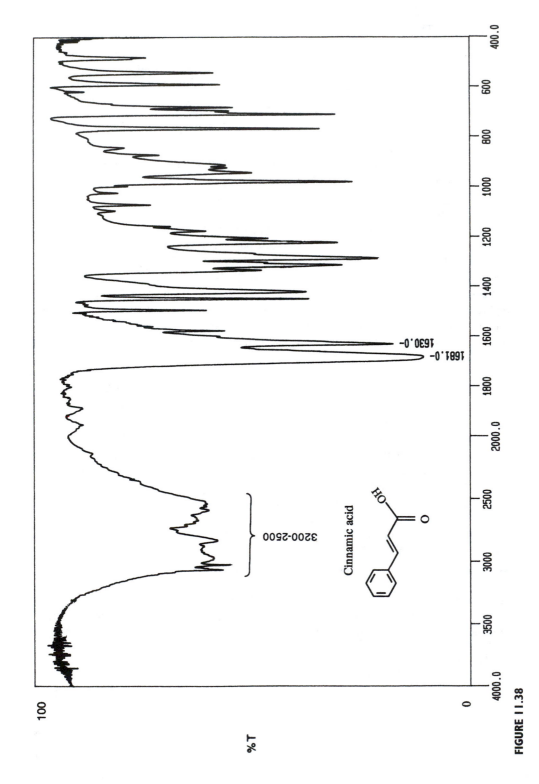

FIGURE II.38

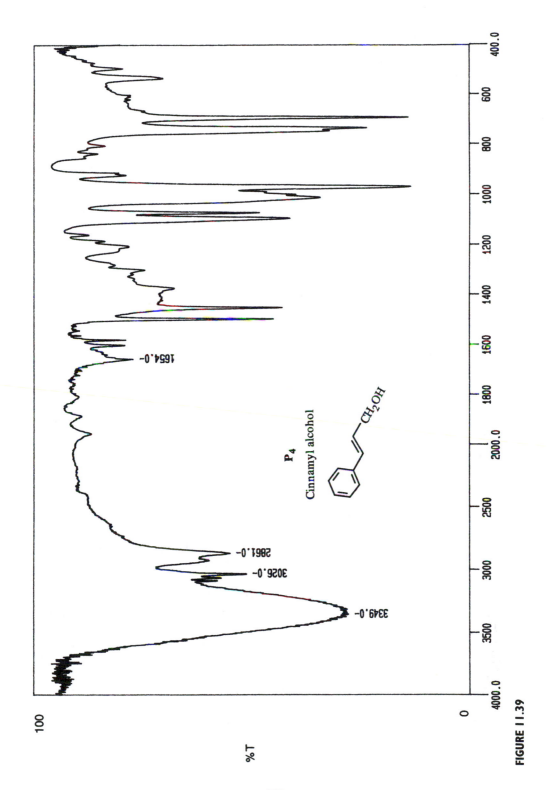

P₄

Cinnamyl alcohol

FIGURE I I.39

293

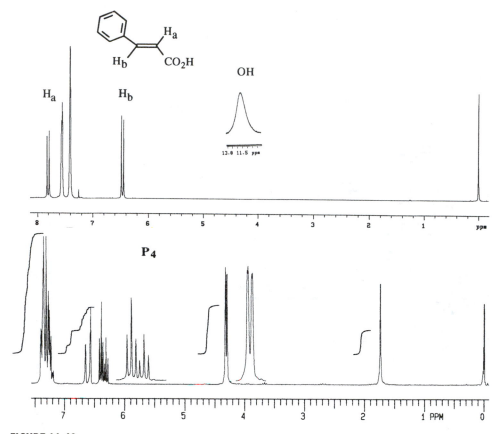

FIGURE 11.40

The spectra of the product are shown in Figures 11.42–11.44. As is clear, all of the appropriate resonances are present. The amide N-methyl group is a doublet ($J = 4$ Hz) at 2.76δ because of weak splitting by the amide N—H proton. The amino N—CH$_3$ group is the sharp singlet at 2.89δ which is not split by the amine N—H proton, due to rapid exchange. The C—H methine proton is a singlet at 4.03δ, and the ethoxy group is evident by the quartet–triplet A$_2$X$_3$ pattern.

However, several pieces of data just do not seem to fit. First, the chemical shift of the amino N-methyl group is at 2.90δ, whereas several known compounds of similar structure had the amino N-methyl group at 2.48δ. The 0.5 ppm shift might be due to the electron-withdrawing properties of the carbethoxy group, but that shift appears to be too large. In fact, the chemical shift of 2.9δ is exactly that expected for an amide N-methyl. However, in this case it is not split and there is already an amide N-methyl signal at 2.76δ, where it should be. Next, the integrated area of the N—H peak at 4.6δ is only 1-H, whereas it should account for two N—H's which exchange. Moreover, the C—H signal at 4.03δ integrates for two protons, rather than one. Although one might discount the discrepancies in the integrated areas as being due to an analytically impure sample and whereas one might force fit the change in chemical

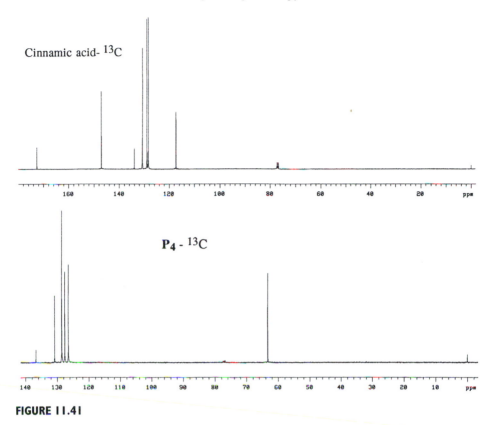

FIGURE 11.41

shift of the amine N—CH$_3$ group, the compound did not dissolve in acid, as was expected for the amine. Adding to the difficulty was the IR spectrum of **F**. In addition to the normal ester C =O stretch at 1745 cm^{-1}, the IR had a C=O stretch at 1636 cm^{-1}, which is at a much lower frequency than a normal secondary amide, $\approx$1685 cm^{-1}.

Based on these discrepancies, the assigned structure had to be discarded. Starting once again from the beginning, the carbethoxy group **a** is set by the A$_2$X$_3$ spin system and the ester C=O stretch. The 2-H singlet at 4.03δ is assigned as a 2-proton CH$_2$ group next to the ester group **b**, since it has the right chemical shift and integrates for two protons. Another fragment which is indicated is the C(O)NHCH$_3$ group, by the chemical shift of the methyl group, its small splitting by the N—H amide proton, and the rather low C=O stretching frequency. This suggests an amide group **c**. What remains is an N—CH$_3$ fragment, which can only be placed between fragments *b* and *c*. That gives the unsymmetric urea as the assigned structure of **F**. This structure fits the data in every way. The methylene group (2-H) and the single N—H proton fit the integration. Both N—CH$_3$ groups are amide types, rather than one amino and one amide type, and both should appear at $\sim$ 2.8-3.0. The urea group has greater resonance stabilization and thus the carbonyl group has more single bond character and comes at a lower IR frequency than an amide. Finally, **F** is *not* an amine and should not behave chemically like one (i.e., it should not dissolve in acid).

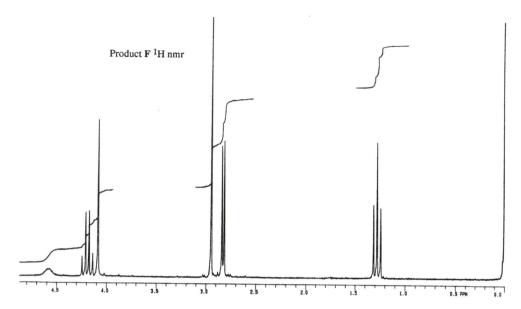

Product F ^{1}H nmr

FIGURE 11.42

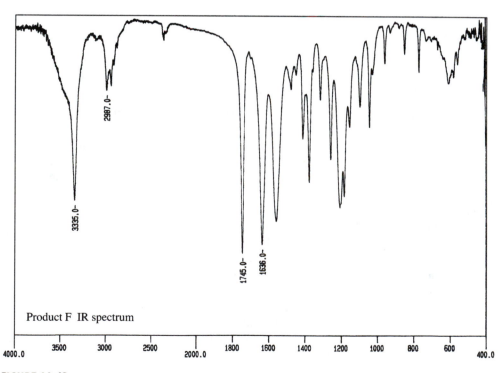

Product F IR spectrum

FIGURE 11.43

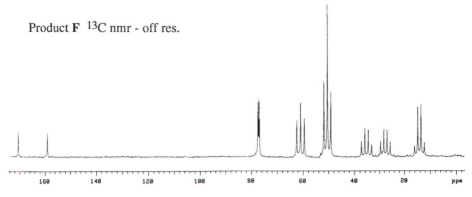

FIGURE 11.44

The structure is confirmed by the off-resonance decoupled ^{13}C spectrum, which shows the CH_2 group as a triplet 61.78, in addition to the other expected carbon resonances.

It is clear that all the data must fit the structure, and vice versa. If not, the assigned structure is probably not correct. It is important to let the data indicate the structure and not to make the data fit a preconceived structure. Often the chemistry will suggest a structure and the data will support that structure. However, that need not always be so, and it is necessary to always check that the data fit the structure. New chemistry is often discovered precisely because expected products are not supported by structural data, and new structures resulting from new chemistry are consequently revealed.

Mass Spectrometry

Mass spectrometry is not a true spectroscopic technique that involves absorption of energy at particular frequencies. Rather, one excites the molecule as a whole and then observes its subsequent reactions. The use of mass spectrometry for structure determination might be described as chemical archeology. A real archeologist picks up fragments of a pot or other vessel, identifies them, fits them back together and can tell what the original object was in great detail. In mass spectrometry, a molecule is purposely broken into pieces, the pieces identified by mass, and the original structure then inferred from the pieces.

To measure the mass spectrum, a molecule is bombarded with a stream of energetic electrons (70 ev) and one of the electrons of the molecule is ejected from one of the orbitals after a collision.

$$M \xrightarrow{e^-} \left[M\overset{\bullet}{+} \right] + 2e^-$$

This produces a charged species with an unpaired electron called a *radical cation*. The first-formed radical cation is called the *molecular ion* because it contains all of the atoms present in the starting molecule. The molecular ion contains a large amount of excess energy deposited by the collision which dislodged the electron. Considering that electrons of 70 ev energy contain 1613.5 kcal/mol of energy, the molecular ion usually contains energies far in excess of bond dissociation energies (80–100 kcal/mol), so it dissociates into fragments.

Fragmentation processes must conserve both charge and spin; thus a radical cation can undergo the following types of cleavages:

$$\left[M\overset{\bullet}{+} \right] \longrightarrow \quad F\overset{\bullet}{+} \quad + \quad \text{neutral molecule}$$

$$F_1^+ \quad + \quad F_2^{\bullet}$$

$$F_1^{\bullet} \quad + \quad F_2^+$$

The fragment products also contain excess energy since fragmentations are adiabatic, and they can themselves undergo further fragmentations to smaller pieces.

Generally, fragmentations occur very rapidly in the region where the initial ionization takes place (called the *source*). The packet of ions, which includes the molecular ion and the fragment ions, is then accelerated by an electric field and focused through slits. Then the beam of ions travels at high speed down a curved tube toward the detector. At this point, one can think of the ion beam as a stream of charged projectiles having different masses. Moving charges are subject to the influences of electric and magnetic fields, so if a magnetic field (or electric field) is applied to the ion beam, the path of the particles will curve. When the trajectory of a particle matches the curve of the tube, the particle will reach the detector. If not, it will hit the wall and will be annihilated. (See Figure 11.45.)

The path of the moving ion will curve according to its speed, mass-to-charge ratio (*m/e*), and the strength of the electric or magnetic field through which it passes. Fragments of low *m/e* will curve more than fragments of higher *m/e*. Thus if the ion packet is accelerated uni-

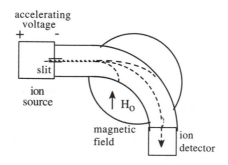

FIGURE 11.45

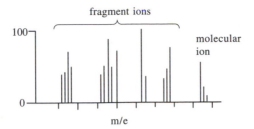

fragment ions

100

molecular
ion

0

m/e

FIGURE 11.46

formly, the magnetic or electric field can be varied so that fragments of different *m/e* ratios can be curved to strike the detector, and be counted, in turn. What results is a series of signals corresponding to different *m/e* values for the various charged fragments produced from the molecular ion (Figure 11.46).

The vast majority of fragments will have only a single positive charge (i.e., $e = 1$); thus the *m/e* ratio of a given ion corresponds to the mass of the ion, in atomic mass units. It is very important to remember that only ions (normally cations or radical cations) are detected. Neutral species (closed shell molecules or radicals) are not detected because they are not accelerated and they are not influenced by the applied field. Thus mass spectrometry yields information about the mass of the molecular ion, and the masses of fragment ions produced from the molecular ion. This so-called cracking pattern provides information about connectivity in the molecule that can be used to reconstruct the intact precursor molecule.

The molecular ion is one of the most important ions in the mass spectrum of a compound for the following reasons:

1. The *m/e* value of the molecular ion is equal to the molecular weight of the compound. This gives a rough estimate of the number of carbon atoms. Furthermore, a knowledge of the history of the sample and the reagents used often permits the molecular formula to be deduced. For example, treatment of *p*-toluic acid with ethyl iodide and potassium carbonate gives an oil whose molecular ion is at *m/e* = 164. This molecular ion corresponds to the addition of 28 mass units to the starting material, consistent with the formation of an ethyl ester by displacement of iodide.

$$H_3C-\langle\bigcirc\rangle-CO_2H \xrightarrow[\substack{K_2CO_3 \\ DMF, \Delta}]{CH_3CH_2I} (M^{\cdot+} \ m/e = 164) \Rightarrow H_3C-\langle\bigcirc\rangle-\overset{\overset{O}{\|}}{C}_{O}CH_2CH_3$$

mw = 136

2. High-resolution mass spectrometers which can measure *m/e* values to four decimal places are capable of confirming the molecular formula of the molecular ions. These so-called exact mass measurements can be used because the atomic weights of the elements are not *exactly* whole numbers (except for ^{12}C, which is the standard at 12.0000 amu). The exact masses of some elements and their most abundant isotopes are given in Table 11.3. To find the exact mass of a molecule, the atomic mass of the most abundant isotope for each element is used to calculate the exact mass of the compound. This is

TABLE 11.3 Exact Masses of Elements and Their Common Isotopes

Element	Isotope	Natural Abundance (%)	Exact Mass
Hydrogen	^1H	100	1.00783
	^2H (deuterium)	0.016	2.01410
Carbon	^{12}C	100	12.0000 (standard)
	^{13}C	1.08	13.0034
Nitrogen	^{14}N	100	14.0031
	^{15}N	0.38	15.0001
Oxygen	^{16}O	100	15.9949
	^{17}O	0.04	16.9991
	^{18}O	0.20	17.9992
Fluorine	^{19}F	100	18.9984
Silicon	^{28}Si	100	27.9769
	^{29}Si	5.10	28.9765
	^{30}Si	3.35	29.9738
Phosphorus	^{31}P	100	30.9738
Sulfur	^{32}S	100	31.9721
	^{33}S	0.78	32.9715
	^{34}S	4.40	33.9679
Chlorine	^{35}Cl	100	34.9689
	^{37}Cl	32.5	36.9659
Bromine	^{79}Br	100	78.9813
	^{81}Br	98	80.9163
Iodine	^{127}I	100	126.9045

compared to the exact mass, measured on a high-resolution mass spectrometer. If the two values agree to the third decimal point, it is certain that the molecular formula used to calculate the exact mass is correct.

Consider the three compounds $C_8H_{16}N_2$, $C_9H_{18}N$, and $C_9H_{16}O$. All would give a molecular ion of $m/e = 140$ in the low resolution mass spectrum. Using the elemental exact masses in Table 11.3, the molecular exact masses are calculated:

$$C_8H_{16}N_2 \Rightarrow (8 \times 12.000) + (16 \times 1.00783) + (2 \times 14.0031) = 140.13148$$

$$C_9H_{18}N \Rightarrow (9 \times 12.000) + (18 \times 1.00783) + 14.0031 \quad = 140.14404$$

$$C_9H_{16}O = (9 \times 12.000) + (16 \times 1.00783) + 15.9949 \quad = 140.12018$$

It is clear that if the mass of the molecular ion can be determined to 0.001 amu, these three compounds can be distinguished clearly. Instead of having to calculate exact masses, one can consult the many published tables of exact masses for any elemental composition, and there are many computer programs that calculate the exact mass after input of the molecule formula.

3. The analysis of isotopic clusters of the molecular ion can be used to infer the presence of elements based on their isotopes. The molecular ion corresponds to the m/e for the

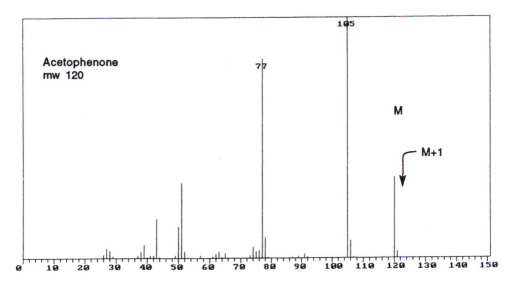

Acetophenone
mw 120

FIGURE 11.47

ion corresponding to some molecular formula. Examination of the molecular ion (see Figure 11.47) reveals that in addition to the expected molecular ion, there is normally a smaller peak at M + 1 and an even smaller one at M + 2. These are due to the fact that there are naturally occurring isotopes of higher mass that, if present in a given molecule, cause its mass to be higher than for the lighter isotopes.

The most obvious example of such behavior is for molecules which contain chlorine or bromine. The two isotopes of chlorine occur naturally in the ratio of $^{35}Cl : {}^{37}Cl = 100 : 32.7$ (3.058 : 1). A molecule such as chlorobenzene would exhibit two distinct molecular ions: one at $m/e = 112$ for those molecules which have the ^{35}Cl isotope, and one at $m/e = 114$ for those

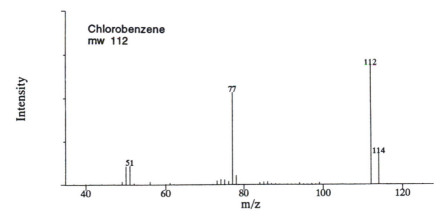

Chlorobenzene
mw 112

FIGURE 11.48

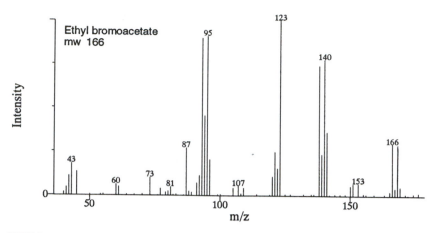

FIGURE 11.49

molecules which contain the ^{37}Cl isotope. The intensities of these peaks should be 3.058 : 1 reflecting the probability that a molecule has one or the other of the isotopes (see Figure 11.48).

A similar situation is seen for molecules containing bromine. The isotopes ^{79}Br and ^{81}Br occur in a ratio ^{79}Br/^{81}Br = 100 : 97.5 (1.026 : 1). Thus, a molecule such as ethyl bromoacetate will exhibit molecular ions at m/e = 166 and m/e = 168 for molecules which contain ^{79}Br and ^{81}Br, respectively. The ratio of peak intensities will be 1.026 : 1 because this is the relative abundance of the two bromine isotopes present in the molecule (see Figure 11.49).

If a molecule contains more than one chlorine atom, the appearance of isotope clusters can be calculated by the probabilities of isotope distributions and the natural abundances of the isotopes. For example, if a molecule contains two chlorine atoms such as *o*-dichlorobenzene, there will be peaks at M, M + 2, and M + 4 for molecules which have two ^{35}Cl, one ^{35}Cl and one ^{37}Cl, and two ^{37}Cl (see Figure 11.50).

The relative intensities of these peaks can be calculated by taking into account the number of combinations that can give the required isotopic substitution and the probability of an isotope being present. An M + 2 peak in the above example will result if either of the chlorine atoms is ^{37}Cl; thus the intensity of an M + 2 peak will be 2 × (1/3.058). An M + 4 peak will occur only if both chlorine atoms are ^{37}Cl thus the intensity of the M + 4 peak will be 2 × (1/3.058)^2. The squared term follows from the necessity that both chlorine atoms must be ^{37}Cl. Thus, the intensities of the peaks in the isotopic cluster of the molecular ion is (approximately):

$$M : M + 2 : M + 4 = 1 : (2 \times 1/3) : 1 \times (1/3)^2 = 1 : 0.667 : 0.111$$

The presence of isotopic clusters is particularly clear for molecules containing chlorine or bromine because of the abundance of two isotopes. The same considerations are applicable, however, for other elements that have smaller abundances of higher isotopes. These natural isotopic abundances are given in Table 11.3. As can be seen, ^{13}C is present to the extent of 1.08% of ^{12}C, whereas ^{2}H is present only to the extent of .016% of ^{1}H. If a molecule such as benzene is examined, the molecular ion is found at m/e = 78, but there is an M + 1 peak and an M + 2 peak with intensities of 6.58% and 0.22%, respectively. The M + 1 peak is due to the

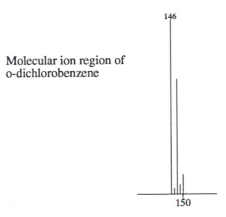

Molecular ion region of
o-dichlorobenzene

FIGURE 11.50

probability that one of the six carbons will be ^{13}C or one of the six hydrogens will be deuterium. The M + 2 peak is due to the probability that two of the carbons in the same molecule will be ^{13}C (the probability of two deuterium atoms in the same molecule is exceedingly small) or that one ^{13}C and one deuterium are present in the same molecule. The intensities of M + 1 and M + 2 peaks can be calculated for various molecular formulas based on these probabilities and they have been tabulated in several texts.

This information can be used to deduce the elemental composition of a compound. For example, the oxidation of 1,2-diazacyclohexane was carried out in cyclohexane. A product was isolated and was found to have a molecular ion of m/e = 84.

$$\text{(structure with } NH, NH) \xrightarrow[C_6H_{12}]{O_2} \text{(structure with } N, N)$$

mw = 84

At this point, the experimenter realized that both the expected product and the reaction solvent have a molecular weight of 84. Measurement of the isotopic cluster of the molecular ion showed an M + 1 peak of 5.30% and an M + 2 peak of 0.15% of the molecular ion. From tables of isotopic abundance ratios, it was found that the expected product $C_4H_8N_2$ should give M + 1 and M + 2 peaks of 5.21% and 0.11%, respectively, and cyclohexane, C_6H_{12}, should give M + 1 and M + 2 peaks of 6.68% and 0.19%, respectively. It is clear that the isolated product is most likely the expected cyclic azo compound and not cyclohexane.

Of course, nowadays exact mass measurement could also distinguish these two molecules, as could a variety of other instrumental techniques. The analysis of isotopic clusters is most useful for detecting the presence of halogens, sulfur, and silicon, all of which have abundant isotopes of two atomic weight units higher, thus leading to relatively large M + 2 peaks.

Fragmentation Processes

Besides the molecular ion, fragmentation processes can be used to infer groups present in the molecule and the connectivity of those groups. The requirement of spin and charge conservation in any fragmentation means that both cations and radical cations can be produced as ions by fragmentation. Because of the great amount of energy deposited in the molecular ion, there is sufficient energy to break any of the bonds in the molecule. It has been found, however, that fragmentations tend not to be random, but occur in such a way that the most stable ions are produced. Normally, the most stable ion is the most abundant ion in the mass spectrum. The most abundant ion is called the *base peak* of the spectrum and is arbitrarily scaled at 100%. The abundances of other ions are given as percentages relative to the base peak. Several examples of very stable ions are shown:

3° carbocation   benzyl cation   allyl cation   acylium ion   oxonium ion

Fragmentations often occur from the molecular ion by loss of neutrals or radicals to give more stable ions or radical ions. The differences in mass correspond to the mass of the un-charged fragment that has been expelled. The mass spectrum of ethane has a molecular ion at $m/e = 30$ and a major peak at $m/e = 15$. This corresponds to the loss of a fragment of 15 amu from the molecular ion. Thus the ethane molecular ion undergoes fragmentation of the C—C bond to give a methyl cation which is detected at $m/e = 15$ and a methyl radical which is not detected, as it is uncharged. This very simple example is indicative of the process:

$$H_3C-CH_3 \xrightarrow{-e^-} \left[H_3C-CH_3 \right]^{+} \longrightarrow CH_3^{\oplus} + CH_3^{\cdot}$$

$$m/e = 30 \qquad\qquad m/e = 15$$

Ethyl benzoate (see Figure 11.51) has a molecular ion at $m/e = 150$ and a base peak at $m/e = 105$ (M − 45) and a smaller peak at $m/e = 77$. The base peak at $m/e = 105$ corresponds to loss of the ethoxy radical from the molecular ion to give the very stable pheny-lacylium ion. Loss of CO from the phenylacylium ion gives the phenyl cation, but due to the instability of the phenyl cation, this pathway is minor. Also, a peak at $m/e = 122$ is observed. This is due to the benzoic acid radical cation, resulting from loss of the neutral ethy-lene molecule from the molecular ion by a different fragmentation process.

1,3-Diphenylpropanone has a molecular ion at $m/e = 210$ and significant fragment ions of $m/e = 119$ (M − 15) and $m/e = 65$. The base peak is $m/e = 91$. In this example, loss of a benzyl radical from the molecular ion produces an acylium ion ($m/e = 119$) which rapidly

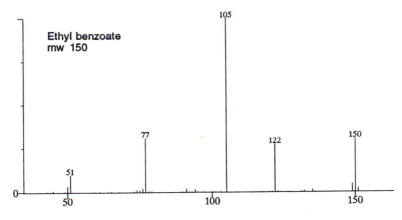

FIGURE 11.51

loses CO because the resulting benzyl cation is extremely stable—one of the most stable ions normally encountered. Examination of the mass spectrum (see Figure 11.52) shows that there are many additional small peaks present other than those just discussed. Their presence is indicative of the high energy deposited in the molecular ion upon ionization. This permits a large number of fragmentations to occur. Nevertheless, the fragmentations which occur most often, and which lead to the most intense peaks, are those that follow common ideas about reactivity and ion stability.

Both ethers and alcohols readily undergo loss of groups next to the oxygen, to produce an oxonium ion.

$$R_2 - \underset{\underset{R_3}{|}}{\overset{\overset{R_1}{|}}{C}} - O - R_4 \xrightarrow{-e^-} R_2 - \underset{\underset{R_3}{|}}{\overset{\overset{R_1}{|}}{C}} \overset{\cdot\,+}{} - O - R_4 \xrightarrow{-R_3\cdot} \underset{R_2}{\overset{R_1}{}} C \overset{\oplus}{=} O - R_4$$

R_4 = H, alkyl, aryl

oxonium ion

Thus tert-butyl ethyl ether $m/e = 102$ and has a very large M − 15 peak due to loss of a methyl radical. The methyl group could be lost from either the *t*-butyl group (path a) or the ethyl group (path b) to give two different oxonium ions with the same m/e value (see Figure 11.53). The base peak at $m/e = 57$ is the *t*-butyl cation and indicates that, at least part of the time, the

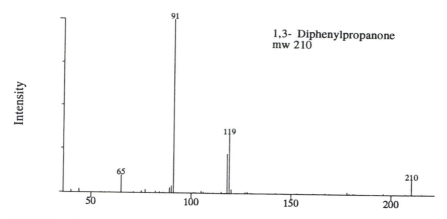

FIGURE 11.52

methyl group is lost from the ethyl group (path b) because subsequent loss of formaldehyde from the oxonium ion gives the *t*-butyl cation. The *t*-butyl cation can also be produced by a single fragmentation of the molecular ion by loss of the ethoxy radical. The stability of the *t*-butyl cation makes it the base peak and ensures its production by a variety of routes. This is not to say that all of the $M - 15$ peak comes from path b. Most likely, there is some contribution to the $m/e = 87$ peak from path a; however, the oxonium ion thus produced is unlikely to fragment into the very unstable ethyl cation.

By working with mass spectral fragmentation patterns, it is possible to develop very keen insight into the ways that molecules disintegrate under high energy conditions. This permits both identification of the structure from the pieces and insight into how they were produced. In conjunction with other structural tools, mass spectrometry provides invaluable insight into molecular formula and connectivity issues in a molecule and is thus an important tool in structure elucidation.

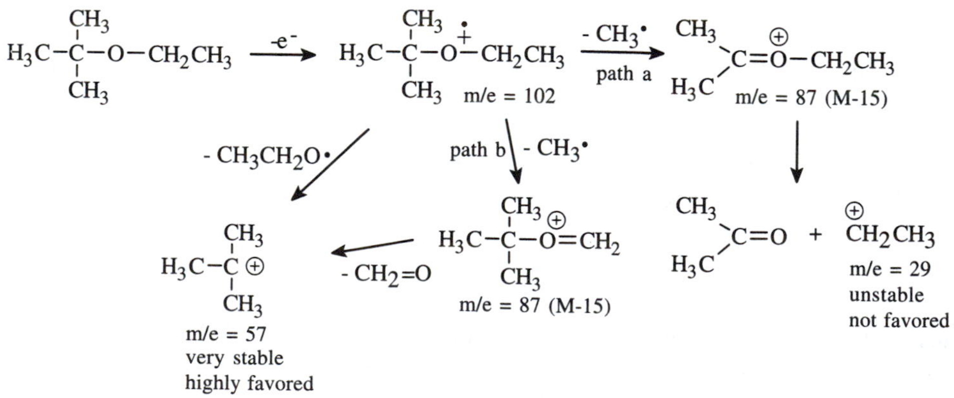

FIGURE 11.53

The foregoing discussion has been a very elementary introduction into mass spectrometry as a tool for structure identification. Advances in sample introduction, methods of ionization, and ion collection and detection have been remarkable, and today the mass spectra of peptides, nucleic acids, proteins, and other biopolymers are routinely obtained. Using known cracking patterns, mass spectrometry is the method of choice for identifying drugs and drug testing since it requires only minute quantities (μg). It has been sent on the Mars probe to look for amino acids as an indication of life forms on Mars. One goal of current research efforts is to use mass spectrometry as a method for sequencing peptides and oligonucleotides by their fragmentation patterns. Mass spectrometry is thus an important analytical and structural tool whose evolution continues at a rapid pace. It remains an important component of structural investigation.

Bibliography

For an alternative discussion of instrumental techniques see J. W. Cooper, *Spectroscopic Techniques for Organic Chemists,* Wiley Interscience, New York, 1980.

A complete discussion of modern instrumental techniques for structure determination is found in R. M. Silverstein, G. C. Bassler, and T. C. Morrill, *Spectrometric Identification of Organic Compounds,* 5th ed., Wiley, New York, 1991.

An excellent collection of nmr and IR spectra for reference purposes is found in *The Aldrich Library of ^{13}C and ^{1}H FT NMR Spectra* and *The Aldrich Library of Infrared Spectra*. Both are available in most libraries, and the former can be purchased from Aldrich Chemical Co. A variety of compilations of mass spectral data are available; see for example F. S. McLafferty and D. B. Stauffer, *The Wiley / NBS Registry of Mass Spectral Data,* 5th ed., Wiley Interscience, New York, 1988.

For a more advanced discussion see A. E. Derome, *Modern NMR Techniques for Chemistry Research,* Pergamon Press, Oxford, UK, 1987.

The classic text on ^{13}C nmr is G. C. Levy, R. L. Lichter, and G. L. Nelson, *Carbon-13 Nuclear Magnetic Resonance Spectroscopy,* 2d ed., Wiley Interscience, New York, 1980.

Problems

1. For the following compounds, label the spin systems present, label the symmetry properties of protons where possible (homotopic, enantiotopic, diastereotopic), and predict the splitting pattern for the proton(s) indicated by an arrow.

2. Tell how you could use ¹H nmr to distinguish the following pairs of compounds. Be specific as to what data you would look for and how you would interpret it. There might be more than one feature in the ¹H nmr that could be used, so give a complete answer.

(a) 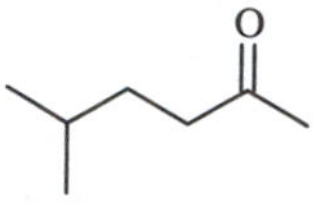 and

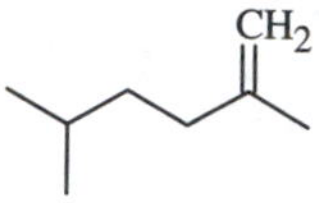

(b) 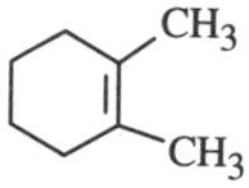 and

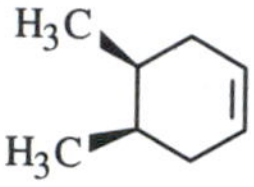

(c) 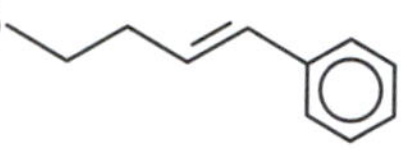 and

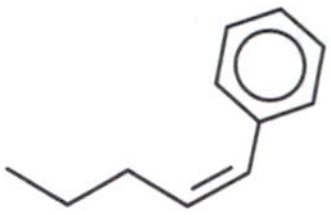

(d) 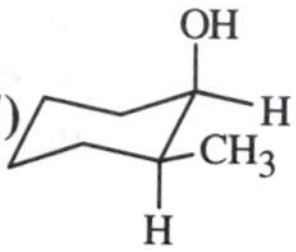 and

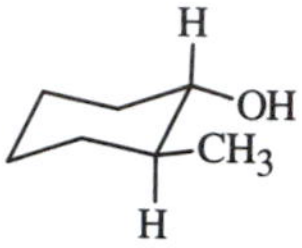

(e) 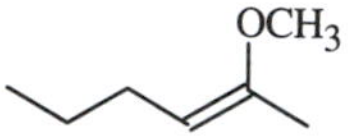 and

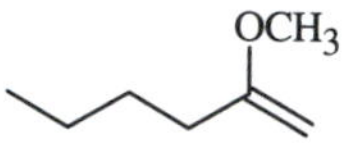

(f) 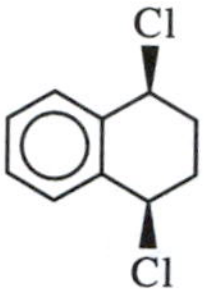 and

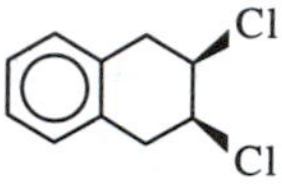

(g)  and

(h) 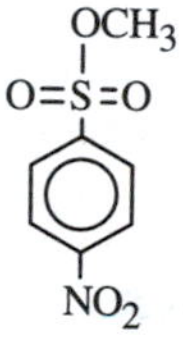 and

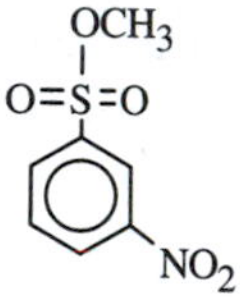

(i) 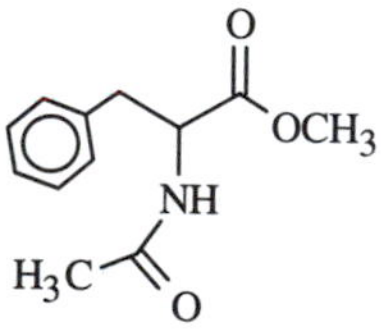 and

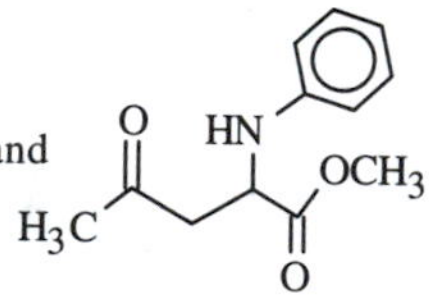

(j) 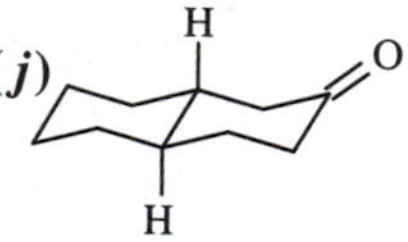 and

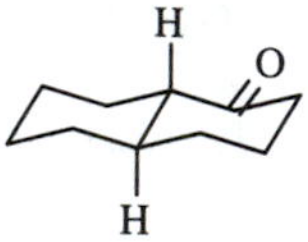

(k) 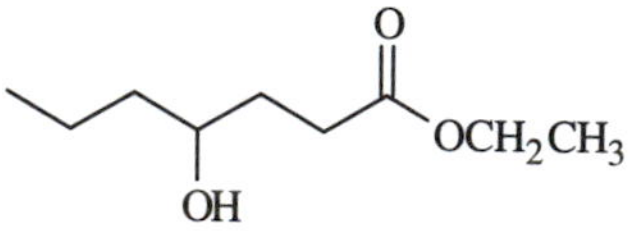 and

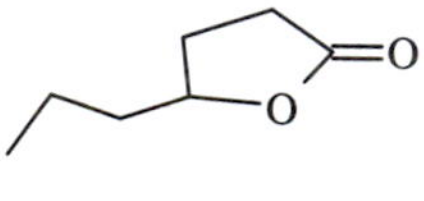

3. Tell how you could use ¹³C nmr to distinguish the following pairs of compounds. Be specific as to what data you would look for and how you would interpret it. There might be more than one feature of ¹³C nmr that could be used, so give a complete answer.

(a) 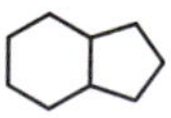 and

(b) 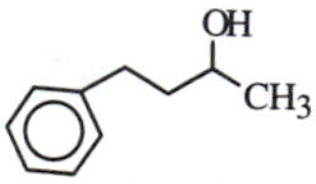 and

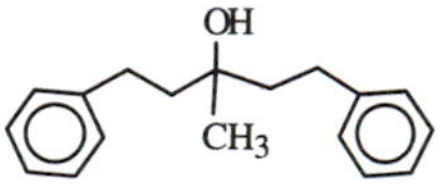

(c) H₃C–CH₂–O–CO–CH₃ and H₃C–CH₂–CO–O–CH₃   (d) [structure: ester with CH₃ branches] and [structure: branched ester with O–CH₃]

(e) [cyclopentane with OCH₂CH₃ substituents] and [cyclopentane with OCH₂CH₃ and OCH₂CH₃]

(f) H₃C–[benzene ring]–CH₂–CH₂–CH₂–CHO and [CH₃–CO–CH₂–benzene ring–CH₂CH₃]

(g) [benzene ring with CO–OCH₂CH₃ and two Cl] and [benzene ring with Cl, CO–OCH₂CH₃, Cl]

(h) [bicyclo structure] and [bicyclo structure]

4. Tell how you could use IR spectroscopy to distinguish the following pairs of compounds. Be specific as to what data you would look for and how you would interpret it. There might be more than one way to distinguish them by IR, so give a complete answer.

(a) [ester: –CO–OCH₂CH₃] and [ketone: –CO–CH₂CH₃]

(b) [cyclohexene with substituent] and [cyclohexane with OH and substituent]

(c) [benzene ring with H₃C and –CH₂–CO–NH₂] and [benzene ring with H₂N and –CH₂–CO–CH₃]

(d) H–C≡C–CH₂CH₂CN and H₃C–C≡C–CH₂CN

(e) [decalin-type ketone with enone] and [decalin-type ketone with enone]

(f) [chain with O–CO–CH₃ ester and N–CH₃ amide] and [chain with OH and N–CH₃ amide, H₃C–CO]

(g) H₃C–[benzene ring]–CH₃ and H₃C–[cyclohexane]–CH₃

(h) [CH₃ branched chain –CO–OH with Cl] and [CH₃ branched chain –CO–Cl with OH]

(i) [bicyclic alkene with CHO] and [bicyclic alkene with CHO]

(j) H₃C–[benzene ring]–C≡C–CO–OCH₃ and H₃C–C≡C–[benzene ring]–O–CO–CH₃

5. Tell how you could use mass spectrometry (MS) to distinguish the following pairs of compounds. Be specific as to what data you would look for and how you would interpret it. There might be more than one way to distinguish them by MS, so give a complete answer.

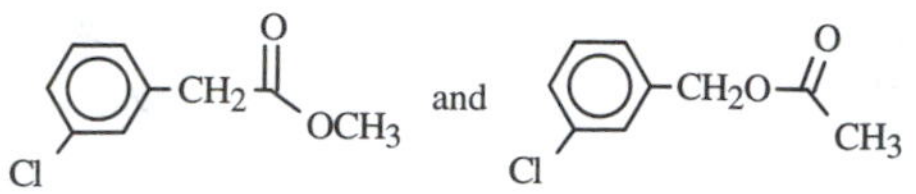

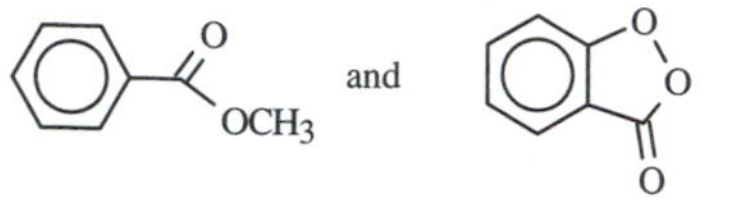

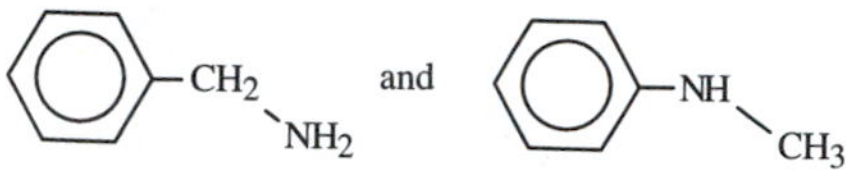

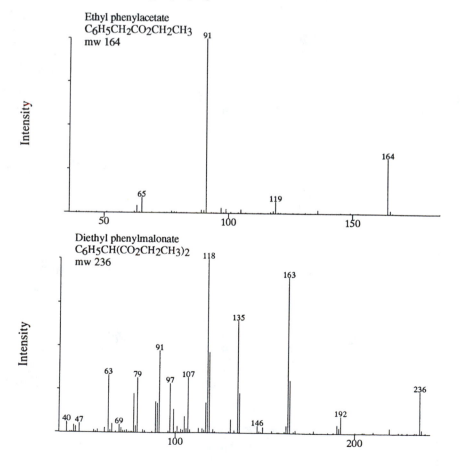

6. Rationalize the major fragmentation pathways observed for ethyl phenylacetate and diethyl phenylmalonate in the accompanying figure.

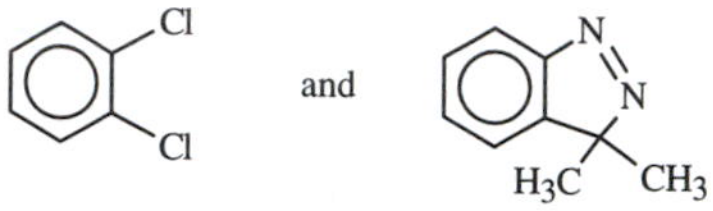

7. Give two instrumental methods that would permit you to distinguish the following. Explain what data you would use and how it would allow you to make the distinction.

(*a*) 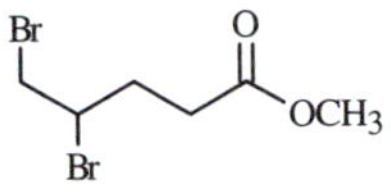 and 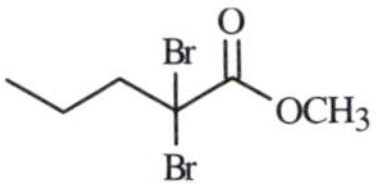

(*b*) 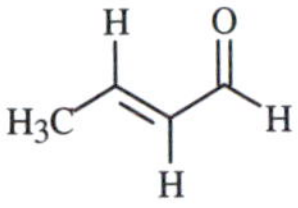 and 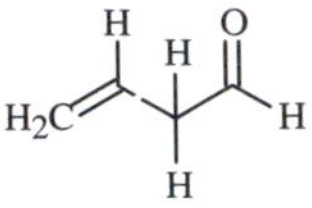

(*c*) 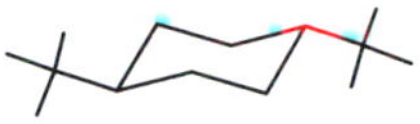 and 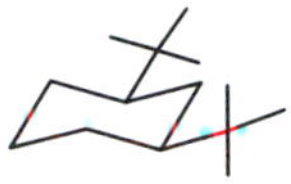

(*d*) 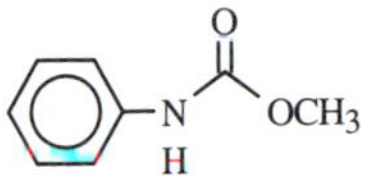 and 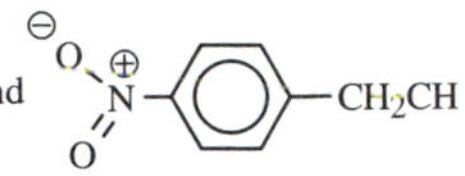

(*e*) 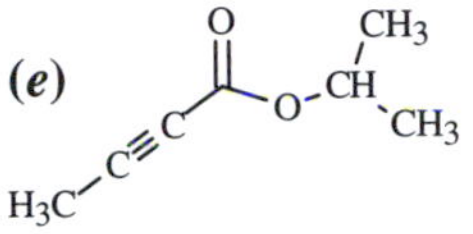 and 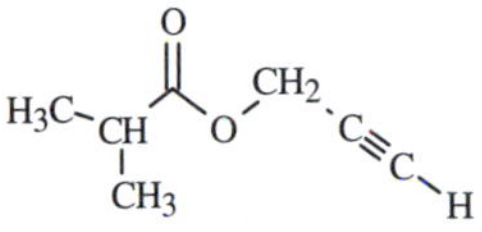

INDEX